AF556100

Digital Control Engineering Analysis and Design

Digital Control Engineering Analysis and Design

Ankit Tiwari

RANDOM PUBLICATIONS
NEW DELHI (INDIA)

Digital Control Engineering Analysis and Design

ISBN 978-93-5111-233-4

Published in 2014 in India by

RANDOM PUBLICATIONS

4376-A/4B, Gali Murari Lal, Ansari Road
New Delhi-110 002
Phone : +91-11-43580356, +91-11-23289044
e-mail: randomexports@gmail.com, sales@randompublications.com, info@randompublications.com

Type Setting by : Keystoneprintads, Delhi-110051
Printed at Thomson Press (India) Ltd

Preface

Digital control is a branch of control theory that uses digital computers to act as system controllers.

Issues in the design and implementation of digital controllers for a buck converter and a boost converter using linear and nonlinear control methods were investigated in this dissertation. The small signal models of the buck and boost converters, obtained using standard state space averaging techniques, were utilized in the dissertation. Analog PID and PI controllers were designed for generic buck and boost converters using standard frequency response techniques. The controllers were then transformed into digital controllers.

Digital controllers are far more convenient to implement on microprocessors than are continuous-time controllers. Continuous-time controllers must be implemented either using analog circuitry (op amps) or using numerical integration routines. Discrete-time controllers, on the other hand, are easily implemented using difference equations, i.e. simple computer software.

Direct digital control (DDC) systems consist of networked microprocessor-based controllers connected to analog and digital devices, which either sense information or control components of a building's heating, ventilating, and air-conditioning (HVAC) system. Control logic initiates and sequences operations programmed with software stored in the hardware. Analog-to-digital (A/D) converters transform analog electrical values into digital information for the microprocessor. The analog sensors may be thought of as transducers that convert a thermodynamic property to an electrical property.

The book also includes the advanced topics on control engineering such as robust control, observer design, optimal control, Lyapunov's stability and digital control system.

I thank all members of my team who have helped in the preparation of the book. My special thanks go to "Random Publications" who have published the book.

– ***Ankit Tiwari***

Contents

1

Digital Control System and Design

DIGITAL CONTROL

Digital control is a branch of control theory that uses digital computers to act as system controllers. Depending on the requirements, a digital control system can take the form of a microcontroller to an ASIC to a standard desktop computer. Since a digital computer is a discrete system, the Laplace transform is replaced with the Z-transform. Also since a digital computer has finite precision (*See quantization*), extra care is needed to ensure the error in coefficients, A/D conversion, D/A conversion, etc. are not producing undesired or unplanned effects.

The application of digital control can readily be understood in the use of feedback. Since the creation of the first digital computer in the early 1940s the price of digital computers has dropped considerably, which has made them key pieces to control systems for several reasons:

- Inexpensive: under $5 for many microcontrollers
- Flexible: easy to configure and reconfigure through software
- Scalable: programs can scale to the limits of the memory or storage space without extra cost
- Adaptable: parameters of the program can change with time (*See adaptive control*)
- Static operation: digital computers are much less prone to environmental conditions than capacitors, inductors, etc.

DIGITAL CONTROLLER IMPLEMENTATION

A digital controller is usually cascaded with the plant in a feedback system. The rest of the system can either be digital or analog.

Typically, a digital controller requires:

- A/D conversion to convert analog inputs to machine readable (digital) format
- D/A conversion to convert digital outputs to a form that can be input to a plant (analog)
- A program that relates the outputs to the inputs

Output Program

- Outputs from the digital controller are functions of current and past input samples, as well as past output samples - this can be implemented by storing relevant values of input and output in registers. The output can then be formed by a weighted sum of these stored values.

The programs can take numerous forms and perform many functions

- A digital filter for low-pass filtering
- A state space model of a system to act as a state observer
- A telemetry system

Stability

Although a controller may be stable when implemented as an analog controller, it could be unstable when implemented as a digital controller due to a large sampling interval. During sampling the aliasing modifies the cutoff parameters. Thus the sample rate characterizes the transient response and stability of the compensated system, and must update the values at the controller input often enough so as to not cause instability. When substituting the frequency into the z operator, regular stability criteria still apply to discrete control systems. Nyquist criteria apply to z-domain transfer functions as well as being general for complex valued functions. Bode stability criteria apply similarly. Jury criterion determines the discrete system stability about its characteristic polynomial.

Design of Digital Controller in S-domain

The digital controller can also be designed in the s-domain (continuous). The Tustin transformation can transform the continuous compensator to the respective digital compensator. The digital compensator will achieve an output which approaches the output of its respective analog controller as the sampling interval is decreased.

$$s = \frac{2(z-1)}{T(z+1)}$$

Tustin Transformation Deduction

Tustin is the $\text{Padé}_{(1,1)}$ approximation of the exponential function $z = e^{sT}$:

$$\begin{aligned} z &= e^{sT} \\ &= \frac{e^{sT/2}}{e^{-sT/2}} \\ &\approx \frac{1+sT/2}{1-sT/2} \end{aligned}$$

And its inverse

$$
\begin{aligned}
s &= \frac{1}{T}\ln(z) \\
&= \frac{2}{T}\left[\frac{z-1}{z+1}+\frac{1}{3}\left(\frac{z-1}{z+1}\right)^3+\frac{1}{5}\left(\frac{z-1}{z+1}\right)^5+\frac{1}{7}\left(\frac{z-1}{z+1}\right)^7+\cdots\right] \\
&\approx \frac{2}{T}\frac{z-1}{z+1} \\
&= \frac{2}{T}\frac{1-z^{-1}}{1+z^{-1}}
\end{aligned}
$$

We must never forget that the digital control theory is the technique to design strategies in discrete time, (and/or) quantized amplitude (and/or) in (binary) coded form to be implemented in computer systems (microcontrollers, microprocessors) that will control the analog (continuous in time and amplitude) dynamics of analog systems.

DIGITAL CONTROL SYSTEMS

DIGITAL SYSTEMS

Digital systems, expressed previously as difference equations or Z-Transform transfer functions can also be used with the state-space representation. Also, all the same techniques for dealing with analog systems can be applied to digital systems, with only minor changes.

DIGITAL SYSTEMS

For digital systems, we can write similar equations, using discrete data sets:

$$x[k+1] = Ax[k] + Bu[k]$$

$$y[k] = Cx[k] + Du[k]$$

Zero-Order Hold Derivation

If we have a continuous-time state equation:

$$x'(t) = Ax(t) + Bu(t)$$

We can derive the digital version of this equation that we discussed above. We take the Laplace transform of our equation:

$$X(s) = (sI - A)^{-1}Bu(s) + (sI - A)^{-1}x(0)$$

Now, taking the inverse Laplace transform gives us our time-domain system, keeping in mind that the inverse Laplace transform of the *(sI - A)* term is our state-transition matrix, ϕ:

$$x(t) = \mathcal{L}^{-1}(X(s)) = \Phi(t - t_0)x(0) + \int_{t_0}^{t} \Phi(t - \tau)Bu(\tau)d\tau$$

Now, we apply a zero-order hold on our input, to make the system digital. Notice that we set our start time $t_0 = kT$, because we are only interested in the behavior of our system during a single sample period:

$$u(t) = u(kT), kT \le t \le (k+1)T$$

$$x(t) = \Phi(t, kT)x(kT) + \int_{kT}^{t} \Phi(t, \tau)Bd\tau u(kT)$$

We were able to remove *u(kT)* from the integral because it did not rely on ô. We now define a new function, Γ, as follows:

$$\Gamma(t, t_0) = \int_{t_0}^{t} \Phi(t, \tau)Bd\tau$$

Inserting this new expression into our equation, and setting *t = (k + 1)T* gives us:

$$x((k+1)T) = \Phi((k+1)T, kT)x(kT) + \Gamma((k+1)T, kT)u(kT)$$

Now ϕ(T) and Γ(T) are constant matrices, and we can give them new names. The *d* subscript denotes that they are digital versions of the coefficient matrices:

$$A_d = \Phi((k+1)T, kT)$$

$$B_d = \Gamma((k+1)T, kT)$$

We can use these values in our state equation, converting to our bracket notation instead:

$$x[k+1] = A_d x[k] + B_d u[k]$$

[H] RELATING CONTINUOUS AND DISCRETE SYSTEMS[EDIT]

Continuous and discrete systems that perform similarly can be related together through a set of relationships. It should come as no surprise that a discrete system and a continuous system will have different characteristics and different coefficient matrices. If we consider that a discrete system is the same as a continuous system, except that it is sampled with a sampling time T, then the relationships below will hold. The process of converting an analog system for use with digital hardware is called discretization. We've given a basic introduction to discretization already, but we will discuss it in more detail here.

Discrete Coefficient Matrices

Of primary importance in discretization is the computation of the associated coefficient matrices from the continuous-time counterparts. If we have the continuous system *(A, B, C, D)*, we can use the relationship *t = kT* to transform the state-space solution into a sampled system:

$$x(kT) = e^{AkT}x(0) + \int_{0}^{kT} e^{A(kT-\tau)}Bu(\tau)d\tau$$

$$x[k] = e^{AkT}x[0] + \int_{0}^{kT} e^{A(kT-\tau)}Bu(\tau)d\tau$$

Now, if we want to analyze the *k+1* term, we can solve the equation again:

$$x[k+1] = e^{A(k+1)T}x[0] + \int_0^{(k+1)T} e^{A((k+1)T-\tau)}Bu(\tau)d\tau$$

Separating out the variables, and breaking the integral into two parts gives us:

$$x[k+1] = e^{AT}e^{AkT}x[0] + \int_0^{kT} e^{AT}e^{A(kT-\tau)}Bu(\tau)d\tau + \int_{kT}^{(k+1)T} e^{A(kT+T-\tau)}Bu(\tau)d\tau$$

If we substitute in a new variable β = *(k + 1)T + T,* and if we see the following relationship:

$$e^{AkT}x[0] = x[k]$$

We get our final result:

$$x[k+1] = e^{AT}x[k] + \left(\int_0^{T} e^{A\alpha}d\alpha\right)Bu[k]$$

Comparing this equation to our regular solution gives us a set of relationships for converting the continuous-time system into a discrete-time system. Here, we will use "d" subscripts to denote the system matrices of a discrete system, and we will use a "c" subscript to denote the system matrices of a continuous system.

Matrix Dimensions:

A: $p \times p$

B: $p \times q$

C: $r \times p$

D: $r \times q$

$$A_d = e^{A_cT}$$

$$B_d = \int_0^{T} e^{A\tau}d\tau B_c$$

$$C_d = C_c$$

$$D_d = D_c$$

This operation can be performed using thisMATLAB command:

c2d

If the A_c matrix is nonsingular, and we can find it's inverse, we can instead define B_d as:

$$B_d = A_c^{-1}(A_d - I)B_c$$

The differences in the discrete and continuous matrices are due to the fact that the underlying equations that describe our systems are different. Continuous-time systems are represented by linear differential equations, while the digital systems are described by difference equations. High order terms in a difference equation are delayed copies of the signals, while high order terms in the differential equations are derivatives of the analog signal.

If we have a complicated analog system, and we would like to implement that system in a digital computer, we can use the above transformations to make our matrices conform to the new paradigm.

Notation

Because the coefficent matrices for the discrete systems are computed differently from the continuous-time coefficient matrices, and because the matrices technically represent different things, it is not uncommon in the literature to denote these matrices with different variables. For instance, the following variables are used in place of A and B frequently:

$$\Omega = A_d$$

$$R = B_d$$

These substitutions would give us a system defined by the ordered quadruple (Ω, R, C, D) for representing our equations.

DIGITAL CONTROLLER DESIGN

The figure below shows the typical continuous feedback system that we have been considering so far in this tutorial. Almost all of the continuous controllers can be built using analog electronics.

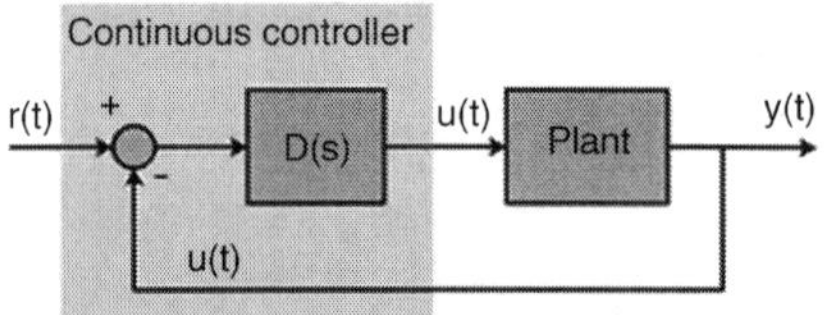

The continuous controller, enclosed in the dashed square, can be replaced by a digital controller, shown below, that performs same control task as the continuous controller. The basic difference between these controllers is that the digital system operates on discrete signals (or samples of the sensed signal) rather than on continuous signals.

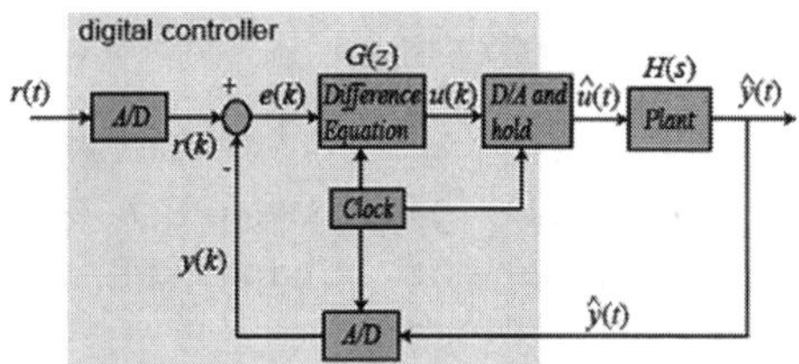

Different types of signals in the above digital schematic can be represented by the following plots.

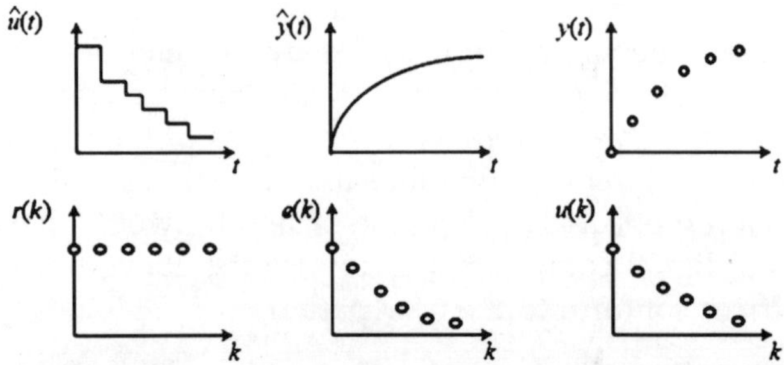

The purpose of this Digital Control Tutorial is to show you how to use MATLAB to work with discrete functions either in transfer function or state-space form to design digital control systems.

ZERO-HOLD EQUIVALENCE

In the above schematic of the digital control system, we see that the digital control system contains both discrete and the continuous portions. When designing a digital control system, we need to find the discrete equivalent of the continuous portion so that we only need to deal with discrete functions.

For this technique, we will consider the following portion of the digital control system and rearrange as follows.

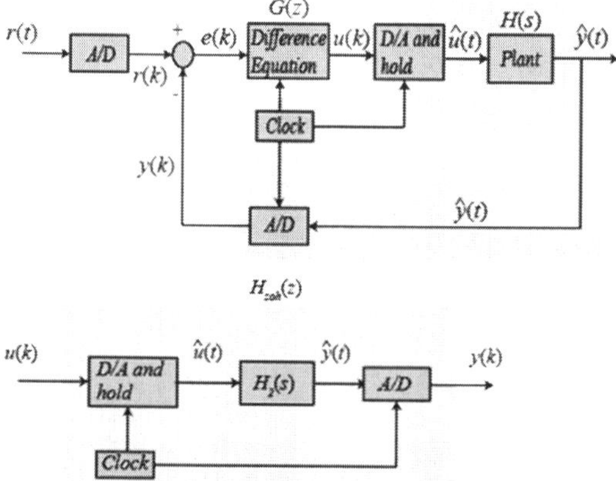

The clock connected to the D/A and A/D converters supplies a pulse every T seconds and each D/A and A/D sends a signal only when the pulse arrives. The purpose of having this pulse is to require that Hzoh(z) have only samples $u(k)$ to work on and produce only samples of output $y(k)$; thus, Hzoh(z) can be realized as a discrete function.

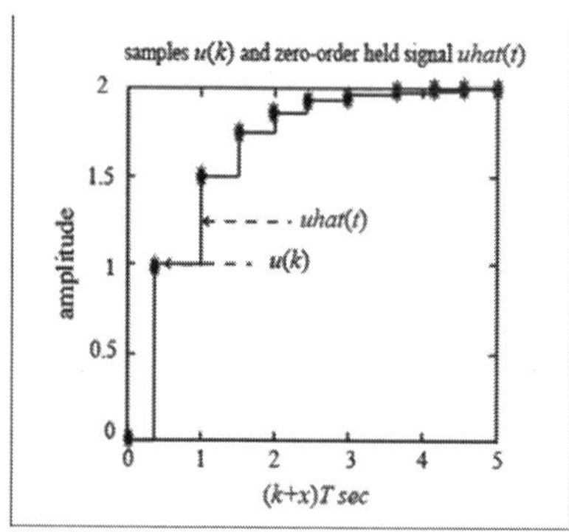

The philosophy of the design is the following. We want to find a discrete function Hzoh(z) so that for a piecewise constant input to the continuous system H(s), the sampled output of the continuous system equals the discrete output. Suppose the signal u(k) represents a sample of the input signal. There are techniques for taking this sample u(k) and holding it to produce a continuous signal uhat(t). The sketch below shows that the uhat(t) is held constant at u(k) over the interval kT to (k + 1)T. This operation of holding uhat(t) constant over the sampling time is called zero-order hold.

The zero-order held signal uhat(t) goes through H2(s) and A/D to produce the output y(k) that will be the piecewise same signal as if the discrete signal u(k) goes through Hzoh(z) to produce the discrete output y(k).

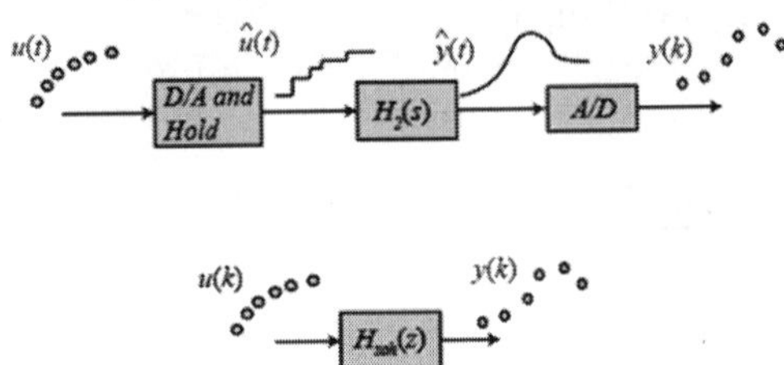

Now we will redraw the schematic, placing Hzoh(z) in place of the continuous portion.

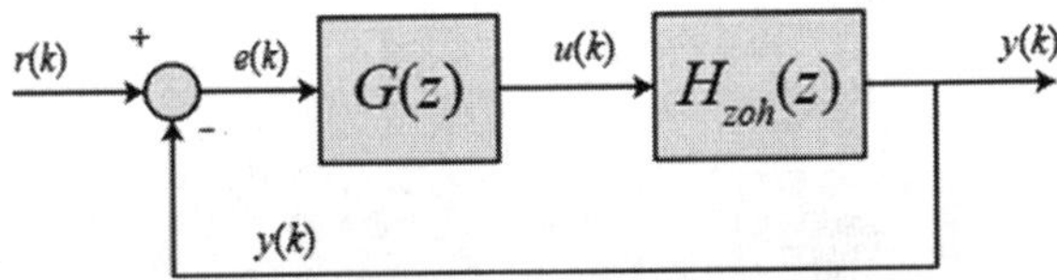

By placing Hzoh(z), we can design digital control systems dealing with only discrete functions.

Note: There are certain cases where the discrete response does not match the continuous response due to a hold circuit implemented in digital control systems. For information, see Lagging effect associated with the hold.

Zero-order Hold Equivalence

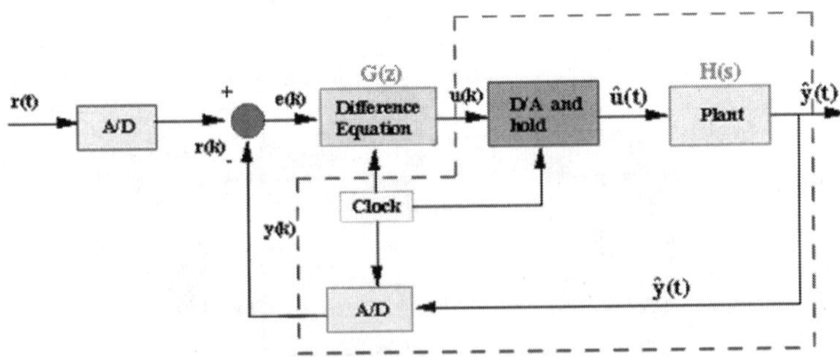

Fig. Discrete Feedback System

In the above schematic of the digital control system, we see that the digital control system contains both discrete and the continuous portions. When designing a digital control system, we need to find the discrete equivalent of the continuous portion so that we only need to deal with discrete functions.

For this technique, we will consider the following portion of the digital control system and rearrange as follows.

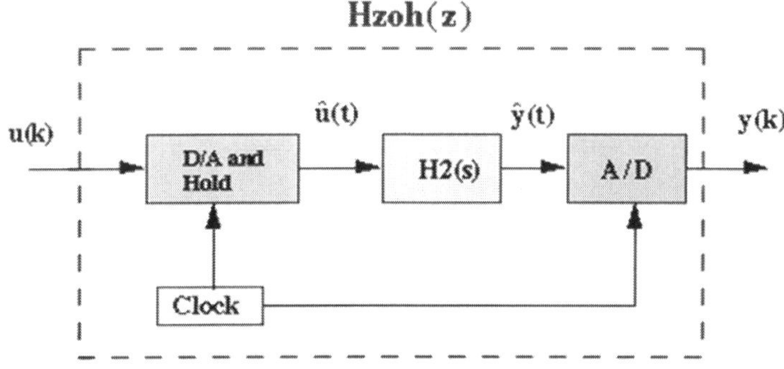

Fig. Discrete Feedback System Rearranged

The clock connected to the D/A and A/D converters supplies a pulse every T seconds, and each D/A and A/D converter sends a signal only when the pulse arrives. The purpose of having this pulse is to require that Hzoh(z) have only samples u(k) to work on and produce only samples of output y(k); thus, Hzoh(z) can be realized as a discrete function.

The philosophy of the design is the following. We want to find a discrete function Hzoh(z) so that for a piecewise constant input to the continuous system H(s), the sampled output of the continuous system equals the discrete output. Suppose the signal u(k) represents a sample of the input signal. There are techniques for taking this sample u(k) and holding it to produce a continuous signal uhat(t). The sketch in Figure 6 below shows that the uhat(t) is held constant at u(k) over the interval kT to (k+1)T. This operation of holding uhat(t) constant over the sampling time is called zero-order hold.

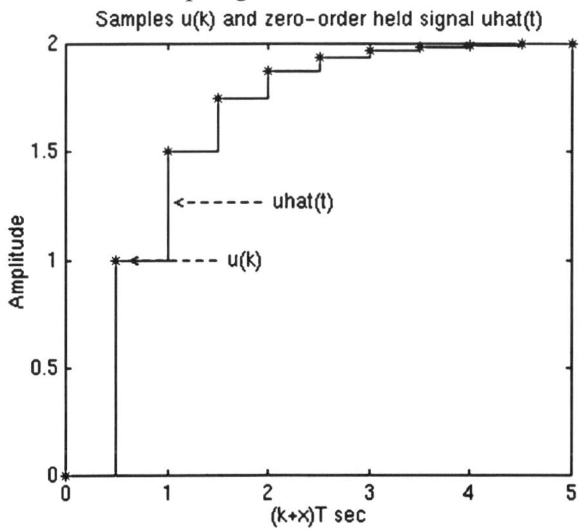

Fig. Zero-order Hold

The zero-order held signal uhat(t) goes through H2(s) and A/D to produce the output y(k) that will be the piecewise same signal as if the discrete signal u(k) goes through Hzoh(z) to produce the discrete output y(k).

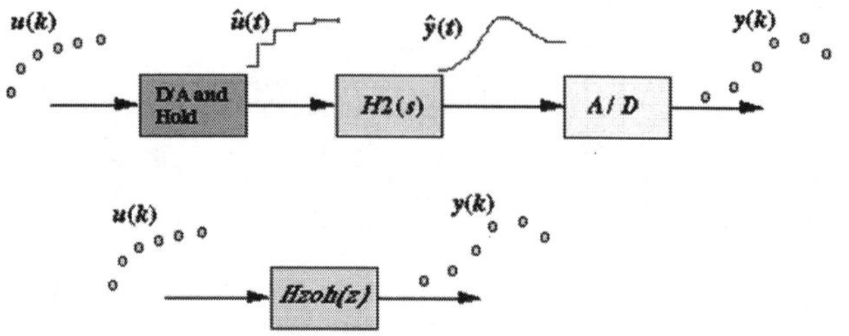

Fig. Output Signal

Now we will redraw the schematic, placing Hzoh(z) in place of the continuous portion.

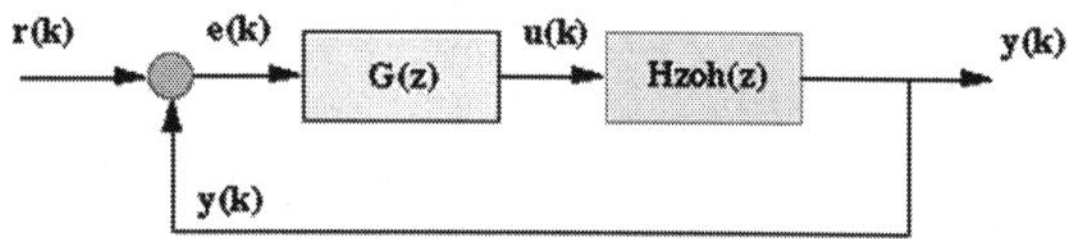

Fig. Redrawn Schematic

By placing Hzoh(z), we can design digital control systems dealing with only discrete functions.

There are certain cases where the discrete response does not match the continuous response due to a hold circuit implemented in digital control systems. In these cases, the hold causes a lagging effect in the discrete response. This effect can be reduced by decreasing the sampling time.

Conversion Using c2d

There is a MATLAB function called c2d that converts a given continuous system (either in transfer function or state-space form) to a discrete system using the zero-order hold operation explained above. The basic command for this in MATLAB is sys_d = c2d(sys,Ts,'zoh') The sampling time (Ts in sec/ sample) should be smaller than 1/(30*BW), where BW is the closed-loop bandwidth frequency.

DESIGN OF SIMPLE DIGITAL CONTROLLERS

Given the flexibility of the digital computer, digital control algorithms need not be restricted to discrete versions of analog designs. In particular, it is possible to formulate controllers that, under ideal conditions, will produce the desired closed loop response. This set of notes introduces some of these digital controller design methodologies. You will need to be familiar with sampled data systems and the associated mathematics to be able to follow the notes. The schematic of a conventional sampled data control system is shown:

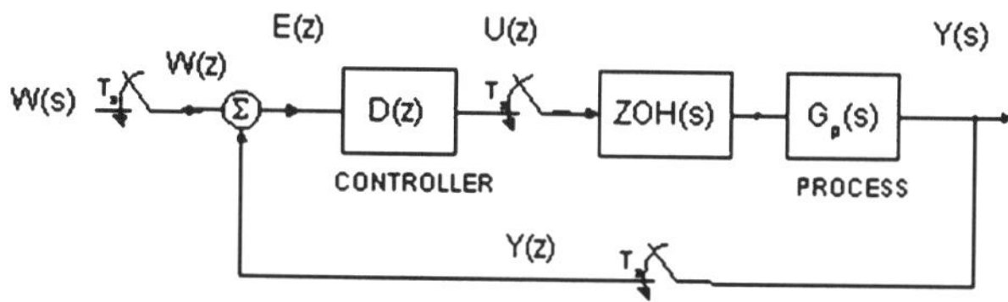

where the position of the samplers are depicted explicitly. Since this is a standard set up, we do not need to show the samplers. Instead, we can make use of the following simplified block diagram

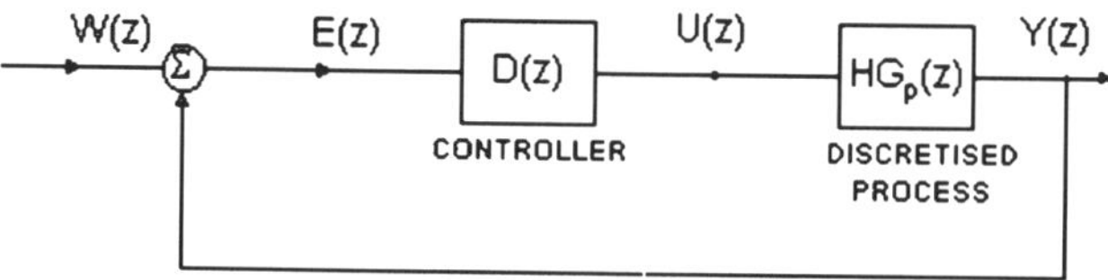

W(z), E(z), U(z) and *Y(z)* are respectively, the set-point, the error, the controller output and the controlled output. *D(z)* represents the digital controller, while the block $HG_p(z)$ refers to the z-transform of the zero-order-hold device in series with the process being controlled.

The zero-order-hold has the Laplace transform

$$L\{ZOH(s)\} = \frac{1-\exp(-sT_s)}{s}$$

Thus,

$$HG_p(z) = Z\left\{\frac{1-\exp(-sT_s)}{s}G_p(s)\right\}$$

Now, we are in a position to start designing digital controllers.

FUNDAMENTALS OF DIRECT DIGITAL CONTROL SYSTEMS

Direct digital control (DDC) systems consist of networked microprocessor-based controllers connected to analog and digital devices, which either sense information or control components of a building's heating, ventilating, and air-conditioning (HVAC) system. Control logic initiates and sequences operations programmed with software stored in the hardware. Analog-to-digital (A/D) converters transform analog electrical values into digital information for the microprocessor. The analog sensors may be thought of as transducers that convert a thermodynamic property to an electrical property. Most systems include stand-alone or remote controllers with software to eliminate the need for continuous communication. A personal computer workstation primarily monitors status and stores back-up copies of the programs, records alarms and trending functions, and archives historical information to operate the energy management system (EMS). Relatively complex strategies and energy management functions are normally available at low levels in the system architecture. Pneumatic actuation can be provided by electric-to-pneumatic transducers.

What Is Control?

Controlling an HVAC system involves three distinct steps:

- Measure a variable and collect data
- Process the data with other information
- Cause a control action

Basic Control Loop

A control loop comprises three main components: sensor, controller, and controlled device. These three components or functions interact to control a medium. Figure 1 shows air temperature as the controlled medium. The sensor measures air temperature and outputs data; the controller processes the data; and the controlled device causes an action.

A pneumatic or electronic control system, where the controller is a separate and distinct piece of hardware. In a DDC system, the controller "function" takes place in software, as shown in figure 2. The double lines indicate the boundary in and out of the DDC controller. Algorithms describing the mathematics of the software handle the controller function.

Sensor

Sensors measure the controlled medium or other control input in an accurate and repeatable manner. HVAC sensors may measure temperature, pressure, or relative humidity. Some sensors may measure other temperatures, time of day, electrical demand condition, or other variables that affect the controller logic. Other sensors input data that influence the control logic or safety, including air flow, water flow, current, fire, smoke, or high/low temperature limit. Sensors are an extremely important part of the control system and can be a weak link in the chain of control.

Controller

Controllers process data that are input from the sensor, apply the logic of control, and cause an output action. The output signal may be transmitted either to the controlled device or to other logical control functions. The controller compares input from sensors with a set of instructions such as set point, throttling range, and action and then produces an output signal. The control logic usually consists of a control response along with other logical decisions that are unique to the specific control application. How the controller functions is referred to as the control response.

Control responses are characterized as:

- Two-position
- Floating
- Proportional (p only)
- Proportional plus integral (pi)
- Proportional plus integral plus derivative (pid)

Controlled Device or Output

A controlled device responds to the signal from a controller or control logic in a way that changes the condition of the controlled medium or the state of the end device. These devices include valve operators, damper operators, electric relays, fans, pumps, compressors, and variable speed drives for fan and pump applications.

Two-Position Control

In a two-position control sequence, the controller compares an analog or variable input with instructions and generates a digital (two-position) output. The instructions involve definitions of upper or lower limits. The output changes value as the input crosses the limits.

There are no standards for defining limits. Common terms for limits include set point and differential. Set point indicates the value where output pulls-in, energizes, or is true. The output changes back or drops out after the input value crosses through the value equal to the difference between the set point and the differential. Two-position control functions as a simple switch. Two-position control can be used in basic control loops for temperature control or for limit control such as freezestats or outside air temperature limits.

Floating Control

A floating control response produces two digital outputs based on changing variable input. One output increases the signal to the controlled device while the other output decreases the signal to the controlled device. The control response also involves an upper and lower limit with the output changing as the variable input crosses these limits. There are no standards for defining these limits, but the terms set point and dead band are common. The set point is a midpoint and the dead band sets the difference between the upper and lower limits. Figure 3 shows an example of control output for floating control of static pressure in a variable air volume system with a variable frequency drive.

When the measured variable is within the dead band or neutral range, neither output is energized-the controlled device does not change mode. For this control response to be stable, the sensor must detect the effect of the controlled device movement very rapidly. Floating control does not function well where there is significant thermodynamic lag in the control loop. Fast air-side control loops respond well to floating control.

Proportional Control

A proportional control response produces an analog or variable output change in proportion to a varying input. A linear relationship exists between input and output (see figure 4). A set point, throttling range, and action normally define this relationship. In proportional control, a unique value of the measured variable corresponds to full travel of the controlled device and a unique value corresponds to zero travel on the controlled device. The change in the measured variable that causes the controlled device to move from fully closed to fully open is called the throttling range. The loop controls within this throttling range, assuming that the system has the capacity to meet the requirements.

The type of action dictates the slope of the control response. In direct-acting proportional control response, the output will rise with an increase in the measured variable. In a reverse-acting response, the output will decrease as the measured variable increases. The set point is an instruction to the control loop and corresponds to a specified value of the controlled device, usually half travel. The value of the measured variable at any given moment is called the control point. Offset is defined as the difference between the control point and the desired condition.

One way to reduce offset is to reduce throttling range. Reducing the throttling range too far will lead to instability. The more quickly the sensor feels the control response, the larger the throttling range has to be to produce stable control.

PI Control

PI control measures offset or error over time. The error is integrated, and a final adjustment is made to the output signal from a proportional part of this model. PI control response (see figure 5) will work the control loop to reduce the offset to zero. A well set-up PI control loop will operate in a narrow band close to the set point and not over the entire throttling range. PI control loops do not perform well when set points are dynamic, sudden load changes occur, or the throttling range is small.

PID Control

PID control adds a predictive element to the control response. In addition to proportional and integral calculation, the controller will compute the derivative or slope of the control response. This calculation dampens a control response that is returning to set point so quickly that it would overshoot the set point. PID is a precision process control response not always required for HVAC applications. PID control should be selectively applied to control loops because it is labor intensive.

Points

The word points is a common term used to describe data storage locations within a DDC system. The data can come from sensors or from software calculations and logic. The data can also be sent to controlled devices or software calculations and logic. Each data storage location has a unique means of identification or addressing. DDC data can be classified three different ways-by type, flow, and source.

Data Classifications

Data types are digital, analog, and accumulating. Digital data may also be called discrete or binary. The value of the data is an integer, a 0 or 1, and usually represents the state or status of a set of contacts. Analog data are represented by a numeric or a decimal number, usually defining a varying electrical input that is a function of temperature, relative humidity, pressure, or some other HVAC variable. Accumulating data are also represented by a numeric or decimal number, where the resulting sum is stored. This type of data is sometimes called a pulse input.

Data flow refers to whether data go to or from the DDC component or logic. Input points describe data used as input information and output points describe data that are output information.

Points can be classified as external when the data are received from an external device or sent to an external device. External points are sometimes referred to as hardware points. Internal points represent data that are created by the logic of the control software. Other terms used to describe these points are:

- Virtual points
- Numeric points
- Data points
- Software points

Global or indirect points are terms used to describe data that are transmitted on the network for use by other controllers.

Common Point Types

Analog input points normally imply an external point and represent a value that varies over time. Common analog inputs for HVAC applications include temperature, pressure, relative humidity, carbon dioxide, or air flow. Analog output points are control signals for modulating valve positions, damper positions, or drive speed.

Digital inputs for HVAC applications are usually status indicators, such as whether or not a motor is running, or for fans, pumps, motors, and lighting contactors. A temperature high limit is considered a digital input because, although it is monitoring an analog value related to temperature, the information that is transmitted to the controller is a digital condition indicating whether or not the temperature has exceeded a threshold. Digital outputs usually control relays that command motors or other devices to turn on or off.

Software Characteristics

There are many different software programs used in DDC systems. DDC systems include operating software, configuration software for the points or system architecture, programming software, and software for alarms. The characteristics of the programming software can be an important difference in various DDC systems. There are three common approaches used to program the logic of DDC systems:

- Line programming
- Template or menu-based programming
- Graphical or block programming

Line programming-based systems use software languages, similar to Basic or FORTRAN with HVAC subroutines. Familiarity with computer programming is helpful in understanding and writing logic for HVAC applications.

Menu-driven, database, or template/tabular programming makes use of templates for common HVAC logical functions. These templates contain the detailed parameters necessary for each logical program block to function. How one block is connected to another or where its data comes from is known as data flow and is programmed in each template.

Graphical or block programming is an extension of tabular programming. Graphical symbols connected by "data flow" lines represent the individual function blocks. Symbols similar to electrical schematics and pneumatic control diagrams depict the process. Graphical diagrams are created and the detailed data are entered in background menus or screens.

Architecture

System architecture is the map or layout of the system used to describe the overall local area network (LAN) structure. This map will show where the operator interfaces with the system and may remotely communicate with the system. The network, or LAN, is the media that connects multiple devices. This network media allows the devices to communicate, share, display and print information, and store data. The most basic task of the system architecture is to connect the DDC controllers so that information can be shared between them.

DDC Controller

A control loop requires a sensor to measure the process variable, the control logic to process data and calculate an instruction, and a controlled device to execute the instruction. A controller (see figura 6) is defined as a device that has inputs (sensors) and outputs (controllable devices) and the ability to execute control logic (software).

LAN Communication

Communications between devices on a network can be characterized as peer-to-peer or polling. On a peer-to-peer LAN, each device can share information with any other device on the LAN without going through a communication manager.

The controllers on the peer-to-peer LAN may be primary controllers, they may be secondary controllers, or they may be a mix. The type of controllers that use the peer-to-peer LAN will vary with different manufacturers.

In a polling controller LAN, the individual controllers cannot pass information directly to one of the other controllers. Data flow from one controller to the interface and then from the interface to the other controller.

The interface device manages communication between the polling LAN controllers and with higher levels in the system architecture. This same device may also supplement the capability of polling LAN controllers by providing the following functions:

- Clock function
- Buffer for trend data, alarms, messages
- Higher order software support

Many systems combine the communications of a peer-to-peer network with a polling network. In this case, the interface communicates in a peer-to-peer fashion with the devices on the peer-to-peer LAN. The polling LAN-based devices can receive data from the peer-to-peer devices but data must flow through the interface.

Controller Classification

Many DDC manufacturers make two distinct levels of controllers. Some make only one. These levels describe where the controller resides within the system architecture on the control network. Knowing the difference between these controllers is important because the appropriate controller is application dependent. Many specifications do not distinguish between the various types of controllers.

Higher-end controllers normally reside on a higher-level network and communicate in a peer-to-peer fashion. These are called primary controllers. Peer-to-peer means that the controllers can share information to other peer-to-peer devices without going through an intermediary device (called a supervisory interface). These primary controllers can range in cost from $1500 to $4000 and more. They have more memory, more sophisticated CPUs, higher resolution A/D converters, more accurate clocks, and can store more complex control strategies as well as trends, schedules, and alarms.

Manufacturers also make lower level controllers that normally reside on a lower level polling network. These controllers have more limited memory and processing capabilities and must use a supervisory interface device to communicate with all other devices. There are many different designs, and they can cost from $100 to $1000. Some are designed for terminal applications like variable air volume boxes or fan coil units. Others may be used for air handling systems with simple to moderately complex sequences of operation. These terminal controllers are usually configured for the number of points required for that application. Some of these controllers use a free form of programming, which requires a complete set of custom programming, while others have application specific programs for typical applications. These programs have selectable parameters that can be set up for each individual application. Since these controllers have more limited memories, they usually do not store historical information (such as trends) and rely on the supervisory interface for this function.

The secondary polling networks are configured such that one supervisory interface can monitor a limited number of controllers. This limitation varies by manufacturer. A large number of controllers on a secondary controller network can negatively affect the number of trends that can be practically used, the amount of data that can be processed, and the speed of transmission over the network. How many is too many on this secondary network? This varies and depends on the manufacturer, the speed of their network, and the application in question.

In HVAC control applications, these lower-level controllers may be adequate for many simple systems, but a primary controller is more appropriate for critical applications. How does one specify these distinctly different controllers? First, the engineer must define requirements of various types of controllers and their corresponding interfaces (both network and operator). Once defined, the engineer can dictate which controller should be used on various applications.

Over the past 20 years, HVAC control systems have undergone dramatic changes. They have evolved from pneumatic controls through various generations of EMS and DDC systems to current generation distributed DDC. At this time the computer industry trend of increasing processing power and memory at lower cost is quickly influencing DDC controllers. The advent of open protocols, the increased availability and use of site/building/campus networks, and the interfacing of DDC systems to the Internet have increased the complexity of these systems. During this relatively short time, we moved from a non-proprietary communication protocol (air pressure) to one that has been very proprietary. In addition, the control logic that was distributed to single function hardware components (receiver controllers, and switching relays) now resides in software. These are significant changes to a critical subsystem of HVAC systems, which is vital to the performance and basic operation of buildings. Historically, HVAC controls get far too little attention in the design, procurement, and installation process. The DDC system is the "brain" of the HVAC system-dictating the position of every damper and valve in a system and determining which fans, pumps, and chillers run and at what speed or capacity. There is a tremendous need for all involved in the process to increase their knowledge of control systems.

For further information: DDC Online (www.DDC-Online.org) is a web site that presents unbiased information about the major DDC product lines. Twenty-one product lines are presented in generically described layers. Vendors' proprietary controllers and interface devices are placed on these layers, allowing a user to compare similar products. The user can penetrate this architecture diagram to arrive at a detailed cut sheet on each product. The cut sheet includes information on the product presented in identical templates for each type of device. One of the goals of this web site is to provide information on products so engineers and owners can do a better comparison of equal products to produce better specifications. The site also includes a basic introduction to controls and a section on input and output devices and criteria.

DIRECT DIGITAL CONTROL SYSTEMS

The various architectures, hardware components and software associated with Direct Digital Control (DDC) systems. To accomplish this goal, a generic framework of the various components and configurations used in current DDC systems has been defined. This framework is used as a yardstick for several DDC manufacturers so readers may compare the relative features and benefits.

Intended Audience

Due to the complexity and proprietary nature of DDC systems, it has become difficult to stay current with the designs, installations, operation and maintenance of DDC systems. This guide was developed specifically to help building owners and consulting/specifying engineers with these issues.

What is an Energy Management System?

For the purposes of this guide, an energy management system (EMS) is defined as a fully functional control system. This includes controllers, various communications devices and the full complement of operational software necessary to have a fully functioning control system. This guide addresses approximately twenty of the DDC vendors who serve the institutional and commercial marketplace in the United States. Vendors who supply a complete line of all the necessary hardware and software are included. This guide does not cover specialty markets (retail grocery, hotels), nor does it cover industrial or process controls.

What is Control?

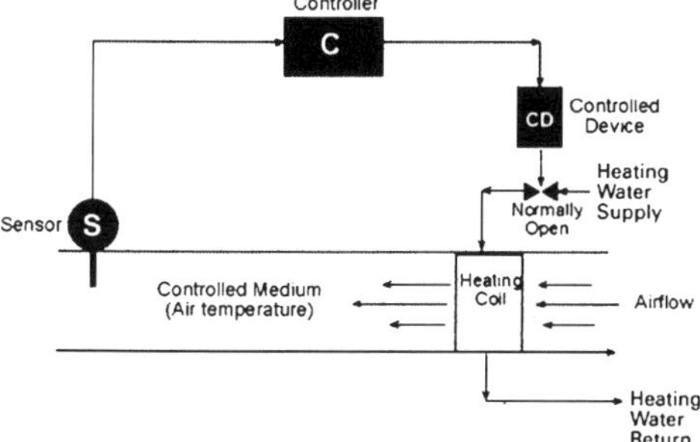

The process of controlling an HVAC system involves three steps. These steps include first measuring data, then processing the data with other information and finally causing a control action. These three functions make up what is known as a control loop. An example of this process is depicted in Figure 1.

Basic Control Loop

The control loop shown in Figure 1 consists of three main components: a sensor, a controller and a controlled device. These three components or functions interact to control a medium. In the example shown in Figure 1, air temperature is the controlled medium. The sensor measures the data, the controller processes the data and the controlled device causes an action.

The Figure 1 could be an example of a pneumatic or electronic control system, where the controller is a separate and distinct piece of hardware. In a DDC system, the controller function takes place in software as shown in Figure 2.

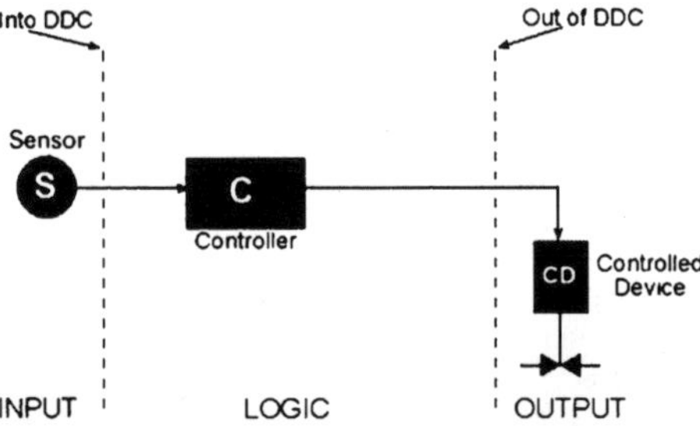

Sensor

The sensor measures the controlled medium or other control input in an accurate and repeatable manner. Common HVAC sensors are used to measure temperature, pressure, relative humidity, airflow stateand carbon dioxide. Other variables may also be measured that impact the controller logic. Examples include other temperatures, time-of-day or the current demand condition. Additional input information (sensed data) that influences the control logic may include the status of other parameters (airflow, water flow, current) or safety (fire, smoke, high/low temperature limit or any number of other physical parameters). Sensors are an extremely important part of the control system and can be the first, as well as a major, weak link in the chain of control.

Controller

The controller processes data that is input from the sensor, applies the logic of control and causes an output action to be generated. This signal may be sent directly to the controlled device or to other logical control functions and ultimately to the controlled device. The controllers function is to compare its input (from the sensor) with a set of instructions such as setpoint, throttling range and action, then produce an appropriate output signal. This is the logic of control. It usually consists of a control response along with other logical decisions that are unique to the specific control application. How the controller functions is referred to as the control response. Control responses are typically one the following:

- Two-Position
- Floating
- Proportional (P only)
- Proportional plus Integral (PI)
- Proportional plus Integral plus Derivative (PID)

Controlled Device or Output

A controlled device is a device that responds to the signal from the controller, or the control logic, and changes the condition of the controlled medium or the state of the end device. These devices include valve operators, damper operators, electric relays, fans, pumps, compressors and variable speed drives for fan and pump applications.

2

Digital Design Procedure and Combinational Logic Circuits

COMBINATIONAL CIRCUITS

Combinational logic is probably the easiest circuitry to design. The outputs from a combinational logic circuit depend only on the current inputs. The circuit has no remembrance of what it did at any time in the past. Much of logic design involves connecting simple, easily understood circuits to construct a larger circuit that performs a much more complicated function. Several simple, often-used combinational logic circuits are the following: Logic circuits for digital systems may be combinational or sequential.

A combinational circuit consists of logic gates whose outputs at any time are determined by combining the values of the applied inputs using logic operations. A combinational circuit performs an operation that can be specified logically by a set of Boolean expression. In addition to using logic gates, sequential circuits employ elements that store bit values. Sequential circuit outputs are a function of inputs and the bit value in storage elements. These values, in turn, are a function of previously applied inputs and stored values. As a consequence, the outputs of a sequential circuit depend not only on the presently applied values of the inputs, but also on pas inputs, and the behaviour of the circuit must be specified by a sequence in time of inputs and internal stored bit values.

A combinational circuit consists of input variables, output variables, logic gates and interconnections. The interconnected logic gates accept signals from the inputs and generate signals at the output. The n input variables come from the environment of the circuit, and the m output variables are available for use by the environment. Each input and output variable exists physically as a binary signal that represents logic 1 or logic 0. For n input variables, there are 2^n possible binary input combinations. For each binary combination of the input variables, there is one possible binary value on each output. Thus, a combinational circuit can be specified by a truth table that lists the output values for each combination of the input variables.

A combinational circuit can also be described by m Boolean function, one for each output variable. Each such function is expressed as function of the n input variables. A combinational circuit performs an operation that can be specified logically by a set of Boolean expression. In addition to using logic gates, sequential circuits employ elements that store bit values. Sequential circuit outputs are a function of inputs and the bit value in storage elements. These values, in turn, are a function of previously applied inputs and stored values.

As a consequence, the outputs of a sequential circuit depend not only on the presently applied values of the inputs, but also on pas inputs, and the behaviour of the circuit must be specified by a sequence in time of inputs and internal stored bit values. A combinational circuit consists of input variables, output variables, logic gates and interconnections. The interconnected logic gates accept signals from the inputs and generate signals at the output. The n input variables come from the environment of the circuit, and the m output variables are available for use by the environment. Each input and output variable exists physically as a binary signal that represents logic 1 or logic 0.

MULTIPLEXERS AND DEMULTIPLEXERS

A multiplexer has a number of inputs (usually a power of two), a number of control signals, and one output. A demultiplexer has one input signal, a number of control signals, and a number of outputs, also usually a power of two. We consider here a 2^N–to–1 multiplexer and a 1–to–2^N demultiplexer.

Circuit	Inputs	ControlSignals	Outputs
Multiplexer	2^N	N	1
Demultiplexer	1	N	2^N

The action of each of these circuits is determined by the control signals. For a multiplexer, the output is the selected input. In a demultiplexer, the input is routed to the selected output. As examples, we show the diagrams for both a four–to–one multiplexer (MUX) and a one–to–four demultiplexer (DEMUX).

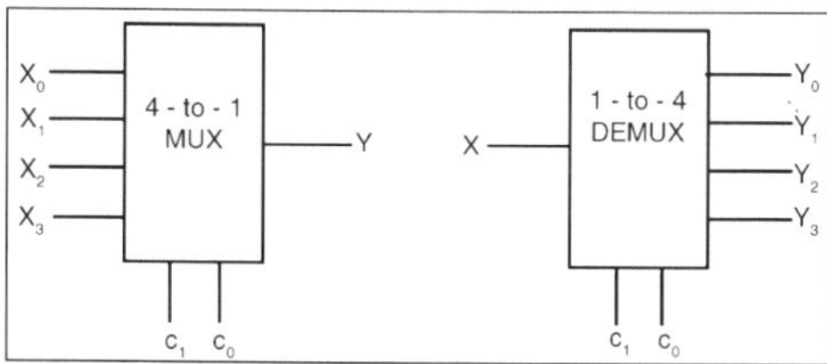

Note that each of the circuits has two control signals. For a multiplexer, the *N* control signals select which of the 2^N inputs will be passed to the output. For a demultiplexer, the *N* control signals select which of the 2^N outputs will be connected to the input.

Multiplexers

The output for a multiplexer can be represented as a Boolean function of the inputs and the control signals. As an example, we consider a 4-input multiplexer, with control signals labeled C_0 and C_1 and inputs labeled I_0, I_1, I_2, and I_3. The output can be described as a truth table or algebraically. Note that each of the truth tables and algebraic expression shows the input that is passed to the output. The truth table is an abbreviated form of the full version, which as a table for independent variables C_0, C_1, I_0, I_1, I_2, and I_3 would have 64 rows.

C_1	C_0	M
0	0	I_0
0	1	I_1
1	1	I_3

$$M = C_1'\bullet C_0'\bullet I_0 + C_1'\bullet C_0\bullet I_1 + C_1\bullet C_0'\bullet I_2 + C_1\bullet C_0\bullet I_3$$

Multiplexers are generally described as 2^N–to–1 devices. These multiplexers have 2^N inputs, one of which is connected to the single output line. The N control lines determine which of the inputs is connected to the output. Here is a circuit for a 4–to–1 multiplexer. Note that the inputs are labeled X_3, X_2, X_1, and X_0 here and I_3, I_2, I_1, and I_0 in the multiplexer equation.

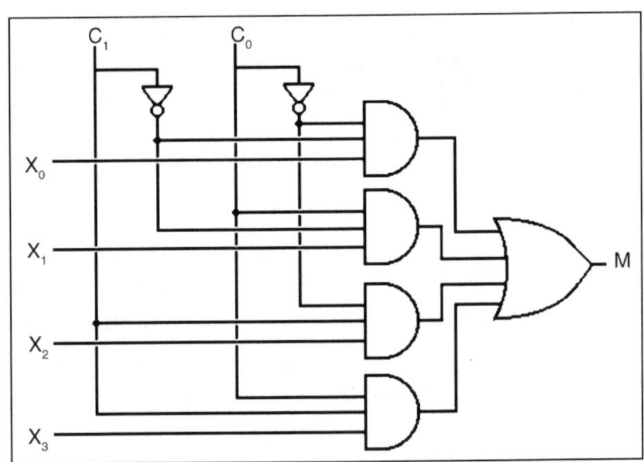

Demultiplexer

Demultiplexers are generally described as 1–to–2^N devices. These multiplexers have one input, which is connected to one of the 2^N output lines.The N control lines determine which of the output line is connected to the input. Here is a circuit for a 1–to–4 demultiplexer, which might be called "active high" in that the outputs not selected are all set to 0. Note that multiplexers do not have the problem of unselected outputs; a MUX has only one output.

Note that, for good measure, we have added an enable–high to the demultiplexer. When this enable is 0, all outputs are 0.

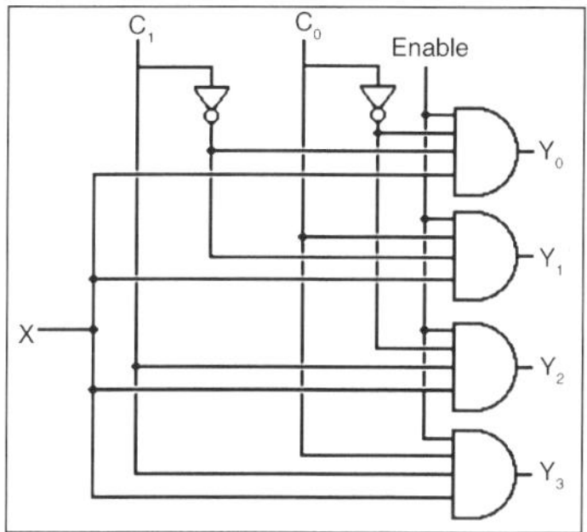

When this enable is 1, the selected output gets the input, X. Remember, that X can have a value of either 0 or 1. We shall return to demultiplexers after we have discussed decoders. At that time, we shall note a similarity between the decoders and demultiplexers, consider the use of a decoder as a demultiplexer, and investigate the possibility of an active–low demultiplexer.

DECODERS AND ENCODERS

We now consider an important class of commercial circuits – encoders and decoders. These perform the functions suggested by the corresponding decimal-binary conversions. In conversion of a decimal number to binary, we obtain the binary equivalent of the number.An encoder has a number of inputs, usually a power of two, and a set of outputs giving the binary code for the "number" of the input.

Encoders

Consider a classic 2^N–to–N encoder. The inputs are labeled $I_0, I_1, ..., I_K,$ where $K = 2^N - 1$. The assumption is that only one of the inputs is active; in our way of thinking only one of the inputs is 1 and the rest are 0. Suppose input J is 1 and the rest are 0. The output of the circuit is the binary code for J. Suppose a 32–to–5 encoder with input 18 active. The output Z is the binary code 10010; $Z_4 = 1$, $Z_3 = 0$, $Z_2 = 0$, $Z_1 = 1$, and $Z_0 = 0$.

Common encoders include 8–to–3, 16–to–4, and 32–to–5. One common exception to the rule of 2^N–to–N is a 10–to–4 encoder, which is used because decimal numbers are so common.Note that three binary bits are not sufficient to encode ten numbers, so we must use four bits and not produce the outputs 1010, 1011, 1100, 1101, 1110, or 1111.

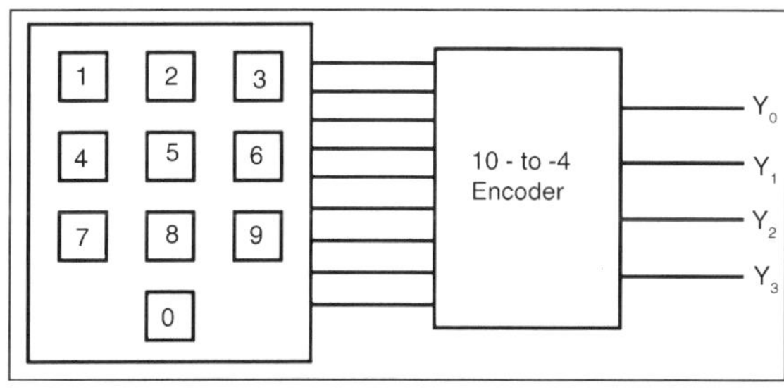

We now present a detailed discussion and a design of a 10-to-4 encoder. We begin with a diagram that might illustrate a possible use of an encoder. In this example, the key pad has ten keys, one for each digit. When a key is pressed, the output line corresponding to that key goes to logic 1 (5 volts) and the other output lines stay at logic 0 (0 volts). Note that there are ten output lines from the key pad, one for each of the keys. These ten output lines form ten input lines into the 10–to–4 encoder.

The 10–to–4 encoder outputs a binary code indicating which of the keys has been pressed. In a complete design, we would require some way to indicate that no key has been pressed. For our discussion, it is sufficient to ignore this common case and assume that a key is active. We first ask why we need four bits for the encoder.N bits will encode 2^N different inputs. As a result, to encode M different items, we need N bits with $2^{N-1} < M \geq 2^N$. To encode 10 inputs, we note that $2^3 < 10 \geq 2^4$, so we need 4 bits to encode 10 items. We now present a table indicating the output of the encoder for each input. In this example, we assume that at any time exactly one input is active.

Input	Y_3	Y_2	Y_1	Y_0
X_0	0	0	0	0
X_1	0	0	0	1
X_2	0	0	1	0
X_3	0	0	1	1
X_4	0	1	0	0
X_5	0	1	0	1
X_6	0	1	1	0
X_7	0	1	1	1
X_8	1	0	0	0
X_9	1	0	0	1

In the table at left, we label the inputs X_0 through X_9, inclusive.

To produce the equations for the outputs, we reason as follows.

Y_3 is 1 when either $X_8 = 1$ or $X_9 = 1$.

Y_2 is 1 when $X_4 = 1$ or $X_5 = 1$ or $X_6 = 1$ or $X_7 = 1$.

Y_1 is 1 when $X_2 = 1$, $X_3 = 1$, $X_6 = 1$, or $X_7 = 1$.

Y_0 is 1 when $X_1 = 1$, $X_3 = 1$, $X_5 = 1$, $X_7 = 1$, or $X_9 = 1$.

These observations lead to the following equations, used to design the encoder.

$$Y_3 = X_8 + X_9$$
$$Y_2 = X_4 + X_5 + X_6 + X_7$$
$$Y_1 = X_2 + X_3 + X_6 + X_7$$
$$Y_0 = X_1 + X_3 + X_5 + X_7 + X_9$$

Here is the circuit for the 10-to-4 encoder.

The student will note that input X_0 is not connected to an output. This gives rise to the following problem for the circuit: how does one differentiate between

X_0 being active and no input being active. That might be a problem for real encoder design.

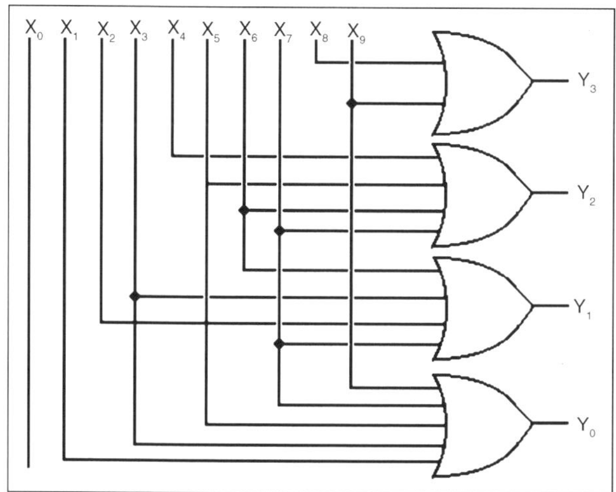

The most straightforward modification of the circuit would be to create the logical OR of the ten inputs and pass that signal as a "key pressed" signal. We mention this only to show that some applications must handle this case; we shall not consider it further in this course. Another issue with encoders is what to do if two or more inputs are active. For 'plain" encoders the output is not always correct; for example, in the above circuit with inputs X_3 and X_5 active would output

$$Y_3 = 0, Y_2 = 1, Y_1 = 1, \text{ and } Y_0 = 1.$$

Priority encoders are designed to avoid the problem of multiple inputs by implementing a priority order on the inputs and producing the output for the input that has priority. For example, in a 32-to-5 priority encoder, having inputs 18 and 29 active would produce either the binary code 10010 (for 18) or 11101 (for 29), depending on the priority policy, and not the output 11111 that a plain encoder would produce.

Decoders

An N–to–2^N decoder does just the opposite, taking an N bit binary code and activating the output labeled with the corresponding number. Consider a 4–to–16 decoder with outputs labeled Z_0, Z_1, ..., Z_{15}. Suppose the input is $I_3 = 1$, $I_2 = 0$, $I_1 = 0$, and $I_0 = 1$ for the binary code 1001. Then output Z_9 is active and the other outputs are not active. Again, the main exception to the N–to–2^N rule for decoders is the 4 – to – 10 decoder, which is a common circuit. Note that it takes 4 bits to encode 10 items, as 3 bits will encode only 8. This author's preference would be to use a 4–to–16 decoder and ignore some of the outputs, but this author does not establish commercial practice. The main advantage is that the 4–to–10 decoder chip would have 6 fewer pins than a 4–to–16 decoder; a 16–pin chip is standard and cheaper to manufacture than a 22–pin chip.

Another issue is whether the signals are active high or active low. Our examples have been constructed for active high circuits. Consider the 4–to–16 decoder as an example. If the input code is 1001, then the output Z_9 is a logic 1 (+5 volts) and all other outputs are logic 0 (0 volts). This approach is active high. In real commercial circuits, we often have outputs as active low, in which case the above decoder would have output Z_9 as a logic 0 (0 volts) and all other inputs as logic 1 (+5 volts). This reflects an issue with design using real TTL circuits or with standard circuit emulators, such as MultiSim or Multi–Media Logic. Decoders have N inputs and 2^N outputs. We consider a 3–to–8 decoder; thus $N = 3$. The design that we present will assume that the decoder is active high; again that the selected output becomes logic 1 (5 volts) and the others remain at logic 0 (ground or 0 volts).

Decimal	Binary
0	000
1	001
2	010
3	011
4	100
5	101
6	110
7	111

The decoder is based on the association of binary numbers to decimal numbers, as is shown in the figure at right. Since this is a 3–to–8 decoder, we have three inputs, labeled X_2, X_1, and X_0; and eight outputs, labeled Y_7, Y_6, Y_5, Y_4, Y_3, Y_2, Y_1, and Y_0.

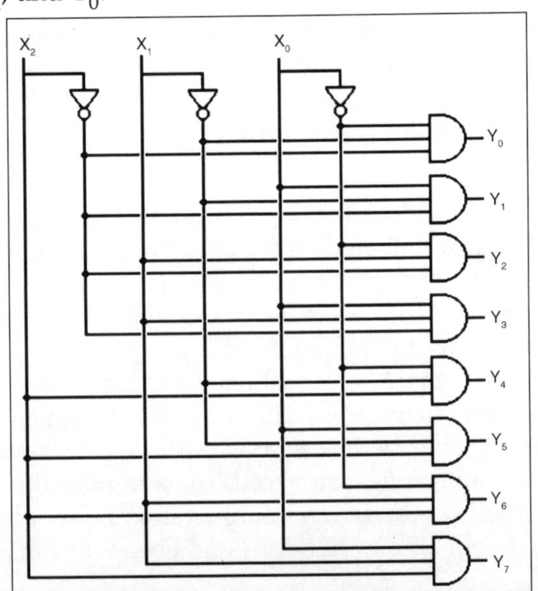

The observation that leads to the design of the decoder is the obvious one that the values of X_2, X_1, and X_0 determine the output selected. For this part of the discussion, I choose to ignore the enable input.

We introduce the Enable output in a later discussion.

The following Boolean equations determine the 3–to–8 decoder.

$$Y_0 = \overline{X_2} \bullet \overline{X_1} \bullet \overline{X_0} \qquad Y_4 = X_2 \bullet \overline{X_1} \bullet \overline{X_0}$$

$$Y_1 = \overline{X_2} \bullet \overline{X_1} \bullet X_0 \qquad Y_5 = X_2 \bullet \overline{X_1} \bullet X_0$$

$$Y_2 = \overline{X_2} \bullet X_1 \bullet \overline{X_0} \qquad Y_6 = X_2 \bullet X_1 \bullet \overline{X_0}$$

$$Y_3 = \overline{X_2} \bullet X_1 \bullet X_0 \qquad Y_7 = X_2 \bullet X_1 \bullet X_0$$

Here is the circuit diagram for the 3–to–8 active–high decoder.

THE ENABLE INPUT

We now consider another important input to the decoder chip. This is the enable input. If the decoder enable signal is active high, then the decoder is active when enable is 1 and not active when enable = 0. We shall consider enabled-high decoders here.

The enable input allows the decoder to be either enabled or disabled.For an active high decoder that is enabled high (Enable = 1 activates it) we have the following.

Enable =0 All outputs of the decoder are 0

Enable = 1 The selected output of the decoder is 1, all other outputs are 0.

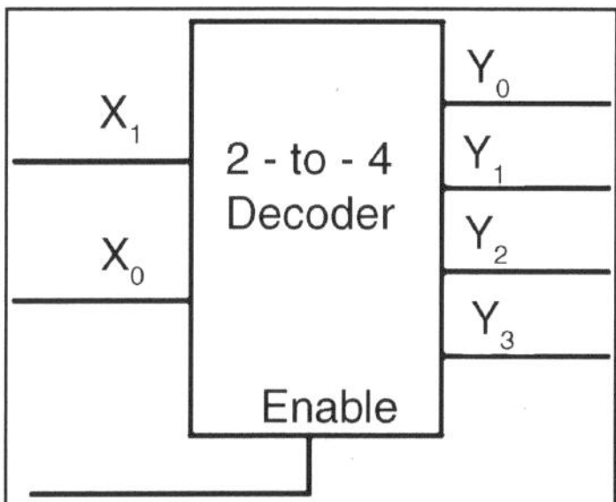

One way to express the effect of the enable input is to use a modified truth table.

Enable	X_1	X_0	Y_0	Y_1	Y_2	Y_3
0	d	d	0	0	0 0	
1	0	0	1	0	0 0	
1	0	1	0	1	0 0	
1	1	0	0	0	1 0	
1	1	1	0	0	0 1	

Thus, all outputs are 0 when the enable input is 0. This is true without regard to the inputs. When the enable input is 1, the outputs correspond to the inputs.

Here is a circuit diagram for a 2–to–4 decoder that is enabled high and active high. The Enable input is passed as an input to all four of the AND gates used to produce the output of the decoder. As a result, when Enable = 0, all of the outputs are 0; the decoder is not active.

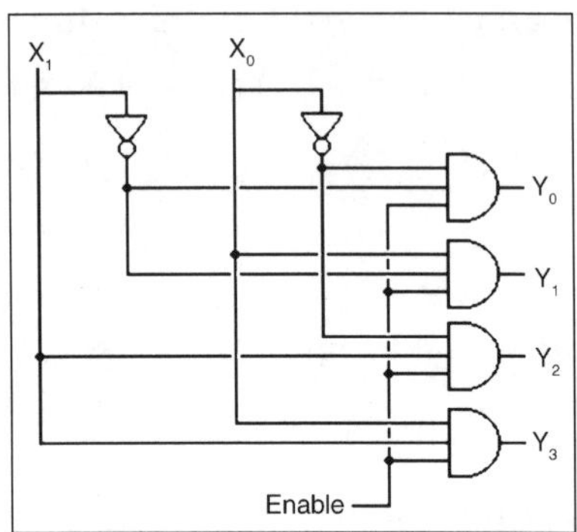

When Enable = 1 in the above circuit, the other two inputs X_1 and X_0 will determine the one of the four outputs is set to 1; the others remaining 0. This is exactly how the active-high decoder should function. The following example illustrates the use of the enable input for decoders. We use two two-to-four decoders to construct a single three-to-eight decoder. The way to do this is to use one of the inputs, conventionally the high-order bit, as an enable signal. This way one of the two-to-four decoders will be enabled and one will not be. Here is the circuit.

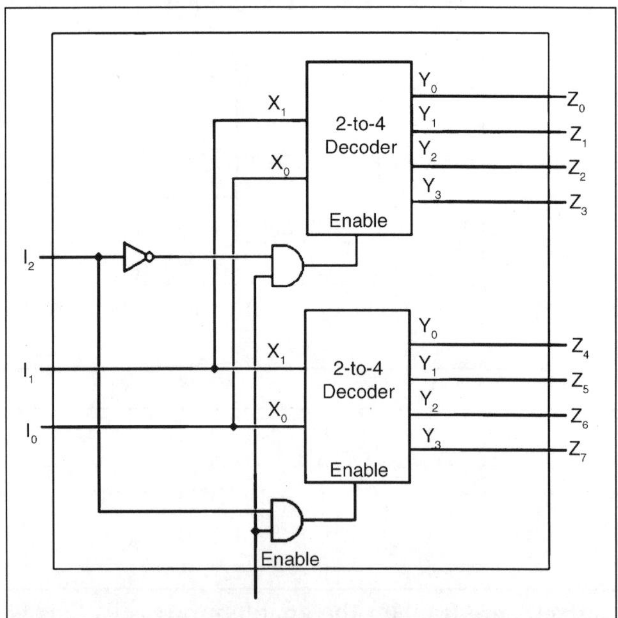

Suppose the three-to-eight decoder is enabled. Under this assumption, the input I_2 selects the two–to–four decoder that is active. It I_2 = 0, then the top decoder is active and the bottom decoder is not active (all its outputs are 0). It $I_2 = 1$, then the bottom decoder is active and the top decoder is not active (all its outputs are 0). In either case, the inputs I_1 and I_0 are passed to both two-to-four decoders. When $I_2 = 0$, the inputs I_1 and I_0 select which of outputs Y_0, Y_1, Y_2, or Y_3 in the top two-to-four decoder is active and thus which of Z_0, Z_1, Z_2, or Z_3 is active.

When $I_2 = 1$, the inputs I_1 and I_0 select which of outputs Y_0, Y_1, Y_2, or Y_3 in the bottom two-to-four decoder is active and thus which of Z_4, Z_5, Z_6, or Z_7 is active. Thus we have constructed the equivalent of a three-to-eight decoder.

ANALYSIS PROCEDURE OF COMBINATIONAL CIRCUITS

The bulk of the Combinational Analysis module is accessed through a single window of that name allowing you to view truth tables and Boolean expressions. This window can be opened in two ways.

VIA THE WINDOW MENU

Select Combinational Analysis, and the current Combinational Analysis window will appear. If you haven't viewed the window before, the opened window will represent no circuit at all. Only one Combinational Analysis window exists within Logisim, no matter how many projects are open. There is no way to have two different analysis windows open at once.

VIA THE PROJECT MENU

From a window for editing circuits, you can also request that Logisim Analyse the current circuit by selecting the Analyse Circuit option from the Project menu. Before Logisim opens the window, it will compute Boolean expressions and a truth table corresponding to the circuit and place them there for you to view. For the analysis to be successful, each input must be attached to an input pin, and each output must be attached to an output pin. Logisim will only Analyse circuits with at most eight of each type, and all should be single-bit pins. Otherwise, you will see an error message and the window will not open.

In constructing Boolean expressions corresponding to a circuit, Logisim will first attempt to construct a Boolean expressions corresponding exactly to the gates in the circuit. But if the circuit uses some non-gate components (such as a multiplexer), or if the circuit is more than 100 levels deep (unlikely), then it will pop up a dialog box telling you that deriving Boolean expressions was impossible, and Logisim will instead derive the expressions based on the truth table, which will be derived by quietly trying each combination of inputs and reading the resulting outputs. After analyzing a circuit, there is no continuing relationship between the circuit and the Combinational Analysis window. That is, changes to the circuit will not be reflected in the window, nor will changes to the Boolean expressions and/or truth table in the window be reflected in the circuit. Of course, you are always free to Analyse a circuit again; and, as we will see later, you can replace the circuit with a circuit corresponding to what appears in the Combinational Analysis window.

LIMITATIONS

Logisim will not attempt to detect sequential circuits: If you tell it to Analyse a sequential circuit, it will still create a truth table and corresponding Boolean expressions, although these will not accurately summarize the circuit behaviour. (In fact, detecting sequential circuits is provably impossible, as it would amount to solving the Halting Problem. Of course, you might hope that Logisim would make at least some attempt - perhaps look for flip-flops or cycles in the wires - but it does not.) As a result, the Combinational Analysis system should not be used indiscriminately: Only use it when you are indeed sure that the circuit you are analyzing is indeed combinational!

Logisim will make a change to the original circuit that is perhaps unexpected: The Combinational Analysis system requires that each input and output have a unique name that conforming to the rules for Java identifiers. (Roughly, each character must either a letter or a digit, and the first character must be a letter. No spaces allowed!) It attempts to use the pins' existing labels, and to use a list of defaults if no label exists. If an existing label doesn't follow the Java-identifier rule, then Logisim will attempt to extract a valid name from the label if at all possible. Incidentally, the ordering of the inputs in the truth table will match their top-down ordering in the original circuit, with ties being broken in left-right order.

LOGIC CIRCUITS FOR DIGITAL SYSTEMS

DESIGN PROCEDURE

Logic circuits for digital systems may be combinational or sequential. A combinational circuit consists of logic gates whose outputs at any time are determined by combining the values of the applied inputs using logic operations. A combinational circuit performs an operation that can be specified logically by a set of Boolean expression. In addition to using logic gates, sequential circuits employ elements that store bit values. Sequential circuit outputs are a function of inputs and the bit value in storage elements.

These values, in turn, are a function of previously applied inputs and stored values. As a consequence, the outputs of a sequential circuit depend not only on the presently applied values of the inputs, but also on pas inputs, and the behaviour of the circuit must be specified by a sequence in time of inputs and internal stored bit values. A combinational circuit consists of input variables, output variables, logic gates and interconnections. The interconnected logic gates accept signals from the inputs and generate signals at the output. The n input variables come from the environment of the circuit, and the m output variables are available for use by the environment. Each input and output variable exists physically as a binary signal that represents logic 1 or logic 0.

For n input variables, there are 2^n possible binary input combinations. For each binary combination of the input variables, there is one possible binary value on each output. Thus, a combinational circuit can be specified by a truth table that lists the output values for each combination of the input variables. A combinational circuit can also be described by m Boolean function, one for each output variable. Each such function is expressed as function of the n input variables.

Combinational Circuit Design

The design of combinational circuit starts from a specification of the problem and culminates in a logic diagram or set of Boolean equations from which the logic diagram can be obtained. The procedure involves the following steps:

- From the specifications of the circuit, determine the required number of inputs and outputs, and assign a letter symbol to each.
- Derive the truth table that defines the required relation ship between inputs and outputs.
- Obtain the simplified Boolean functions of each outputs as function of the input variables.
- Draw the logic diagram.
- Verify the correctness of the design.

SUBTRACTOR OF BINARY ADDER

HALF ADDER/SUBTRACTOR

This figure shows the configuration for a Half Adder/Subtractor. The first logic gate which is a XOR allows the circuit to do the complement of the binary bit when we want to do a subtraction. In the case that is needed to implement a addition the XOR keep the number the same.

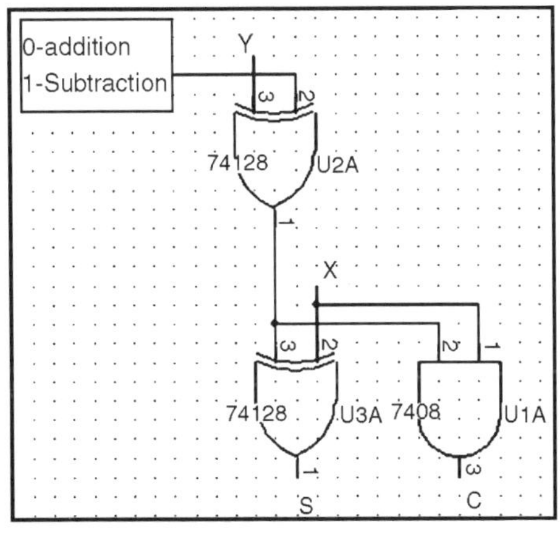

Truth Table

A	B	C	S
0	0	0	0
0	1	0	1
1	0	0	1
1	1	1	0

FULLADDER/SUBTRACTOR

Full Adder Circuit

This circuit adds two binary one bit numbers. Also, it manages a carry that could come from another circuit.

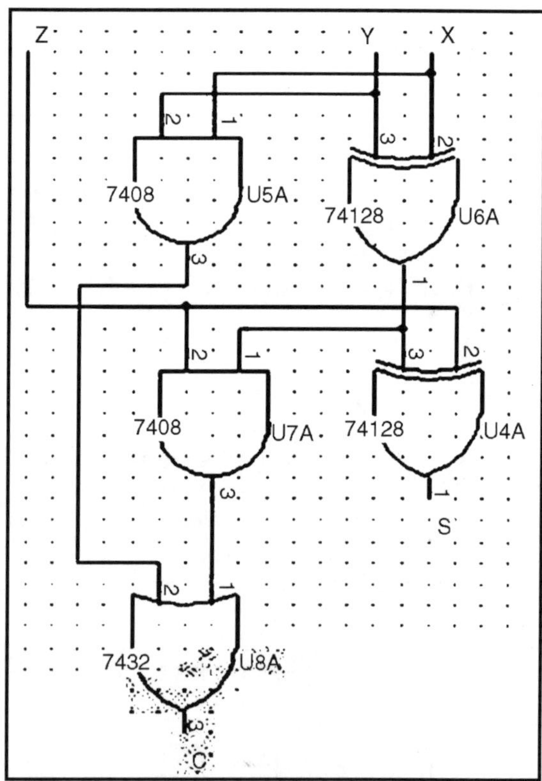

[(AA)'*(BB)']' = A+B

X	Y	Z	C	S
0	0	0	0	0
0	0	1	0	1
0	1	0	0	1

0	1	1	1	0
1	0	0	0	1
1	0	1	1	0
1	1	0	1	0
1	1	1	1	1

This figure shows the configuration for a Full Adder/Subtractor. The box with the name of"Full Adder" has all the logic to do the addition and subtraction depending of the inputs that it receives. The XOR gate that is outside the box allows the circuit to do the complement of the binary bit when we want to do a subtraction. In the case that is needed to implement an addition the XOR keep the number the same. In the previous graph the internal logic for the box"Full Adder" is shown.

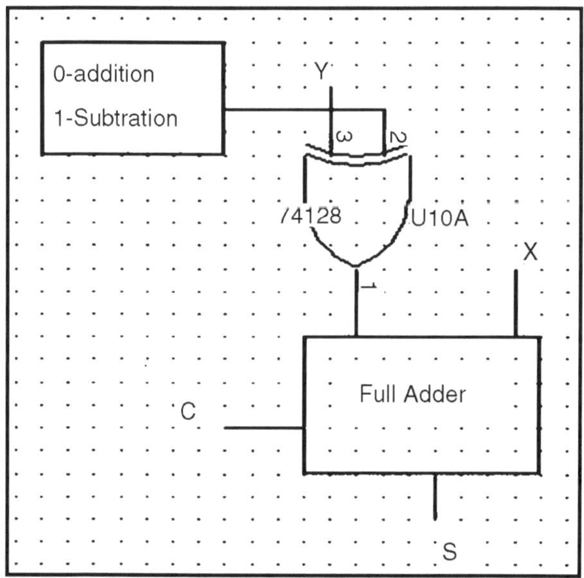

3-BIT ADDER/SUBTRACTOR

The 3-bit adder/Subtractor was implemented with three Full adder circuits and three XOR gates outside which implemented the operation (addition/ subtraction) selected by the user. In this circuit, we have three inputs for the first three bits binary number and three inputs for the second three bits binary number.

When the addition is selected the XOR gates keep the binary numbers the same and add them together. On the other hand, when a subtraction is selected the XOR gates complement the second number which is represented with a"Y" and after that, the Full Adder circuit adds both numbers together.

The 3-bit Adder/Subtractor circuit has four outputs. The first three outputs represent the three bits that were the result of the subtraction or addition. The last bit represents a carry.

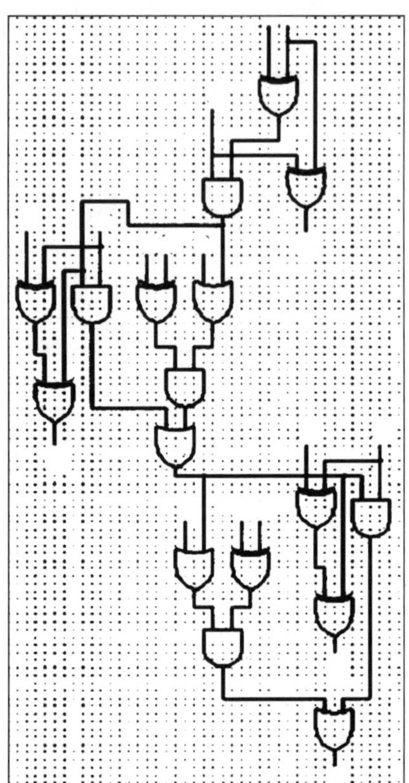

BINARY MULTIPLICATION TECHNIQUE

Binary multiplication uses the same technique as decimal multiplication. In fact, binary multiplication is much easier because each digit we multiply by is either zero or one. Consider the simple problem of multiplying 1102 by 102. We can use this problem to review some terminology and illustrate the rules for binary multiplication.

- First, we note that 1102 is our multiplicand and 102 is our multiplier.

$$\begin{array}{r} 110 \\ \times\ 10 \\ \hline \end{array}$$

- We begin by multiplying 1102 by the rightmost digit of our multiplier which is 0. Any number times zero is zero, so we just write zeros below. 110

$$\begin{array}{r} 110 \\ \times\ 10 \\ \hline 000 \end{array}$$

- Now we multiply the multiplicand by the next digit of our multiplier which is 1. To perform this multiplication, we just need to copy the multiplicand and shift it one column to the left as we do in decimal multiplication.

```
   110
 × 10
 -----
   000
  110
```

- Now we add our results together. The product of our multiplication is 11002. 110

```
   110
 × 10
 -----
   000
  110
  ----
  1100
```

When performing binary multiplication, remember the following rules:

- Copy the multiplicand when the multiplier digit is 1. Otherwise, write a row of zeros.
- Shift your results one column to the left as you move to a new multiplier digit.
- Add the results together using binary addition to find the product.

DECODERS MULTIPLEXER

In both the multiplexer and the demultiplexer, part of the circuits decode the address inputs, i.e. it translates a binary number of n digits to 2n outputs, one of which (the one that corresponds to the value of the binary number) is 1 and the others of which are 0. It is sometimes advantageous to separate this function from the rest of the circuit, since it is useful in many other applications. Thus, we obtain a new combinatorial circuit that we call the decoder. It has the following truth table (for n = 3):

```
a2 a1 a0 | d7 d6 d5 d4 d3 d2 d1 d0
----------------------------------
 0 0 0 | 0 0 0 0 0 0 0 1
 0 0 1 | 0 0 0 0 0 0 1 0
 0 1 0 | 0 0 0 0 0 1 0 0
 0 1 1 | 0 0 0 0 1 0 0 0
 1 0 0 | 0 0 0 1 0 0 0 0
 1 0 1 | 0 0 1 0 0 0 0 0
 1 1 0 | 0 1 0 0 0 0 0 0
 1 1 1 | 1 0 0 0 0 0 0 0
```

Here is the circuit diagram for the decoder:

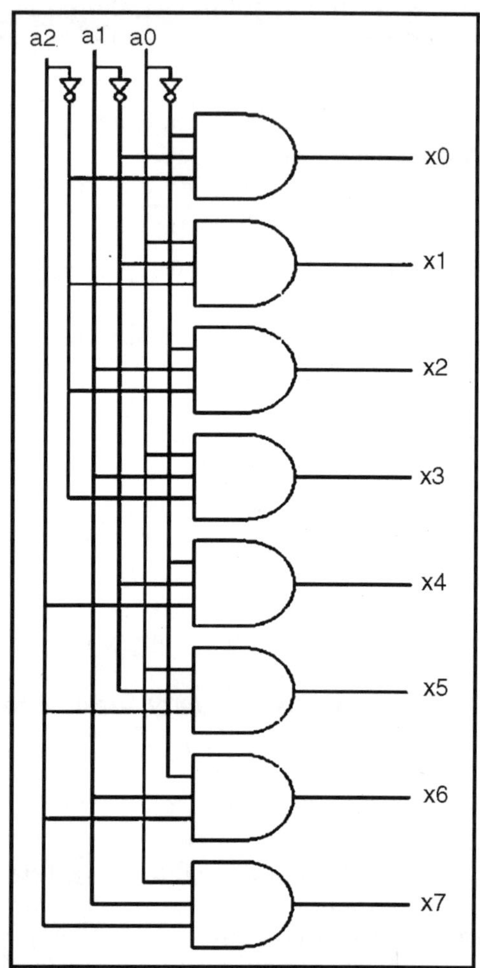

DECODER GENERATION

X86's decoder is generated using an ISA description like all the other ISAs, although how it does that is a bit different. Most of the instructions for most of the other ISAs are defined by passing chunks of code that perform the instruction into an instruction format.

The format is basically a template which puts wraps that bit of code in the structure needed to support it and you have your instruction object. Because almost all of the instructions in x86 are microcoded and many can be encoded in multiple ways and hence appear in the decoder more than once, and because the same non-trivial decoding rules apply to many different instructions, X86 uses the decoder as a layer of indirection and defines the majority of its instructions elsewhere.

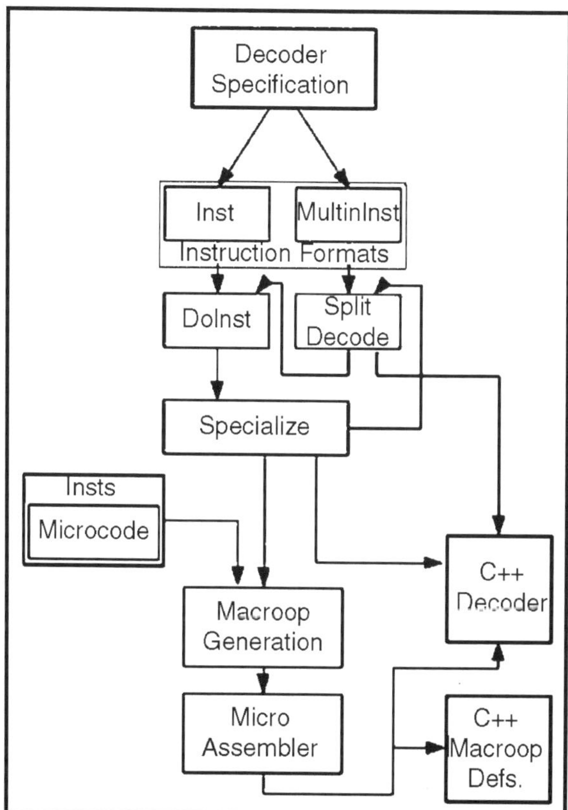

X86 almost exclusively uses only two different instruction formats, Inst and MultiInst. MultiInst is just a compact way of describing multiple related Insts. An Inst essentially selects an instruction like XOR and provides a specification for its operands.

Inside the instruction format, the instruction name and operand specification are passed to a python function called"specializeInst" which figures out what to do with it. If the operand specification describes more than one version of the instruction, for instance one that uses memory and one that uses registers, the instruction's information is passed into another funciton,"doSplitDecode", which separates out those versions and passes each individually back through the same system.

This goes on until the instructions have been fully split out and code has been generated to figure out what version to use. As a nice bonus, the MultiInst format doesn't add much complexity to this model since it can simply jump right into doSplitDecode and continue as normal.

There is one additional format for string instructions that works similar to MultiInst, except instead of specializing the instruction based on its operands, it specializes it based on its prefixes.

There are also a few instructions that don't use this system, but because there are so few and because they work like the instructions in other ISAs I won't describe them here.

At this point, the code for selecting the right version of an instruction is put into the C++ decoder function.

Almost all of this function is built this way, with the minor exception of small bits of logic that glue everything together and make large scale distinctions the number of opcode bytes.

3 TO 8 DECODER

The 3 to 8 decoder unit takes 3 address lines as input and outputs 8 address enable lines. Which of the 8 output is enabled is dependant upon the configuration of the 3 address line.

If A0, A1, and A2 are all 0, then address 0 is enabled. If A0=0, A1=1, A2=0, then address 3 is enables. With 3 address lines, the number of words that can be addressed is 8 (2^3=8).

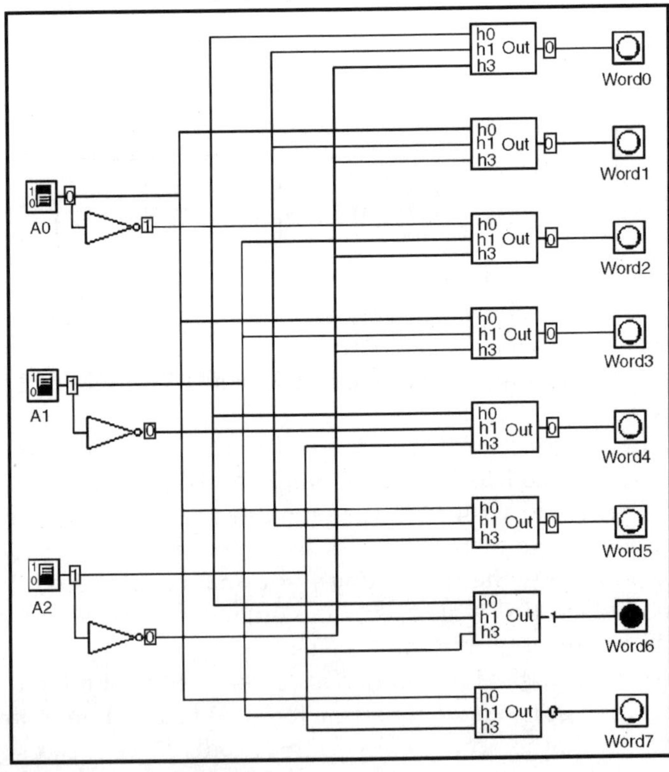

ENCODERS

An encoder is a circuit that changes a set of signals into a code. Lets begin making a 2-to-1 line encoder truth table by reversing the 1-to-2 decoder truth table.

D_1	D_0	A
0	1	0
1	0	1

This truth table is a little short. A complete truth table would be

D_1	D_0	A
0	0	
0	1	0
1	0	1
1	1	

One question we need to answer is what to do with those other inputs? Do we ignore them? Do we have them generate an additional error output? In many circuits this problem is solved by adding sequential logic in order to know not just what input is active but also which order the inputs became active. A more useful application of combinational encoder design is a binary to 7-segment encoder. The seven segments are given according

D_0
D_1 D_2
D_3
D_4 D_5
D_6

Our truth table is:

I_3	I_2	I_1	I_0	D_6	D_5	D_4	D_3	D_2	D_1	D_0
0	0	0	0	1	1	1	0	1	1	1
0	0	0	1	0	0	1	0	0	1	0
0	0	1	0	1	0	1	1	1	0	1
0	0	1	1	1	0	1	1	0	1	1
0	1	0	0	0	1	1	1	0	1	0
0	1	0	1	1	1	0	1	0	1	1
0	1	1	0	1	1	0	1	1	1	1
0	1	1	1	1	0	1	0	0	1	0
1	0	0	0	1	1	1	1	1	1	1
1	0	0	1	1	1	1	1	0	1	1

Deciding what to do with the remaining six entries of the truth table is easier with this circuit. This circuit should not be expected to encode an undefined combination of inputs, so we can leave them as"don't care" when we design the circuit. The boolean equations are

$$D_0 = I_3 + I_1 + \bar{I}_3\bar{I}_2\bar{I}_1\bar{I}_0 + \bar{I}_3 I_2\bar{I}_1 I_0$$
$$D_1 = I_3 + \bar{I}_2\bar{I}_1 + I_2\bar{I}_1 + I_2\bar{I}_0$$
$$D_2 = I_2 + \bar{I}_3\bar{I}_2\bar{I}_1\bar{I}_0 + \bar{I}_3 I_2 I_1 I_0$$
$$D_3 = I_3 + I_1\bar{I}_0 + I_2\bar{I}_1$$
$$D_4 = I_1\bar{I}_0 + \bar{I}_2\bar{I}_1\bar{I}_0$$
$$D_5 = I_3 + I_2 + I_0$$
$$D_6 = I_3 + I_1\bar{I}_0 + \bar{I}_3\bar{I}_2 I_1 + \bar{I}_3\bar{I}_2\bar{I}_1\bar{I}_0 + \bar{I}_3 I_2\bar{I}_1 I_0$$

and the circuit is

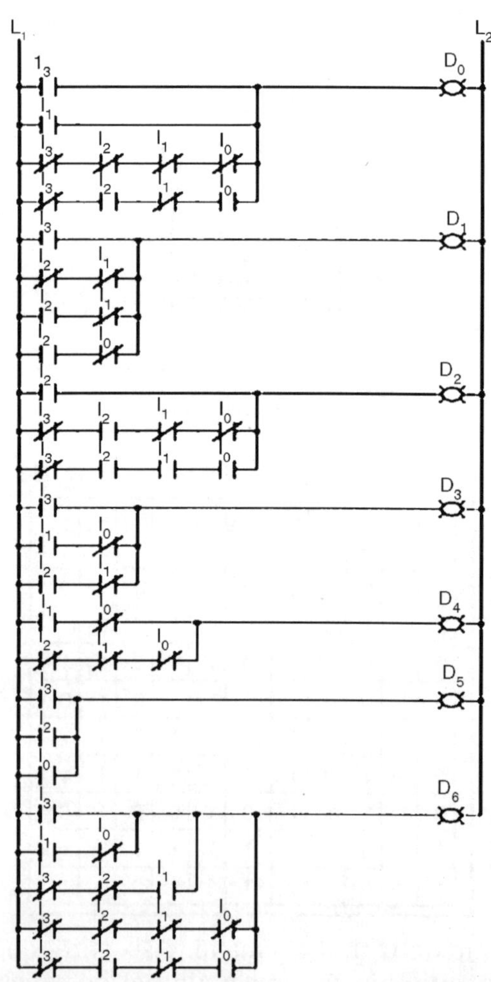

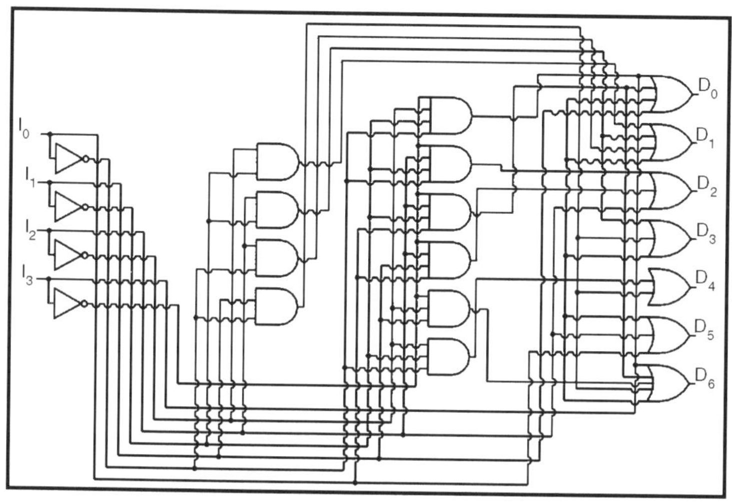

MAGNITUDE COMPARATOR

74L85 4-BIT MAGNITUDE COMPARATOR

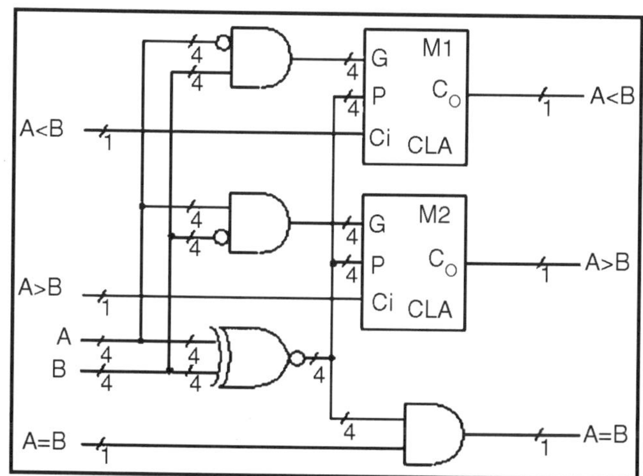

STATISTICS

11 inputs; 3 outputs; 33 gates;

FUNCTION

The 74L85 magnitude comparator can be functionally modeled as above. This is a simplification of implementing a magnitude comparator by a carry function with an inverted input bus as shown. Using this concept, common elements of the three comparator functions A < B, A > B, and A = B are combined to construct the model shown above, which maps directly onto the gate-level realization of the 74L85.

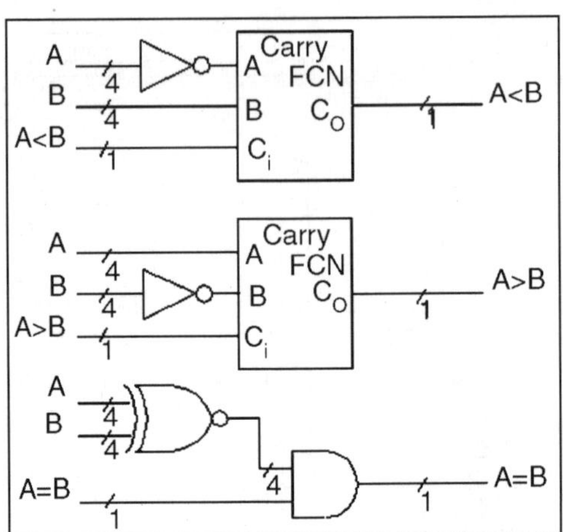

MULTIPLEXES COMBINATIONAL CIRCUIT

A multiplexer is a combinatorial circuit that is given a certain number (usually a power of two) data inputs, let us say 2n, and n address inputs used as a binary number to select one of the data inputs.

The multiplexer has a single output, which has the same value as the selected data input. In other words, the multiplexer works like the input selector of a home music system. Only one input is selected at a time, and the selected input is transmitted to the single output. While on the music system, the selection of the input is made manually, the multiplexer chooses its input based on a binary number, the address input. The truth table for a multiplexer is huge for all but the smallest values of n. We therefore use an abbreviated version of the truth table in which some inputs are replaced by `-' to indicate that the input value does not matter. Here is such an abbreviated truth table for n = 3. The full truth table would have 2(3 + 23) = 2048rows.

a_2	a_1	a_0	d_7	d_6	d_5	d_4	d_3	d_2	d_1	d_0	I	x
-	-	-	-	-	-	-	-	-	-	-	---	-
0	0	0	-	-	-	-	-	-	-	0	I	0
0	0	0	-	-	-	-	-	-	-	1	I	1
0	0	1	-	-	-	-	-	-	0	-	I	0
0	0	1	-	-	-	-	-	-	1	-	I	1
0	1	0	-	-	-	-	-	0	-	-	I	0
0	1	0	-	-	-	-	-	1	-	-	I	1
0	1	1	-	-	-	-	0	-	-	-	I	0
0	1	1	-	-	-	-	1	-	-	-	I	1
1	0	0	-	-	-	0	-	-	-	-	I	0
1	0	0	-	-	-	1	-	-	-	-	I	1
1	0	1	-	-	0	-	-	-	-	-	I	0
1	0	1	-	-	1	-	-	-	-	-	I	1

1	1	0	-	0	-	-	-	-	-	-	|	0
1	1	0	-	1	-	-	-	-	-	-	|	1
1	1	1	0	-	-	-	-	-	-	-	|	0
1	1	1	1	-	-	-	-	-	-	-	|	1

We can abbreviate this table even more by using a letter to indicate the value of the selected input, like this:

a_2	a_1	a_0	d_7	d_6	d_5	d_4	d_3	d_2	d_1	d_0	|	x
-	-	-	-	-	-	-	-	-	-	-	---	-
0	0	0	-	-	-	-	-	-	-	c	|	c
0	0	1	-	-	-	-	-	-	c	-	|	c
0	1	0	-	-	-	-	-	c	-	-	|	c
0	1	1	-	-	-	-	c	-	-	-	|	c
1	0	0	-	-	-	c	-	-	-	-	|	c
1	0	1	-	-	c	-	-	-	-	-	|	c
1	1	0	-	c	-	-	-	-	-	-	|	c
1	1	1	c	-	-	-	-	-	-	-	|	c

The same way we can simplify the truth table for the multiplexer, we can also simplify the corresponding circuit. Indeed, our simple design method would yield a very large circuit. The simplified circuit looks like this:

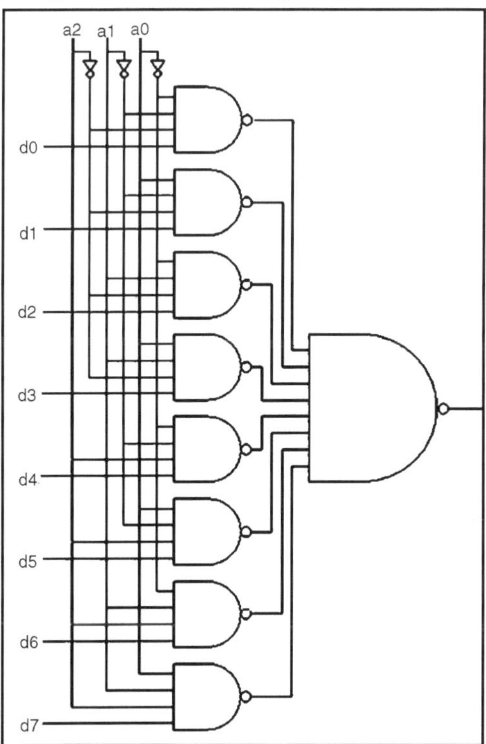

A multiplexer performs the function of selecting the input on any one of'n' input lines and feeding this input to one output line. Multiplexers are used as one method of reducing the number of integrated circuit packages required by a particular circuit design. This in turn reduces the cost of the system. Assume that we have four lines, C0, C1, C2 and C3, which are to be multiplexed on a single line, Output (f). The four input lines are also known as the Data Inputs. Since there are four inputs, we will need two additional inputs to the multiplexer, known as the Select Inputs, to select which of the C inputs is to appear at the output. Call these select lines A and B. The gate implementation of a 4-line to 1-line multiplexer is shown below:

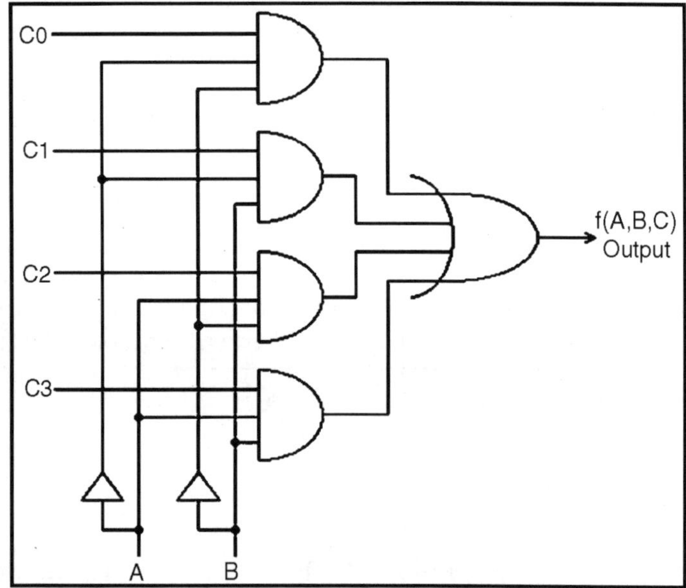

The circuit symbol for the above multiplexer is:

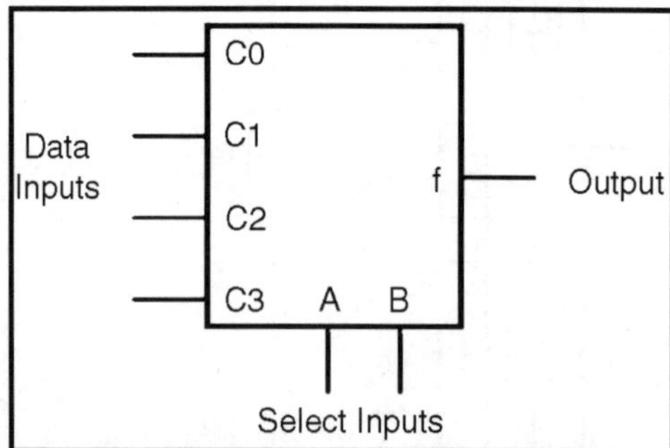

4 INPUT MULTIPLEXER

The multiplexer concept is not limited to two data inputs. If we add a second addressing input, B, we can control as many as four data inputs, as shown to the left. A third and fourth addressing input will allow the multiplexer to control eight or sixteen inputs, respectively.

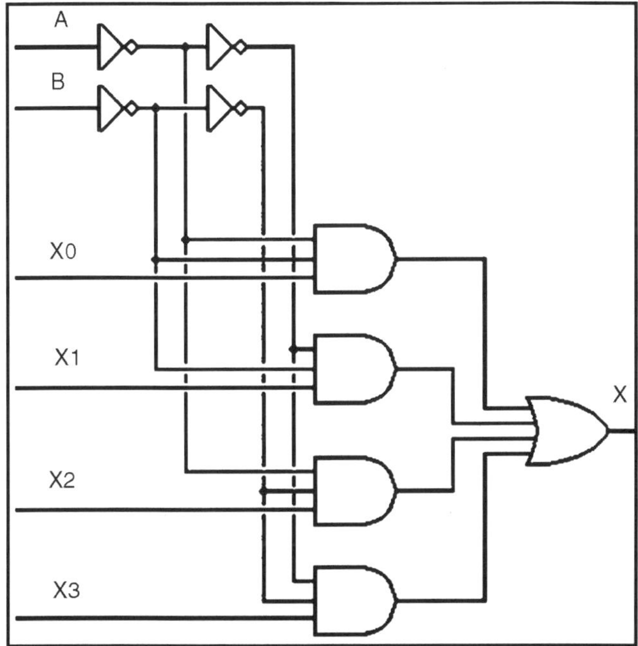

Inputs A and B are the addressing inputs to this multiplexer. They select which of the four data inputs will be transmitted to the final output, X. If the data inputs are to be multiplexed for transmission to a distant location, the inputs must cycle through all four possible addresses more than twice for each single cycle of each of the data inputs. Otherwise the input data cannot be reconstructed accurately at the receiving end.

COMBINATIONAL LOGIC CIRCUITS

Unlike Sequential Logic Circuits whose outputs are dependant on both their present inputs and their previous output state giving them some form of Memory, the outputs of Combinational Logic Circuits are only determined by the logical function of their current input state, logic "0" or logic "1", at any given instant in time as they have no feedback, and any changes to the signals being applied to their inputs will immediately have an effect at the output. In other words, in a Combinational Logic Circuit, the output is dependant at all times on the combination of its inputs and if one of its inputs condition changes state so does the output as combinational circuits have "no memory", "timing" or "feedback loops".

COMBINATIONAL LOGIC

Combinational logic is used in computer circuits to do boolean algebra on input signals and on stored data. Practical computer circuits normally contain a mixture of combinational and sequential logic. For example, the part of an arithmetic logic unit, or ALU, that does mathematical calculations is constructed using combinational logic. Other circuits used in computers, such as half adders, full adders, half subtractors, full subtractors, multiplexers, demultiplexers, encoders and decoders are also made by using combinational logic.

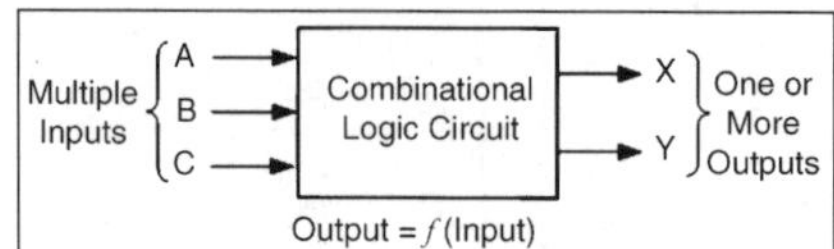

Combinational Logic Circuits are made up from basic logic NAND, NOR or NOT gates that are "combined" or connected together to produce more complicated switching circuits. These logic gates are the building blocks of combinational logic circuits. An example of a combinational circuit is a decoder, which converts the binary code data present at its input into a number of different output lines, one at a time producing an equivalent decimal code at its output. Combinational logic circuits can be very simple or very complicated and any combinational circuit can be implemented with only NAND and NOR gates as these are classed as "universal" gates.

The three main ways of specifying the function of a combinational logic circuit are:

- *Boolean Algebra*: This forms the algebraic expression showing the operation of the logic circuit for each input variable either True or False that results in a logic "1" output.
- *Truth Table:* A truth table defines the function of a logic gate by providing a concise list that shows all the output states in tabular form for each possible combination of input variable that the gate could encounter.
- *Logic Diagram:* This is a graphical representation of a logic circuit that shows the wiring and connections of each individual logic gate, represented by a specific graphical symbol, that implements the logic circuit.

and all three of these logic circuit representations are shown below.

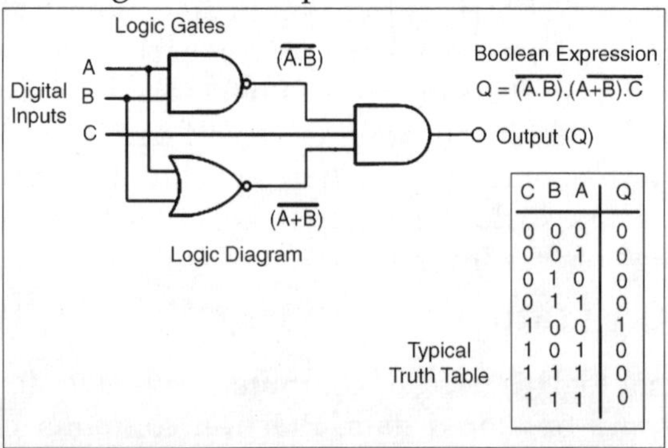

C	B	A	Q
0	0	0	0
0	0	1	0
0	1	0	0
0	1	1	0
1	0	0	1
1	0	1	0
1	1	1	0
1	1	1	0

As combinational logic circuits are made up from individual logic gates only, they can also be considered as "decision making circuits" and combinational logic is about combining logic gates together to process two or more signals in order to produce at least one output signal according to the logical function of each logic gate. Common combinational circuits made up from individual logic gates that carry out a desired application include Multiplexers, De-multiplexers, Encoders, Decoders, Fulland Half Adders, etc.

Classification of Combinational Logic

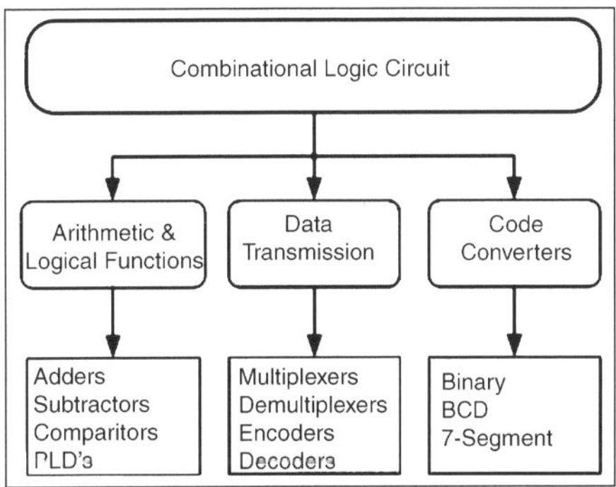

One of the most common uses of combinational logic is in Multiplexer and De-multiplexer type circuits. Here, multiple inputs or outputs are connected to a common signal line and logic gates are used to decode an address to select a single data input or output switch. A multiplexer consist of two separate components, a logic decoder and some solid state switches, but before we can discuss multiplexers, decoders and de-multiplexers in more detail we first need to understand how these devices use these "solid state switches" in their design.

FUNCTIONS OF COMBINATIONAL LOGIC CIRCUIT

The term "combinational" comes to us from mathematics. In mathematics a combination is an unordered set, which is a formal way to say that nobody cares which order the items came in. Most games work this way, if you rolled dice one at a time and get a 2 followed by a 3 it is the same as if you had rolled a 3 followed by a 2. With combinational logic, the circuit produces the same output regardless of the order the inputs are changed. There are circuits which depend on the when the inputs change, these circuits are called sequential logic. Even though you will not find the term "sequential logic". Practical circuits will have a mix of combinational and sequential logic, with sequential logic making sure everything happens in order and combinational logic performing functions like arithmetic, logic, or conversion.

You have already used combinational circuits. Each logic gate discussed previously is a combinational logic function. Let's follow how two NAND gate works if we provide them inputs in different orders.

We begin with both inputs being 0.

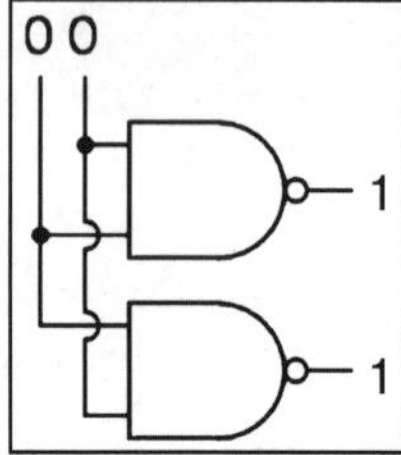

We then set one input high.

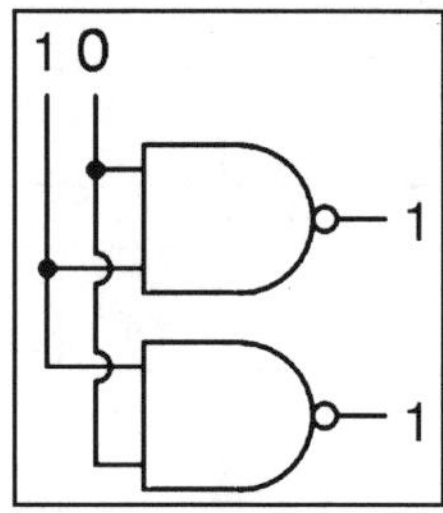

We then set the other input high.

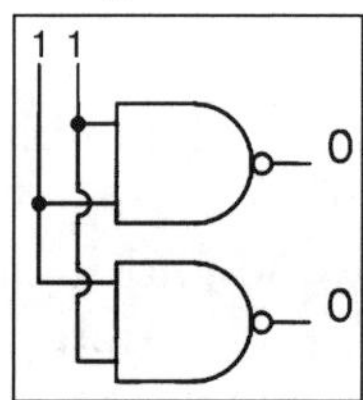

So NAND gates do not care about the order of the inputs, and you will find the same true of all the other gates covered up to this point (*AND, XOR, OR, NOR, XNOR,* and *NOT*).

A HALF-ADDER

As a first example of useful combinational logic, let's build a device that can add two binary digits together. We can quickly calculate what the answers should be:

$0 + 0 = 0$ $0 + 1 = 1$ $1 + 0 = 1$ $1 + 1 = 10_2$

So we well need two inputs (a and b) and two outputs. The low order output will be called Σ because it represents the sum, and the high order output will be called C_{out} because it represents the carry out.

Table. 6.2 The Truth Table is

A	B	Σ	C_{OUT}
0	0	0	0
0	1	1	0
1	0	1	0
1	1	0	1

Simplifying boolean equations or making some Karnaugh map will produce the same circuit shown below, but start by looking at the results. The Σ column is our familiar *XOR* gate, while the C_{out} column is the AND gate. This device is called a half-adder for reasons that will make sense in the next section.

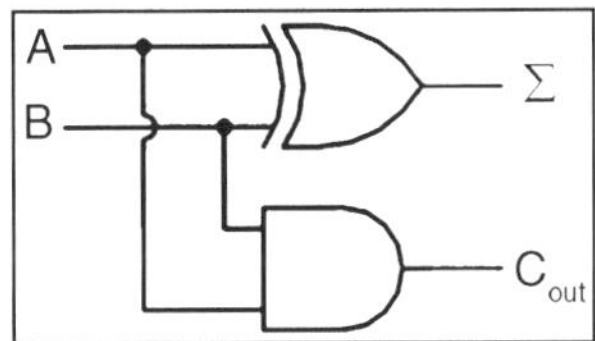

or in ladder logic

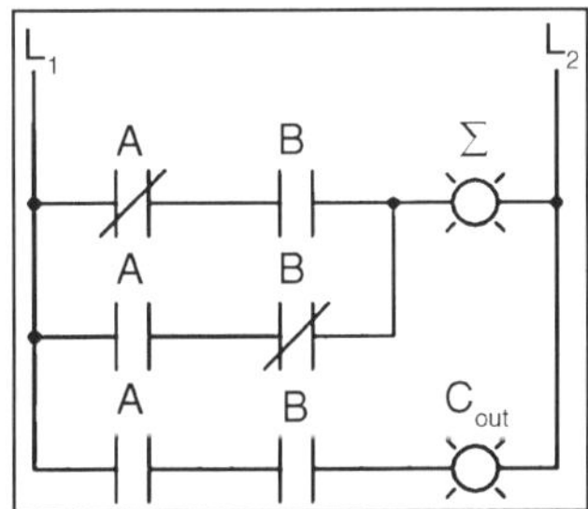

A FULL-ADDER

The half-adder is extremely useful until you want to add more that one binary digit quantities. The slow way to develop a two binary digit adders would be to make a truth table and reduce it. Then when you decide to make a three binary digit adder, do it again. Then when you decide to make a four digit adder, do it again. Then when... The circuits would be fast, but development time would be slow. Looking at a two binary digit sum shows what we need to extend addition to multiple binary digits.

11
11
11
—
110

Look at how many inputs the middle column uses. Our adder needs three inputs; *a*, *b*, and the carry from the previous sum, and we can use our two-input adder to build a three input adder.

Σ is the easy part. Normal arithmetic tells us that if $\Sigma = a + b + C_{in}$ and $\Sigma_1 = a + b$, then $\Sigma = \Sigma_1 + C_{in}$.

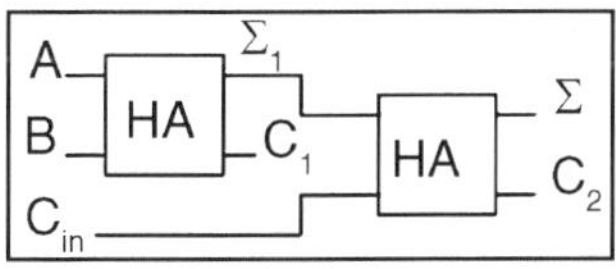

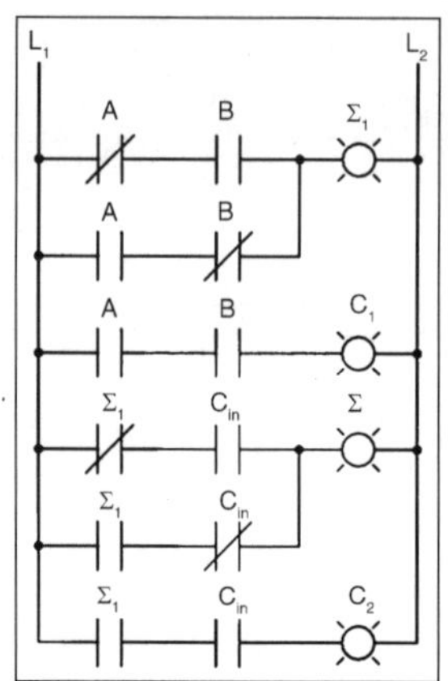

What do we do with C_1 and C_2? Let's look at three input sums and quickly calculate:

C_{in} + a + b = ?

0 + 0 + 0 = 0   0 + 0 + 1 = 1   0 + 1 + 0 = 1   0 + 1+ 1 = 10

1 + 0 + 0 = 1   1 + 0 + 1 = 10   1 + 1 + 0 = 10   1 + 1+ 1 = 11

If you have any concern about the low order bit, please confirm that the circuit and ladder calculate it correctly.

In order to calculate the high order bit, notice that it is 1 in both cases when $a + b$ produces a C_1. Also, the high order bit is 1 when a + b produces a Σ_1 and C_{in} is a 1. So We will have a carry when C_1 OR (Σ_1 AND C_{in}). Our complete three input adder is:

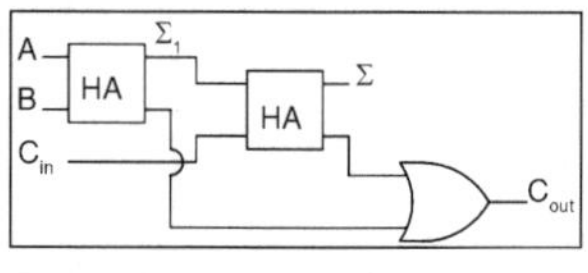

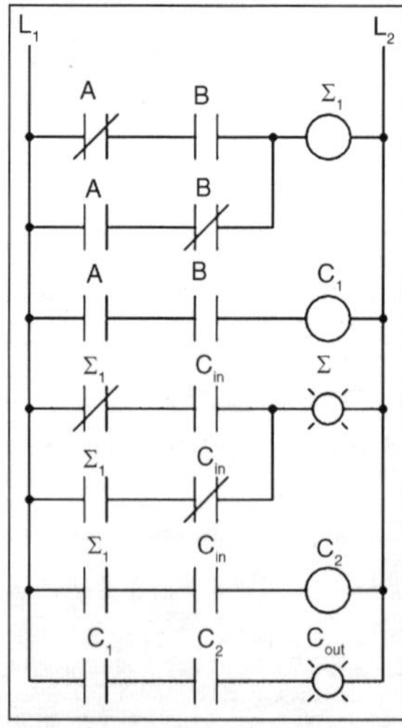

For some designs, being able to eliminate one or more types of gates can be important, and you can replace the final *OR* gate with an *XOR* gate without changing the results. We can now connect two adders to add 2 bit quantities.

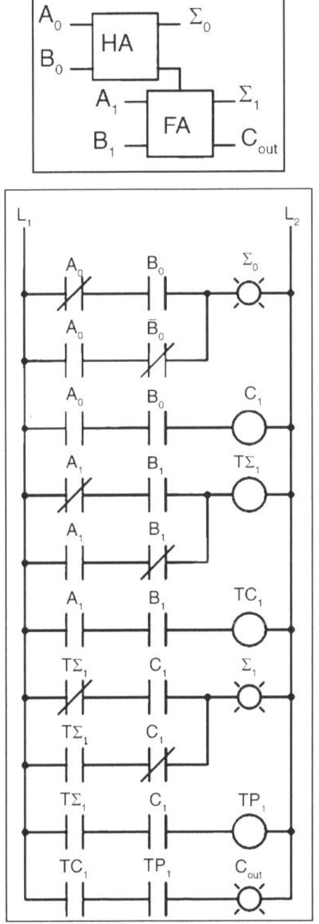

A_0 is the low order bit of *A*, A_1 is the high order bit of *A*, B_0 is the low order bit of *B*, B_1 is the high order bit of *B*, $\Sigma_0$is the low order bit of the sum, Σ_1 is the high order bit of the sum, and C_{out} is the Carry. A two binary digit adder would never be made this way. Instead the lowest order bits would also go through a full adder.

Cin
A0 FA Σ0
B0
A1 FA Σ1
B1 Cout

There are several reasons for this, one being that we can then allow a circuit to determine whether the lowest order carry should be included in the sum. This allows for the chaining of even larger sums. Consider two different ways to look at a four bit sum.

111 1< –+ 11<+ –
0110 | 01 | 10
1011 | 10 | 11
— — – | — | —
10001 1 +–100 +-101

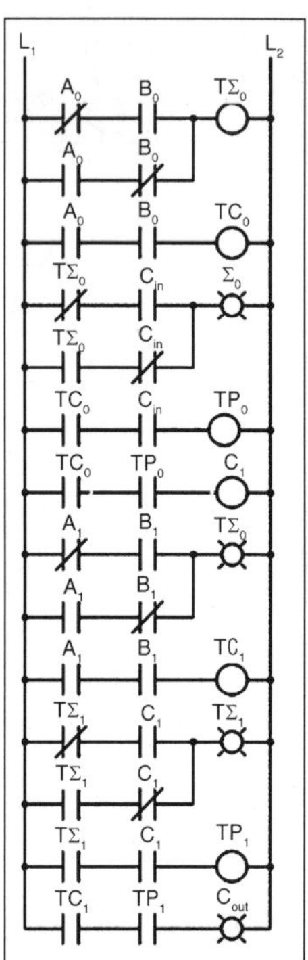

If we allow the programme to add a two bit number and remember the. carry for later, then use that carry in the next sum the programme can add any number of bits the user wants even though we have only provided a two–bit adder. Small *PLCs* can also be chained together for larger numbers.

These full adders can also can be expanded to any number of bits space allows. As an example, here's how to do an 8 bit adder.

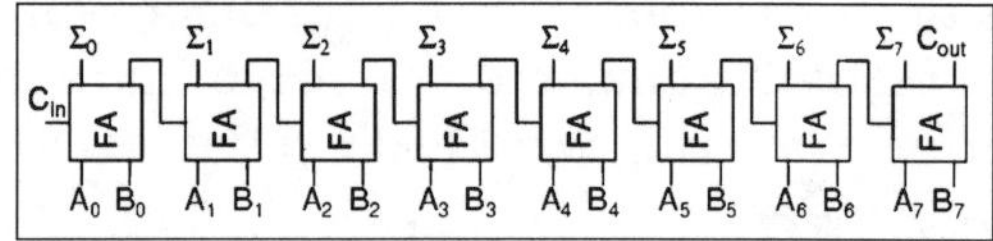

This is the same result as using the two 2-bit adders to make a 4-bit adder and then using two 4-bit adders to make an 8-bit adder or re-duplicating ladder logic and updating the numbers.

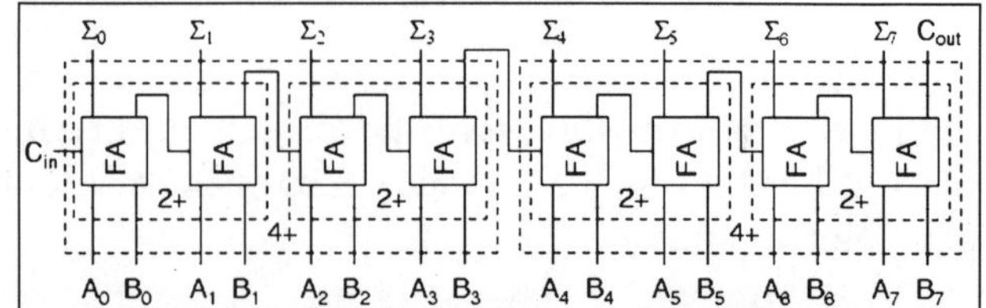

Each "2+" is a 2-bit adder and made of two full adders. Each "4+" is a 4-bit adder and made of two 2-bit adders. And the result of two 4-bit adders is the same 8-bit adder we used full adders to build.

For any large combinational circuit there are generally two approaches to design: you can take simpler circuits and replicate them; or you can design the complex circuit as a complete device.

Using simpler circuits to build complex circuits allows a you to spend less time designing but then requires more time for signals to propagate through the transistors. The 8-bit adder design above has to wait for all the C_{xout}signals to move from $A_0 + B_0$ up to the inputs of Σ_7.

If a designer builds an 8-bit adder as a complete device simplified to a sum of products, then each signal just travels through one *NOT* gate, one *AND* gate and one *OR* gate. A seventeen input device has a truth table with 131,072 entries, and reducing 131,072 entries to a sum of products will take some time. When designing for systems that have a maximum allowed response time to provide the final result, you can begin by using simpler circuits and then attempt to replace portions of the circuit that are too slow. That way you spend most of your time on the portions of a circuit that matter.

3

Digital Engineering and Circuits Analysis

INTRODUCTION

Computers understand only two numbers, 0 and 1, and do all their arithmetic operations in this binary mode. Many electrical and electronic devices have two states: they are either off or on. A light switch is a familiar example, as are vacuum tubes and transistors. Because computers have been a major application for integrated circuits from their beginning, digital integrated circuits have become commonplace. It has thus become easy to design electronic systems that use digital language to control their functions and to communicate with other systems.

A major advantage in using digital methods is that the accuracy of a stream of digital signals can be verified, and, if necessary, errors can be corrected. In contrast, signals that vary in proportion to, say, the sound of an orchestra can be corrupted by "noise," which once present cannot be removed. An example is the sound from a phonograph record, which always contains some extraneous sound from the surface of the recording groove even when the record is new. The noise becomes more pronounced with wear.

Contrast this with the sound from a digital compact disc recording. No sound is heard that was not present in the recording studio. The disc and the player contain error-correcting features that remove any incorrect pulses (perhaps arising from dust on the disc) from the information as it is read from the disc. As electronic systems become more complex, it is essential that errors produced by noise be removed; otherwise, the systems may malfunction. Many electronic systems are required to operate in electrically noisy environments, such as in an automobile. The only practical way to assure immunity from noise is to make such a system operate digitally. In principle it is possible to correct for any arbitrary number of errors, but in practice this may not be possible. The amount of extra information that must be handled to correct for large rates of error reduces the capacity of the system to handle the desired information, and so trade-offs are necessary.

A consequence of the veritable explosion in the number and kinds of electronic systems has been a sharp growth in the electrical noise level of the environment. Any electrical system generates some noise, and all electronic systems are to some degree susceptible to disturbance from noise. The noise may be conducted along wires connected to the system, or it may be radiated through the air. Care is necessary in the design of systems to limit the amount of noise that is generated and to shield the system properly to protect it from external noise sources.

DIGITAL DESIGN IN DIGITAL CIRCUITS ANALYSIS

Digital circuits are made from analog components. The design must assure that the analog nature of the components doesn't dominate the desired digital behaviour. Digital systems must manage noise and timing margins, parasitic inductances and capacitances, and filter power connections.

Bad designs have intermittent problems such as "glitches", vanishingly-fast pulses that may trigger some logic but not others, "runt pulses" that do not reach valid "threshold" voltages, or unexpected ("undecoded") combinations of logic states. Additionally, where clocked digital systems interface to analog systems or systems that are driven from a different clock, the digital system can be subject to metastability where a change to the input violates the set-up time for a digital input latch. This situation will self-resolve, but will take a random time, and while it persists can result in invalid signals being propagated within the digital system for a short time.

Since digital circuits are made from analog components, digital circuits calculate more slowly than low-precision analog circuits that use a similar amount of space and power. However, the digital circuit will calculate more repeatably, because of its high noise immunity. On the other hand, in the high-precision domain (for example, where 14 or more bits of precision are needed), analog circuits require much more power and area than digital equivalents.

CONSTRUCTION

A digital circuit is often constructed from small electronic circuits called logic gates that can be used to create combinational logic. Each logic gate represents a function of boolean logic. A logic gate is an arrangement of electrically controlled switches, better known as transistors.

Each logic symbol is represented by a different shape. The actual set of shapes was introduced in 1984 under IEEE\ANSI standard 91-1984. "The logic symbol given under this standard are being increasingly used now and have even started appearing in the literature published by manufacturers of digital integrated circuits." The output of a logic gate is an electrical flow or voltage, that can, in turn, control more logic gates.

Logic gates often use the fewest number of transistors in order to reduce their size, power consumption and cost, and increase their reliability. Integrated circuits are the least expensive way to make logic gates in large volumes. Integrated circuits are usually designed by engineers using electronic design automation software.

Another form of digital circuit is constructed from lookup tables, (many sold as "programmable logic devices", though other kinds of PLDs exist). Lookup tables can perform the same functions as machines based on logic gates, but can be easily reprogrammed without changing the wiring. This means that a designer can often repair design errors without changing the arrangement of wires. Therefore, in small volume products, programmable logic devices are often the preferred solution. They are usually designed by engineers using electronic design automation software.

When the volumes are medium to large, and the logic can be slow, or involves complex algorithms or sequences, often a small microcontroller is programmed to make an embedded system. These are usually programmed by software engineers. When only one digital circuit is needed, and its design is totally customised, as for a factory production line controller, the conventional solution is a programmable logic controller, or PLC. These are usually programmed by electricians, using ladder logic.

STRUCTURE OF DIGITAL SYSTEMS

Engineers use many methods to minimize logic functions, in order to reduce the circuit's complexity. When the complexity is less, the circuit also has fewer errors and less electronics, and is therefore less expensive.

The most widely used simplification is a minimization algorithm like the Espresso heuristic logic minimizer within a CAD system, although historically, binary decision diagrams, an automatedQuine–McCluskey algorithm, truth tables, Karnaugh maps, and Boolean algebra have been used. Representations are crucial to an engineer's design of digital circuits. Some analysis methods only work with particular representations.

The classical way to represent a digital circuit is with an equivalent set of logic gates. Another way, often with the least electronics, is to construct an equivalent system of electronic switches (usually transistors). One of the easiest ways is to simply have a memory containing a truth table. The inputs are fed into the address of the memory, and the data outputs of the memory become the outputs. For automated analysis, these representations have digital file formats that can be processed by computer programmes. Most digital engineers are very careful to select computer programmes ("tools") with compatible file formats.

To choose representations, engineers consider types of digital systems. Most digital systems divide into "combinational systems" and "sequential systems." A combinational system always presents the same output when given the same inputs. It is basically a representation of a set of logic functions. A sequential system is a combinational system with some of the outputs fed back as inputs. This makes the digital machine perform a "sequence" of operations. The simplest sequential system is probably a flip flop, a mechanism that represents a binary digit or "bit". Sequential systems are often designed as state machines. In this way, engineers can design a system's gross behaviour, and even test it in a simulation, without considering all the details of the logic functions.

Sequential systems divide into two further subcategories. "Synchronous" sequential systems change state all at once, when a "clock" signal changes state. "Asynchronous" sequential systemspropagate changes whenever inputs change. Synchronous sequential systems are made of well-characterised asynchronous circuits such as flip-flops, that change only when the clock changes, and which have carefully designed timing margins.

The usual way to implement a synchronous sequential state machine is to divide it into a piece of combinational logic and a set of flip flops called a "state register." Each time a clock signal ticks, the state register captures the feedback generated from the previous state of the combinational logic, and feeds it back as an unchanging input to the combinational part of the state machine. The fastest rate of the clock is set by the most time-consuming logic calculation in the combinational logic.

The state register is just a representation of a binary number. If the states in the state machine are numbered (easy to arrange), the logic function is some combinational logic that produces the number of the next state.

In comparison, asynchronous systems are very hard to design because all possible states, in all possible timings must be considered. The usual method is to construct a table of the minimum and maximum time that each such state can exist, and then adjust the circuit to minimize the number of such states, and force the circuit to periodically wait for all of its parts to enter a compatible state (this is called "self-resynchronisation"). Without such careful design, it is easy to accidentally produce asynchronous logic that is "unstable", that is, real electronics will have unpredictable results because of the cumulative delays caused by small variations in the values of the electronic components. Certain circuits (such as the synchroniser flip-flops, switchdebouncers, arbiters, and the like which allow external unsynchronised signals to enter synchronous logic circuits) are inherently asynchronous in their design and must be analysed as such.

As of 2005, almost all digital machines are synchronous designs because it is much easier to create and verify a synchronous design—the software currently used to simulate digital machines does not yet handle asynchronous designs. However, asynchronous logic is thought to be superior, if it can be made to work, because its speed is not constrained by an arbitrary clock; instead, it runs at the maximum speed of its logic gates. Building an asynchronous circuit using faster parts makes the circuit faster.

Many digital systems are data flow machines. These are usually designed using synchronous register transfer logic, using hardware description languages such as VHDL or Verilog. In register transfer logic, binary numbers are stored in groups of flip flops called registers. The outputs of each register are a bundle of wires called a "bus" that carries that number to other calculations. A calculation is simply a piece of combinational logic. Each calculation also has an output bus, and these may be connected to the inputs of several registers. Sometimes a register will have a multiplexer on its input, so that it can store a number from any one of several buses. Alternatively, the outputs of several items may be connected to a bus through buffers that can turn off the output of all of the devices except one. A sequential state machine controls when each register accepts new data from its input.

In the 1980s, some researchers discovered that almost all synchronous register-transfer machines could be converted to asynchronous designs by using first-in-first-out synchronisation logic. In this scheme, the digital machine is characterised as a set of data flows. In each step of the flow, an asynchronous "synchronisation circuit" determines when the outputs of that step are valid, and presents a signal that says, "grab the data" to the stages that use that stage's inputs. It turns out that just a few relatively simple synchronisation circuits are needed.

The most general-purpose register-transfer logic machine is a computer. This is basically an automatic binary abacus. The control unit of a computer is usually designed as a microprogram run by a microsequencer. A microprogram is much like a player-piano roll. Each table entry or "word" of the microprogram commands the state of every bit that controls the computer. The sequencer then counts, and the count addresses the memory or combinational logic machine that contains the microprogram. The bits from the microprogram control the arithmetic logic unit, memory and other parts of the computer, including the microsequencer itself.

In this way, the complex task of designing the controls of a computer is reduced to a simpler task of programming a collection of much simpler logic machines. Computer architecture is a specialised engineering activity that tries to arrange the registers, calculation logic, buses and other parts of the computer in the best way for some purpose. Computer architects have applied large amounts of ingenuity to computer design to reduce the cost and increase the speed and immunity to programming errors of computers. An increasingly common goal is to reduce the power used in a battery-powered computer system, such as a cell-phone. Many computer architects serve an extended apprenticeship as microprogrammers. "Specialised computers" are usually a conventional computer with a special-purpose microprogram.

AUTOMATED DESIGN TOOLS

To save costly engineering effort, much of the effort of designing large logic machines has been automated. The computer programmes are called "electronic design automation tools" or just "EDA." Simple truth table-style descriptions of logic are often optimised with EDA that automatically produces reduced systems of logic gates or smaller lookup tables that still produce the desired outputs. The most common example of this kind of software is the Espresso heuristic logic minimizer.

Most practical algorithms for optimising large logic systems use algebraic manipulations or binary decision diagrams, and there are promising experiments with genetic algorithms and annealing optimisations. To automate costly engineering processes, some EDA can take state tables that describe state machines and automatically produce a truth table or a function table for the combinational logic of a state machine. The state table is a piece of text that lists each state, together with the conditions controlling the transitions between them and the belonging output signals.

It is common for the function tables of such computer-generated state-machines to be optimised with logic-minimization software such as Minilog. Often, real logic systems are designed as a series of sub-projects, which are combined using a "tool flow." The tool flow is usually a "script," a simplified computer language that can invoke the software design tools in the right order.

Tool flows for large logic systems such as microprocessors can be thousands of commands long, and combine the work of hundreds of engineers.

Writing and debugging tool flows is an established engineering specialty in companies that produce digital designs. The tool flow usually terminates in a detailed computer file or set of files that describe how to physically construct the logic. Often it consists of instructions to draw the transistors and wires on an integrated circuit or a printed circuit board.

Parts of tool flows are "debugged" by verifying the outputs of simulated logic against expected inputs. The test tools take computer files with sets of inputs and outputs, and highlight discrepancies between the simulated behaviour and the expected behaviour.

Once the input data is believed correct, the design itself must still be verified for correctness. Some tool flows verify designs by first producing a design, and then scanning the design to produce compatible input data for the tool flow. If the scanned data matches the input data, then the tool flow has probably not introduced errors.

The functional verification data are usually called "test vectors." The functional test vectors may be preserved and used in the factory to test that newly constructed logic works correctly. However, functional test patterns don't discover common fabrication faults. Production tests are often designed by software tools called "test pattern generators". These generate test vectors by examining the structure of the logic and systematically generating tests for particular faults. This way the fault coverage can closely approach 100 per cent, provided the design is properly made testable.

Once a design exists, and is verified and testable, it often needs to be processed to be manufacturable as well. Modern integrated circuits have features smaller than the wavelength of the light used to expose the photoresist. Manufacturability software adds interference patterns to the exposure masks to eliminate open-circuits, and enhance the masks' contrast.

DESIGN FOR TESTABILITY

There are several reasons for testing a logic circuit. When the circuit is first developed, it is necessary to verify that the design circuit meets the required functional and timing specifications. When multiple copies of a correctly designed circuit are being manufactured, it is essential to test each copy to ensure that the manufacturing process has not introduced any flaws.

A large logic machine (say, with more than a hundred logical variables) can have an astronomical number of possible states. Obviously, in the factory, testing every state is impractical if testing each state takes a microsecond, and there are more states than the number of microseconds since the universe began. Unfortunately, this ridiculous-sounding case is typical. Fortunately, large logic machines are almost always designed as assemblies of smaller logic machines. To save time, the smaller sub-machines are isolated by permanently-installed "design for test" circuitry, and are tested independently. One common test scheme known as "scan design" moves test bits serially (one after another) from external test equipment through one or more serial shift registers known as "scan chains". Serial scans have only one or two wires to carry the data, and minimize the physical size and expense of the infrequently-used test logic.

After all the test data bits are in place, the design is reconfigured to be in "normal mode" and one or more clock pulses are applied, to test for faults (*e.g.*, stuck-at low or stuck-at high) and capture the test result into flip-flops and/or latches in the scan shift register(s). Finally, the result of the test is shifted out to the block boundary and compared against the predicted "good machine" result. In a board-test environment, serial to parallel testing has been formalised with a standard called "JTAG" (named after the "Joint Test Action Group" that proposed it).

Another common testing scheme provides a test mode that forces some part of the logic machine to enter a "test cycle." The test cycle usually exercises large independent parts of the machine.

TRADE-OFFS

Several numbers determine the practicality of a system of digital logic: cost, reliability, fanout and speed. Engineers explored numerous electronic devices to get an ideal combination of these traits.

Cost

The cost of a logic gate is crucial. In the 1930s, the earliest digital logic systems were constructed from telephone relays because these were inexpensive and relatively reliable. After that, engineers always used the cheapest available electronic switches that could still fulfill the requirements. The earliest integrated circuits were a happy accident. They were constructed not to save money, but to save weight, and permit the Apollo Guidance

Computer to control an inertial guidance system for a spacecraft. The first integrated circuit logic gates cost nearly $50 (in 1960 dollars, when an engineer earned $10,000/year). To everyone's surprise, by the time the circuits were mass-produced, they had become the least-expensive method of constructing digital logic. Improvements in this technology have driven all subsequent improvements in cost.

With the rise of integrated circuits, reducing the absolute number of chips used represented another way to save costs. The goal of a designer is not just to make the simplest circuit, but to keep the component count down. Sometimes this results in slightly more complicated designs with respect to the underlying digital logic but nevertheless reduces the number of components, board size, and even power consumption. For example, in some logic families, NAND gates are the simplest digital gate to build. All other logical operations can be implemented by NAND gates. If a circuit already required a single NAND gate, and a single chip normally carried four NAND gates, then the remaining gates could be used to implement other logical operations like logical and. This could eliminate the need for a separate chip containing those different types of gates.

Reliability

The "reliability" of a logic gate describes its mean time between failure (MTBF). Digital machines often have millions of logic gates. Also, most digital machines are "optimised" to reduce their cost. The result is that often, the failure of a single logic gate will cause a digital machine to stop working. Digital machines first became useful when the MTBF for a switch got above a few hundred hours. Even so, many of these machines had complex, well-rehearsed repair procedures, and would be non-functional for hours because a tube burned-out, or a moth got stuck in a relay. Modern transistorised integrated circuit logic gates have MTBFs greater than 82 billion hours (8.2×10^{10}) hours, and need them because they have so many logic gates.

Fanout

Fanout describes how many logic inputs can be controlled by a single logic output without exceeding the current ratings of the gate.[6] The minimum practical fanout is about five. Modern electronic logic using CMOS transistors for switches have fanouts near fifty, and can sometimes go much higher.

Speed

The "switching speed" describes how many times per second an inverter (an electronic representation of a "logical not" function) can change from true to false and back. Faster logic can accomplish more operations in less time. Digital logic first became useful when switching speeds got above fifty hertz, because that was faster than a team of humans operating mechanical calculators. Modern electronic digital logic routinely switches at five gigahertz

(5×10^9 hertz), and some laboratory systems switch at more than a terahertz (1×10^{12} hertz).

Logic families

Design started with relays. Relay logic was relatively inexpensive and reliable, but slow. Occasionally a mechanical failure would occur. Fanouts were typically about ten, limited by the resistance of the coils and arcing on the contacts from high voltages.

Later, vacuum tubes were used. These were very fast, but generated heat, and were unreliable because the filaments would burn out. Fanouts were typically five to seven, limited by the heating from the tubes' current. In the 1950s, special "computer tubes" were developed with filaments that omitted volatile elements like silicon. These ran for hundreds of thousands of hours.

The first semiconductor logic family was resistor-transistor logic. This was a thousand times more reliable than tubes, ran cooler, and used less power, but had a very low fan-in of three. Diode-transistor logic improved the fanout up to about seven, and reduced the power. Some DTL designs used two power-supplies with alternating layers of NPN and PNP transistors to increase the fanout. Transistor transistor logic (TTL) was a great improvement over these. In early devices, fanout improved to ten, and later variations reliably achieved twenty. TTL was also fast, with some variations achieving switching times as low as twenty nanoseconds. TTL is still used in some designs. Emitter coupled logic is very fast but uses a lot of power. It was extensively used for high-performance computers made up of many medium-scale components (such as the Illiac IV). By far, the most common digital integrated circuits built today use CMOS logic, which is fast, offers high circuit density and low-power per gate. This is used even in large, fast computers, such as the IBM System z.

DIGITAL INTEGRATED CIRCUITS

Digital integrated circuits, sometimes called an IC, chip or microchip, are the building blocks for nearly every type of modern electronics and electrical devices, whether for home or office, work or pleasure.

ORIGIN OF DIGITAL CIRCUITS

Digital integrated circuit development was necessary to resolve the fundamental issues of vacuum tubes, the leading technology of the first half of the 20th century. Vacuum tubes are bulky, fragile, power hungry, hot and failed frequently.

History

On Sept. 12, 1958, Jack Kilby successfully demonstrated the first IC with a single transistor and all its components on a single piece of germanium semiconductor material. In the first 50 years, ICs have advanced from single transistors to complex microprocessors with multiple CPU cores. On Dec. 3, 2009, the Helsinki University of Technology announced that a team of researchers had succeeded in building the first single atom transistor, the theoretical minimum component size. Technological alternatives such as quantum computing evolved as semiconductor technology approached theoretical limits.

Features

Digital integrated circuits operate based on binary mathematics using electrical signals to represent "ones" and "zeros." Digital integrated circuits can perform any number of operations, from a simple switch to a complex microprocessor.

HISTORY OF INTEGRATED CIRCUITS

The experiments performed on semi conductors revealed that they can perform the variety of vacuum tube functions. The incorporation of bulk of puny transistors on a small chip was a major leap in manually assembling electronic components on a really small circuit. Mass production and efficiency of Integrated Circuits greatly replaced the discrete transistors from many devices.

The two main features of integrated circuits like low cost and high performance gave it edge over the discrete transistors. The low cost is associated to the bulk production and printing using photolithography. The material used in constructing an IC is less than that used in discrete circuits. The high performance can be contributed to the quick switching and low power consumption.

The idea of IC was developed by the scientists working for Royal Radar. The idea propagated further when Jack Kilby and Robert Noyce working separately brought a remarkable invention. The chips made by both of them were meant to perform same functions whereas the material used was not the same. Noyce made a silicone chip and the Kilby manufactured a germanium chip. The development in this field was not new it was in 1949 when the first idea related to integrated chip came to view. From then onwards the development and growth is continued.

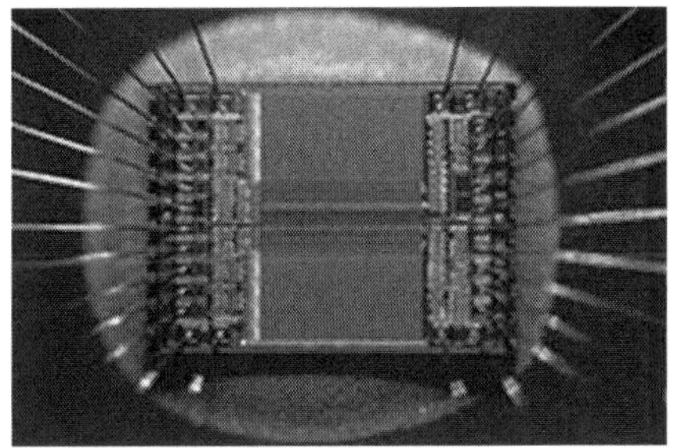

Applications of Integrated Circuts

Integrated circuits today have become ubiquitous. It seems that it was this invention which has led to the growth in technology. This remarkable invention has helped companies and manufacturers of electronic devices to design and introduce new and advance products. It can be said that it was due to the micro chips that we can use portable digital devices today.

The integrated chip is an important component of laptops, palm tops, MP3s, playstaions, mobile phones, net books and an unlimited array of electronic devices. The existence of all modern communication and electronic devices is based today on the use of integrated circuits. This chip is small but really efficient and swift when it comes to the performance. It is due to this desirable feature that it is part of almost all cell phones and we carry them anywhere in the world without being worrying about performance decline.

Taxonomy of Integrated Circuits

The integrated circuits can be divided into three broad categories like digital ICs, analog ICs and mixed signal ICs. Digital integrated circuits contain many flip flops, multiplexers and logic gates in a single chip comprising only few millimeters. This small size has provided its edge over all other technologies. Therefore, the manufactures of digital devices now prefer it over discrete transistors. These small chips consume less power and they are efficient. The manufacturing of these circuits help in reducing manufacturing cost due to the wide scale integration.

The working of the digital microchips is guided by the binary mathematical process in the form of 1 and 0 signals. On the contrary the analog micro circuits work by continuous flow of signals. The various examples of analog circuits include power management circuits, sensors and operational amplifiers.

These integrated circuits are used for performing functions like active filtering, demodulation, amplification and mixing. The expertly designed analog circuits remove the need for manufacturing an entirely new circuit from abrasion. This provides the facility to the mobile phone manufactures to buy and install the IC instead of preparing one. The digital and analog circuits can be combined in on one chip to enhance the performance.

TYPES OF INTEGRATED CIRCUITS

Integrated circuits, often called semiconductors, chips or simply ICs, fall into numerous categories, or types. Leading semiconductor trade organisations and market research firms use these classifications to report on the health of the various chip market segments. Although many classifications exist, the three primary classifications are analog, digital and mixed-signal integrated circuits.

DIGITAL INTEGRATED CIRCUITS

Digital integrated circuits, primarily used to build computer systems, also occur in cellular phones, stereos and televisions. Digital integrated circuits include microprocessors, microcontrollers and logic circuits. They perform mathematical calculations, direct the flow of data and make decisions based on Boolean logic principles. The Boolean system used centers on on two numbers: 0 and 1. On the other hand, the base 10 system, the number system we learn in elementary school, is based on 10 numbers: 0, 1, 3, 4, 5, 6, 7, 8 and 9.

Analog Integrated Circuits

Analog integrated circuits most commonly make up a part of power supplies, instruments and communications. In these applications, analog integrated circuits amplify, filter and modify electrical signals. In cellular phones, they amplify and filter the incoming signal from the phone's antenna. The sound encoded into that signal has a low amplitude level; after the circuit filters the sound signal from the incoming signal, the circuit amplifies the sound signal and sends it to the speaker in your cell phone, allowing you to hear the voice on the other end.

Mixed-Signal Integrated Circuits

Mixed-signal circuits occur in cellular phones, instrumentation, motor and industrial control applications. These circuits convert digital signals to analog signals, which in turn set the speed of motors, the brightness of lights and the temperature of heaters, for example. They also convert digital signals to sound waveforms, allowing for the design of digital musical instruments such as electronic organs and computer keyboards capable of playing music.Mixed-signal integrated circuits also convert analog signals to digital signals. They will convert analog voltage levels to digital number representations of the voltage level of the signals. Digital integrated circuits then perform mathematical calculations on these numbers.

Memory-Integrated Circuits

Though primarily used in computer systems, memory-integrated circuits also occur in cellular phones, stereos and televisions. A computer system may include 20 to 40 memory chips, while other types of electronic systems may contain just a few. Memory circuits store information, or data, as two numbers: 0 and 1. Digital integrated circuits will often retrieve these numbers from memory and perform calculations with them, then save the calculation result a memory chip's data storage locations. The more data it accesses—pictures, sound and text—the more memory an electronic system will require.

CONSIDER OF ASYNCHRONOUS CIRCUITS AND SYSTEMS

The circuit is considered to be *asynchronous* if it does not employ a periodic clock signal *C* to synchronise its internal changes of state. Therefore the state changes occur in direct response to signal changes on primary (data) input lines, and different memory elements can change state at different times.

In asynchronous sequential circuits the inputs are levels and there are no clock pulses; the inputs events drive the circuit.

In general, an asynchronous circuit does not need the precise timing control supported by flip-flops. It may therefore contain latches rather than flip-flops. In many cases, an asynchronous circuit simply relies on the propagation delays of its component gates and connections, combined with the circuit's feedback structure, to implement its memory functions.

ASYNCHRONOUS SYSTEMS

We now begin our discussion of a different type of system named Asynchronous system. A block diagram is shown of an asynchronous system. The first major difference compared to the synchronous system is that there is no global clock. This is the reason why this system is called asynchronous. Like the synchronous system, it has a combinational block, which performs a given function.

However, this system does not have an edge dependant element like the flip-flop. Instead it has delay elements, which have the duty of taking the output of the combinational block and passing the next state assignments to the combinational block through the feedback loop. Since there is no global clock, as soon the delay elements receive data, output information is passed on the combinational block after a known finite time delay.

Unlike the synchronous system where state transitioning is regulated by a clock, the asynchronous system's output is continuously changing. This means that output of the combinational block, which contains next state information and function results of the combinational block, is always being sampled by the delay element. Hence any type of unwanted transient called glitch on the input could cause an error at the output of the system. The transient at the input of the asynchronous system could also cause erroneous state assignment, which may cause the system to fail.

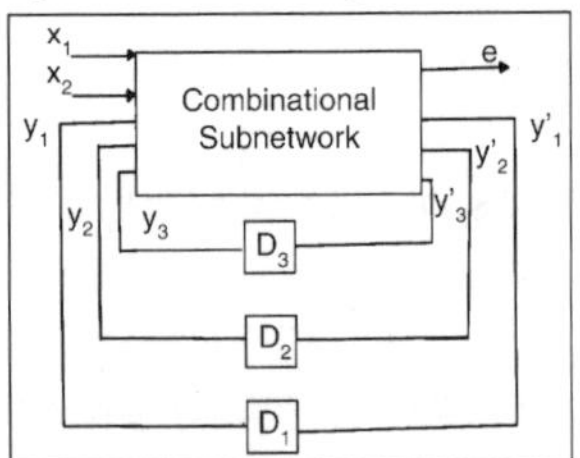

Fig. Combinational Sub Network

There must be restrictions put on asynchronous systems in order for correct operation. The first restriction we put on asynchronous systems is called "fundamental mode ". In fundamental mode operation, our inputs to figure above are restricted to change only when our next state has stabilised.

Since there is no clock in our system, a given input pattern might cause the machine to go through unstable states until it reaches a stable state where the machine will stop. When the machine has reached a stable state, it will be ready for the next input change.

Another restriction that must be put on asynchronous systems is single input change. This means when the output of the combinational block has reached a stabilised state, the input bits can only change one bit at a time. Single input bit change restriction is used so that asynchronous systems can avoid race conditions between states. An example of a race condition could be if a transition has to be made from state 01 to state 10. This creates a race because 01 can go to 11 then to 10 or 01 can go to 00 then 10. Hence a race is created between state 11 and 00.

There are two types of important race conditions that could occur in asynchronous systems. The race condition in the previous example was a non-critical race because the destination (10) was eventually reached. The counter part is a critical race in which the state can go to 11 or 00 and stay at that particular state. This means that the final destination was not reached. The result of a critical race is that the machine might not function since the system is stuck at some unwanted state resulting in an erroneous result.

Fundamental mode and single input restrictions are needed since they allow the asynchronous system to avoid state races, but unwanted glitches at the output of the function block can still occur. When a machine is given a single input change, the system will go through a series of transient states before a stable state is reached. In doing so, the output of the combinational block might toggle a few times before it reaches the final output value. These unwanted transient outputs are called glitches. Potential glitches are called hazards. In asynchronous systems there are many types of hazards. There are hazards associated with the combinational block, and there are also hazards due to the overall behaviour of the asynchronous sequential system.

Hazards associated with asynchronous sequential systems can cause the machine to oscillate between two states before a final stable state is reached. In some cases, the final destination can never be reached since the machine is oscillating between the two. When doing hazard analysis for asynchronous sequential systems, all relevant transitions (single input change) must be analysed for the next state and output variables.

In addition, the combinational block must also be inspected for possible combinational hazards. Combinational hazard can be due to implementation of the SOP or POS system or the hazard can be due to the function itself. Hazards due to implementations can be fixed by altering the implementation, but functional hazards can only be fixed by the insertion of delay elements or the addition of a global clock. Another important characteristic of asynchronous systems that must be addressed is computation time or the speed of the system. In the case of synchronous systems, the maximum speed that could be achieved is a function of the worst-case delay through the combinational circuit. As analysed in earlier sections of the paper, the combinational block has to have a stabilised output before a clock transition occurs. This means that the worst-case delay of the combinational block must be small enough so that the output of the block stabilises before the next clock cycle.

However in the case of asynchronous systems, the computation time is not dependant on the worst-case delay rather the speed of the system depends on the average delay of the combinational block. This is a very important characteristic when computational time is an important factor in the design of asynchronous systems.

BENEFITS OF ASYNCHRONOUS SYSTEMS

There are many benefits in Asynchronous systems. Some of these benefits are actual solutions to problems faced by synchronous or clocked systems. An important characteristic in pure asynchronous systems is that there is no dynamic power consumption. Having a global clock means that the entire systems will consume dynamic power because most of the components in the synchronous system dependant on clock edges.

However in asynchronous systems, there is no global clock hence there is no dynamic power consumption. As noted in: *"VLSI Digital Signal Processing Systems"*, by Parhi, asynchronous systems do not have dynamic power consumption since they do not have global clocks. It is also indicated that asynchronous systems have the characteristic of going into power saving mode when they are not involved with computation. What this means is that power consumption only occurs at blocks within the whole system where computation is involved. These characteristics make asynchronous designs very attractive for low power applications.

In pure asynchronous systems, which are called self-timed systems, additional circuitry has to be built into the overall systems to place order and regulation in the network. Strict regulations help with most of the problems that plague asynchronous systems. Problems like race conditions between states and certain types of glitches at the output of the system can be better regulated so that the system can operate more efficiently. In order to better understand these additional circuits, asynchronous pipeline datapath theory must be addressed.

According to *"Digital Integrated circuits,* Rabaey, asynchronous pipeline systems have the characteristic of indicating when a given computation is completed. The system also has the ability to initiate a new computation. A block diagram of an asynchronous pipeline datapath. As an input pattern arrives at the input of the system, a "request" signal to block A is raised. If block A is not active at that time, the block transfers the data and acknowledges this information to the input block. Block A is enabled by raising the "start" signal and after a finite time the "done" signal is activated indicating the completion of the data computation. It is at this point that the "request" signal is given to block B.

If the function is available the "acknowledge" signal goes high. This is when the output from block A is transferred to block B and block A is permitted to execute the next computation. The "done" signal indicates that all outputs are stabilised and ready for input change. Also, the "acknowledge" and "request" scheme, which are reffered to as "handshaking protocols" allow for logical ordering of operation. The benefit of asynchronous pipelined systems is that they provide order and regulation for the entire system. These regulations allow the asynchronous system to avoid unwanted races and some potential hazards.

Drawbacks of Asynchronous Systems

As indicated by the previous section, asynchronous pipeline systems can be built to aid the system in regulating computations and other vital activities. However these extra circuits have to be included throughout the chip to ensure that the system is well regulated. This means that the area of the system will be large due to the addition of extra circuitry. Larger area leads to an increase in the cost of the entire system. An example of this drawback is discussed in an article titled *"A Comparison of Synchronous and Asynchronous FSMD Designs"* by Auletta, etal.

In the article, a 16-bit factoring chip is implemented in both synchronous and asynchronous technologies. The key comparisons are made in the field of area. The article indicates that when the chip is optimised for area, the asynchronous type is 5 times larger than the synchronous version. When the chip was optimised for speed once again the asynchronous type was 3 times larger in area than the synchronous implementation. Hence even if the synchronous systems have distributed clock lines throughout a given chip, the overall area might still be smaller than an asynchronous implementation with pipelines throughout the system.

Another drawback to asynchronous systems is there are few CAD tools available for asynchronous system analysis. Most of the tools in industry are catered towards synchronous designs because most of the products in industry are based on synchronous system technology. The most recent asynchronous CAD tools include: *PETRIFY, TANGRAM, COSPAN, OCCAM, SIS, MEAT, ALCYON.*

According to *"Design Flow and CAD Tools for Asynchronous Cells"* by Cuche etal, Alcyon has the ability to perform a variety of different type's analysis. For example, this tool can indicate if there are races in a given system. If there is a race in the system, Alcyon can even indicate if it's a critical or non-critical race. Even though CAD tools exist for asynchronous systems, there is only a handful when compared to the large variety of synchronous system based CAD tools.

SYNCHRONOUS SYSTEMS

In synchronous systems, a global clock is used to regulate states and outputs of a given circuit. A synchronous sequential circuit. The circuit is composed of a combinational circuit, which has the duty of performing the function of interest. In addition to the combinational circuit, there is an edge triggered memory device in the feedback loop of the synchronous system. The task of the memory device, also called a flip-flop, is to store information until it senses an edge of a clock where it will then pass the stored information on to the combinational circuit.

The output port of the synchronous system consists of the output of the combinational circuit along with the next state information. In order for the system to operate correctly, the output bits of the combinational block must stabilise before the next clock cycle. Therefore, the delay of the combinational circuit must be less than the period of the global clock. The inputs to the flip-flops must stabilise before the clock edge occurs because the data will be stored once the memory device detects the edge of the clock. If the data stored is not stabilised, a junk state will be passed on to the combinational circuit. This junk data will cause the circuit to give erroneous outputs. Hence, the input to the clocked flip-flop must stabilise before the clock period, in order for correct information to be passed on to the combinational circuit.

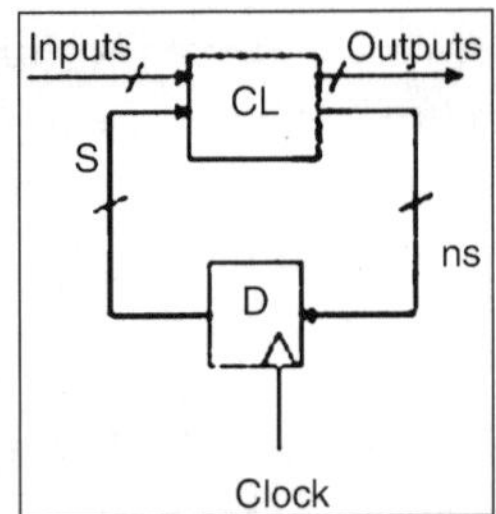

One of the most important delays in a synchronous system is that of the combinational circuit. The delay of the combinational network must be less than the period of the global clock. In other words, careful delay analysis of the combinational circuit must be considered in order to determine the worst-case path in the combinational circuit, which attributes to the maximum delay.

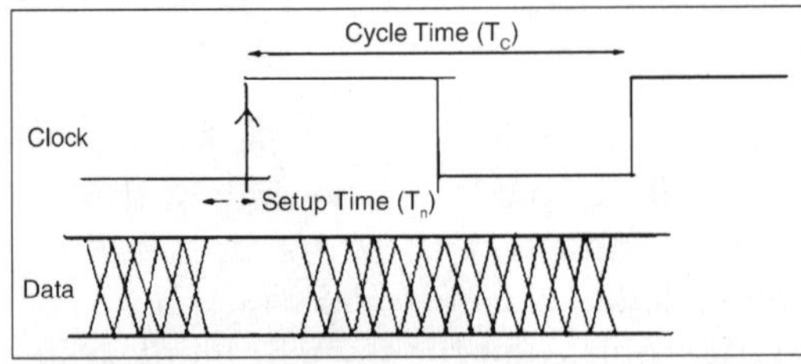

Fig. Cycle Time

Another type of delay that is very important to clocked circuits is skew. Skew can be classified as a delay in sequential synchronous systems. In synchronous systems, clock waveform relationships are very crucial to functionality of the system. Parasitic capacitances and resistances on clock lines can cause unwanted skews in a given waveform. These unwanted skews could alter clock relationships by misaligning synchronised clock waveforms. Incorrect clock waveform relationships could alter or deteriorate the performance of the synchronous system.

Flip-flops, which store and pass data, are activated by clock edges. It is during clock transitions that we can further analyse synchronous sequential systems. Clock transitions are not infinitely fast; rather they are finite in rise and fall times. This means that when a clock is transitioning from high to low or low to high, the flip-flops will then spend time in regions of the rise or fall time where the transistors inside the flip-flop might be all "on".

In most cmos designs the circuit is composed of a p-channel network and an n-channel network. The pmos device is activated "on" for a voltage low or "0", where the nmos device is activated for a voltage high or "1". When both the p and n-channels are "on", current travels from the positive supply to the negative supply giving rise to power consumption. This phenomenon is referred to as dynamic power consumption because it only occurs during clock transitions in synchronous systems.

Benefits of Synchronous Systems

Having a global clock in a system allows for flexibility. One of the most important benefits of having a global clock is that it can make testing simpler. The global clock can be designed such that it could be activated or deactivated. Having the flexibility of activating or deactivating the global clock allows synchronous designs to be easily characterised and or debugged. An example of a flexible global clock circuit is called the Timing or clock generator. These circuits are located on chip and they produce adjustable clocks. Adjustable clock edges can remedy unwanted skews, alter or correct duty-cycles and even adjust rise/fall times.

Another benefit in synchronous systems is that a technique called pipelining can be used to correct the problem noted earlier regarding the maximum delay of the combinational clock being the limiting factor in the speed of the synchronous system. According to *"VLSI Digital Signal Processing Systems"* by Parhi, a synchronous system can be altered in such a way as to reduce the maximum delay of the combinational block resulting in a possible reduced clock period. The combinational block is broken up into two parts, with a flip-flop in between the combinational blocks.

This architecture allows each computation to occur in two clock cycles. Therefore, the results of the combinational block are available at the flip-flop output at every clock period. The insertion of the flip-flop between the combinational blocks reduces its maximum delay. This means that the speed of the synchronous system is not limited by the maximum delay of the combinational circuit; rather the speed of the synchronous system is a direct function of the clock period. A smaller clock period suggests that we can operate the synchronous system at higher speeds.

Lastly an important advantage in synchronous system is that most of simulation software available to designers is geared towards synchronous systems. Since most of industry uses synchronous based technologies, CAD tools have been designed to cater towards clocked systems. Sophisticated CAD tools, obviously allow the given designer to have the ability to simulate and guarantee that a design is fully functional without any design errors. Some examples of synchronous based CAD tools are: Cadence's *Verilog XL* and *spectre, LogicWorks and HSPICE.*

Drawbacks of Synchronous Systems

The drawbacks of synchronous system include the design of additional circuitry, which control timing waveforms and or skew related problems. However, if a timing generator is not used, the alternative is a clock distribution network that must be laid out in such a way as to counteract skews and unwanted delays. Using an on chip timing generator allows for many flexibilities, but at the same time additional circuitry can increase the chip size. This means that the overall cost of the chip goes up due to the increase in area. However if a timing generator is not used, the cost of the chip can be reduced since control circuitry is limited. For synchronous systems without timing generators, an on chip clock distribution network must be used in which all the clock lines must have equal delays.

The lines represent clock routes with finite delays associated with them. Also shown in the figure are blocks, which represent circuits. H-tree networks allow all clock delays to be equal since the geometry of the network forces equal line lengths for all possible clock paths. This means that each of the circuit blocks sense clock edges at the same time. Although all the clock lines have equal amount of delay, which prevents unwanted skews between clock waveforms, parasitic capacitances and resistances have also been increased. These parasitic capacitances and resistances will degrade rise/fall times by slowing them down.

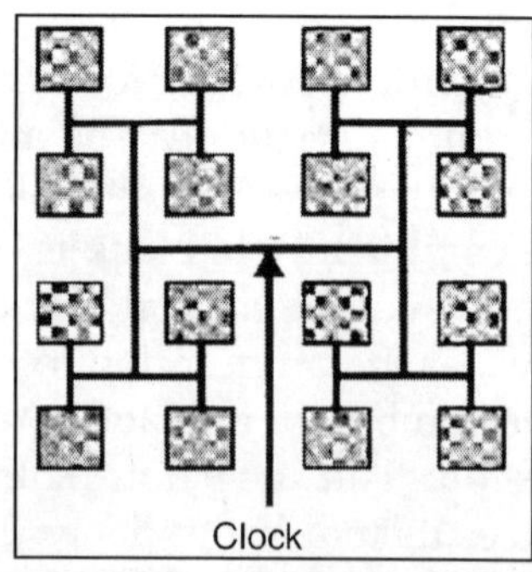

Fig. H-Tree Network

Additional clock buffers might be needed amongst the H-tree network in order to achieve higher rise/fall times. However the size and power consumption of the clock redistribution section of the entire chip can grow. This again can lead to higher cost.

INFORMATION OF DIGITAL AND ANALOG SIGNALS

Signals carry information and are defined as any physical quantity that varies with time, space, or any other independent variable. For example, a sine wave whose amplitude varies with respect to time or the motion of a particle with respect to space can be considered as signals. A system can be defined as a physical device that performs an operation on a signal. For example, an amplifier is used to amplify the input signal amplitude. In this case, the amplifier performs some operation(s) on the signal, which has the effect of increasing the amplitude of the desired information-bearing signal.

Signals can be categorised in various ways; for example discrete and continuous time domains. Discrete-time signals are defined only on a discrete set of times. Continuous-time signals are often referred to as continuous signals even when the signal functions are not continuous; an example is a square-wave signal.

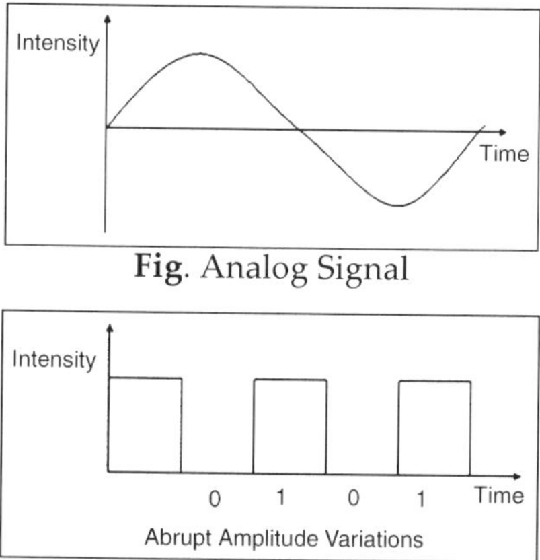

Fig. Analog Signal

Fig. Digital Signal

Another category of signals is discrete-valued and continuous-valued or otherwise known as digital and analog signals. Digital signals are discrete-valued and analog signals are continuous electrical signals that vary in time as shown in Figures. Analog devices and systems process signals whose voltages or other quantities vary in a continuous manner. They can take on any value across a continuous range of voltage, current, or other metric.

The analog signals can have an infinite number of values. Analog systems can be called wave systems. They have a value that changes steadily over time and can have any one of an infinite set of values in a range. Analog signals represent some physical quantity and they can be a model of the real quantity.

Most of the time, the variations corresponds to that of the non-electric (original) signal. For example, the telephone transmitter converts the sounds into an electrical voltage signal. The intensity of the voice causes electric current variations. Therefore, the two are analogous hence the name analog. At the receiving end, the signal is reproduced in the same proportion. Hence the electric current is a model and is an electrical representation of one's voice.

Not all analog signals vary as smoothly as the waveform shown in Figure. Digital signals are non-continuous, they change in individual steps. They consist of pulses or digits with discrete levels or values. The value of each pulse is constant, but there is an abrupt change from one digit to the next. Digital signals have two amplitude levels. The value of which are specified as one of two possibilities such as 1 or 0, HIGH or LOW, TRUE or FALSE and so on. In reality, the values are anywhere within specific ranges and we define values within a given range.

A digital system is the one that handles only discrete values or signals. Any set that is restricted to a finite number of elements contains discrete information. The word digital describes any system based on discontinuous data or events. Digital is the method of storing, processing and transmitting information through the use of distinct electronic pulses that represent the binary digits 0 and 1. Examples of discrete sets are the 10 decimal digits, the 26 letters of the alphabet, etc. A digital system would be to flick the light switch on and off. There's no 'in between' values.

ADVANTAGES OF DIGITAL SIGNALS

The usual advantages of digital circuits when compared to analog circuits are:

- *Noise Margin (resistance to noise/robustness):* Digital circuits are less affected by noise. If the noise is below a certain level (the noise margin), a digital circuit behaves as if there was no noise at all. The stream of bits can be reconstructed into a perfect replica of the original source. However, if the noise exceeds this level, the digital circuit cannot give correct results.
- *Error Correction and Detection:* Digital signals can be regenerated to achieve lossless data transmission, within certain limits. Analog signal transmission and processing, by contrast, always introduces noise.
- *Easily Programmable:* Digital systems interface well with computers and are easy to control with software. It is often possible to add new features to a digital system without changing hardware, and to do this remotely, just by uploading new software. Design errors or bugs can be worked-around with a software upgrade, after the product is in customer hands. A digital system is often preferred because of (re-)programmability and ease of upgrading without requiring hardware changes.

- Cheap Electronic Circuits: More digital circuitry can be fabricated per square millimeter of integrated-circuit material. Information storage can be much easier in digital systems than in analog ones. In particular, the great noise-immunity of digital systems makes it possible to store data and retrieve it later without degradation. In an analog system, aging and wear and tear will degrade the information in storage, but in a digital system, as long as the wear and tear is below a certain level, the information can be recovered perfectly. Theoretically, there is no data-loss when copying digital data. This is a great advantage over analog systems, which faithfully reproduce every bit of noise that makes its way into the signal.

DISADVANTAGES

The world in which we live is analog, and signals from this world such as light, temperature, sound, electrical conductivity, electric and magnetic fields, and phenomena such as the flow of time, are for most practical purposes continuous and thus analog quantities rather than discrete digital ones. For a digital system to do useful things in the real world, translation from the continuous realm to the discrete digital realm must occur, resulting in quantisation errors. This problem can usually be mitigated by designing the system to store enough digital data to represent the signal to the desired degree of fidelity. The Nyquist-Shannon sampling theorem provides an important guideline as to how much digital data is needed to accurately portray a given analog signal.

Digital systems can be fragile, in that if a single piece of digital data is lost or misinterpreted, the meaning of large blocks of related data can completely change. This problem can be diminished by designing the digital system for robustness. For example, a parity bit or other error-detecting or error-correcting code can be inserted into the signal path so that minor data corruptions can be detected and possibly corrected.

Digital circuits use more energy than analog circuits to accomplish the same calculations and signal processing tasks, thus producing more heat as well. In portable or battery-powered systems this can be a major limiting factor.

Digital circuits are made from analog components, and care has to be taken to all noise and timing margins, to parasitic inductances and capacitances, to proper filtering of power and ground connections, to electromagnetic coupling amongst data lines. Inattention to these can cause problems such as "glitches", pulses do not reach valid switching (threshold) voltages, or unexpected ("undecoded") combinations of logic states.

A corollary of the fact that digital circuits are made from analog components is the fact that digital circuits are slower to perform calculations than analog circuits that occupy a similar amount of physical space and consume the same amount of power. However, the digital circuit will perform the calculation with much better repeatability, due to the high noise immunity of digital circuitry.

THERMAL OXIDATION OF SILICON

The oxide of silicon, or silicon dioxide (SiO_2), is one of the most important ingredients in integrated circuits. Thermal SiO_2 is amorphous. It has a density of 2.2 gm/cm^3 and molecular Density of $2.3x10^{22}$ molecules/cm^3. The crystalline SiO_2 is also possible and known as Quartz has a density of 2.65 gm/cm^3. SiO_2 has excellent properties which makes them necessary in every part of the integrated circuits.

It is an excellent electrical insulator having energy gap ~ 9 eV with a resistivity greater than 10 20 ohm-cm and breakdown electric field greater than 10MV/cm. Si technology became popular because of the stable and reproducible Si/SiO_2 interface. Conformal oxide growth on exposed Si surface is easily possible. SiO_2 is a good diffusion mask for common dopants such as. B, P, As, Sb. In addition there exists good etching selectivity between Si and SiO_2. The formation of SiO_2 on a silicon surface is most often accomplished through a process called thermal oxidation. As its name implies, is a technique that uses extremely high temperatures (usually between 700-1200°C) to promote the growth rate of oxide layers whose thicknesses range from 20 to 10000 angstroms. During the process, silicon substrate is exposed to a high purity oxidising species like oxygen gas (dry oxidation) or water vapour (wet oxidation). The chemical reaction at the silicon surface for dry and wet oxidation is given as,

$$Si + O_2 \rightarrow SiO_2$$

$$Si + 2H_2O \rightarrow SiO_2 + 2H_2$$

Oxidation of silicon is not difficult, since silicon has a natural inclination to form a stable oxide even at room temperature, as long as an oxidising ambient is present. In both cases the oxidising species diffuses through the growing oxide and reacts with the silicon surface. These oxidation reactions occur at the Si-SiO_2 interface, *i.e.*, silicon at the interface is consumed as oxidation takes place. As the oxide grows the Si-SiO_2 interface moves into the silicon substrate. As a result, the Si-SiO_2 interface will always be below the original Si wafer surface. The SiO_2 surface, on the other hand, is always above the original Si surface. SiO_2 formation therefore proceeds in two directions relative to the original wafer surface as shown in Figure below.

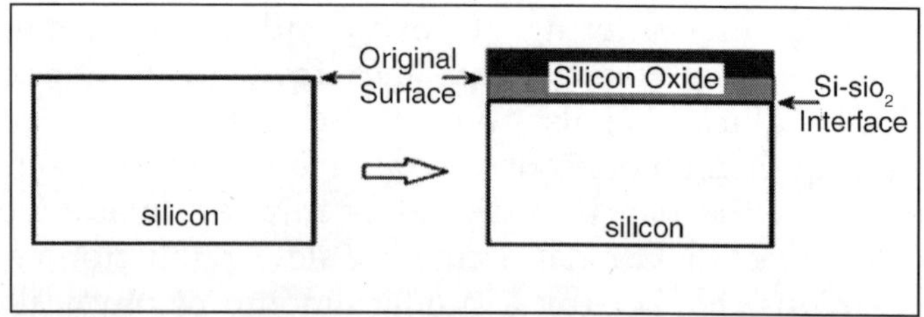

Fig. The Silicon-Silicon Dioxide Interface in Thermal Oxides

The amount of silicon consumed by the formation of silicon dioxide is also fairly predictable from the relative densities and molecular weights of Si and SiO_2, *i.e.*, the thickness of silicon consumed is 44 per cent of the final thickness of the oxide formed. Thus, an oxide that is 1000 angstroms thick will consume about 440 angstroms of silicon from the substrate. In another words, 1μm thick Si oxidises to 2.17 μm thick SiO_2.

Thermal oxidation is accomplished using an oxidation furnace which provides the heat needed to elevate the oxidising ambient temperature. A furnace typically consists of a temperature controlled heating system, fused quartz process tubes, arrange for controlled flow of various gases. The heating system usually consists of several heating coils that control the temperature around the furnace tubes. The wafers are placed in quartz glassware known as boats. A boat can contain many wafers. The oxidising agent (oxygen or steam) then enters the process tube through its source end, subsequently diffusing to the wafers where the oxidation occurs.

During dry oxidation, the silicon wafer reacts with the ambient oxygen, forming a layer of silicon dioxide on its surface. In wet oxidation, the water is heated in the 40-80°C range and oxygen or nitrogen carrier gases are used for the flow of water vapors to the chamber. Alternatively hydrogen and oxygen gases are introduced into a torch chamber where they react to form water molecules, which are then made to enter the reactor where they diffuse towards the wafers. The water molecules react with the silicon to produce the oxide and another byproduct, *i.e.*, hydrogen gas.

Kinetics of SiO_2 Growth

The Linear and Parabolic growth laws were developed by Deal and Grove, and are known as the Linear Parabolic Model. This oxide growth model has been empirically proven to be accurate over a wide range of temperatures (700-1300°C), oxide thicknesses (300-20,000 angstroms), and oxidant partial pressures (0.2-25 atmospheres). Figures below pictures various diffusions possible and the concentration of species during thermal oxidation and is the basis for Deal and Grove model.

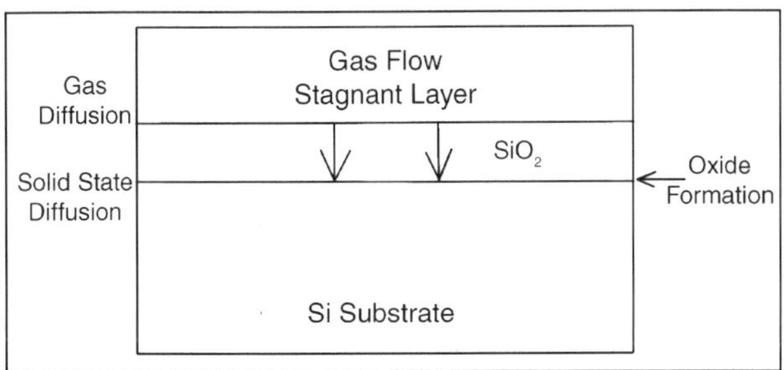

Fig. A Model for Thermal Oxidation of Silicon Indicating Various Diffusions Possible

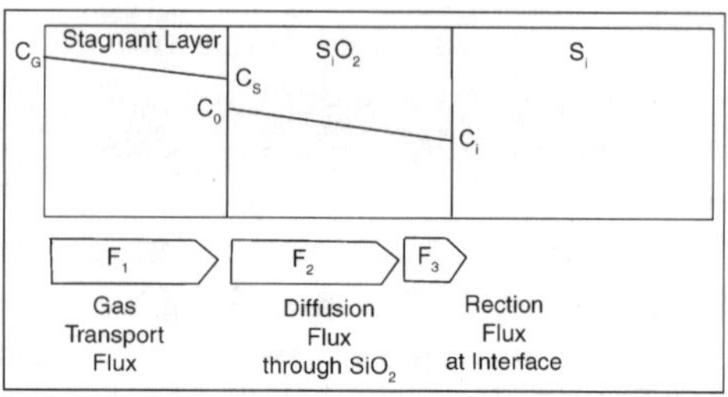

Fig. A Model for Thermal Oxidation of Silicon Indicating the Concentration of Species

The gas-phase flux F_1 is proportional to the difference between the oxidant concentration in the bulk of the gas (C_G) and the oxidant concentration adjacent to the oxide surface (C_S).

$$F_1 = h_G\left(C_G - C_S\right)$$

The Fick's law of solid sate diffusion states that,

$$F_2 = -D\frac{dC}{dx} \cong D\frac{\left(C_0 - C_i\right)}{x}$$

$$F_3 = K_5 C_i$$

h_G is the mass transfer constant (cm/s),D is the diffusion coefficient (cm^2/s),k_S is the surface reaction transfer constant (cm/s), Using $C_s = \frac{N}{V}$ and $PV = NkT$, Henry's law to relate C_0 and C_S,

$$C_0 = HP_S = H\left(kT\right)C_S$$

Where H is the Henry's constant and P_S is the partial pressure of oxidant gases at surface,

$$C_S = \frac{C_0}{HkT_0}$$

If the equilibrium concentration of oxidising species in oxide is $C_A = HkTC_G$,

$$F_1 = \frac{h_G}{HkT}\left(C_A - C_0\right)$$

Taking $\frac{h_G}{HkT}$ as h, at steady state,

$$F_1 = F_2 = F_3$$

Solving these two equations for two unknowns C_0 and C_i

$$C_i = \frac{C_A}{1 + \frac{k_S}{h} + \frac{k_S x}{D}}$$

$$C_0 = C_i\left(1+\frac{k_S x}{D}\right)$$

$$F = F_1 = F_2 = F_3 = KsC_i = \frac{k_S C_A}{1+\frac{k_S}{h}+\frac{k_S x}{D}}$$

To convert *F* into oxide thickness growth rate,

$$F = N_1\frac{dx}{dt} = \frac{k_S C_A}{1+\frac{k_S}{h}+\frac{k_S x}{D}}$$

Where N_1 is the oxidant molecules per unit volume required to form a unit volume of SiO_2. N_1 is 2.3 x 10^{22} cm^{-3} for dry oxidation and 4.6 x10^{22} cm^{-3} for wet oxidation. Taking boundary conditions as $x = x_0$ at $t=0$. The oxide thickness grown at any point of time *t* is modeled as $x^2 + Ax = B(t+\tau)$ where A and B constants.

$$A = 2D\left(\frac{1}{k_S}+\frac{1}{h}\right)$$

$$B = 2D\frac{C_A}{N_1}$$

$$\tau = \frac{x_i^2 + Ax_i}{B}$$

The time displacement τ is included to account for the oxide layer (at $t = 0$) formed by the accelerated growth in the initial phase of oxidation.

$$x = \frac{A}{2}\left[\sqrt{1+\left(\frac{t+\tau}{A^2/4B}\right)}-1\right]$$

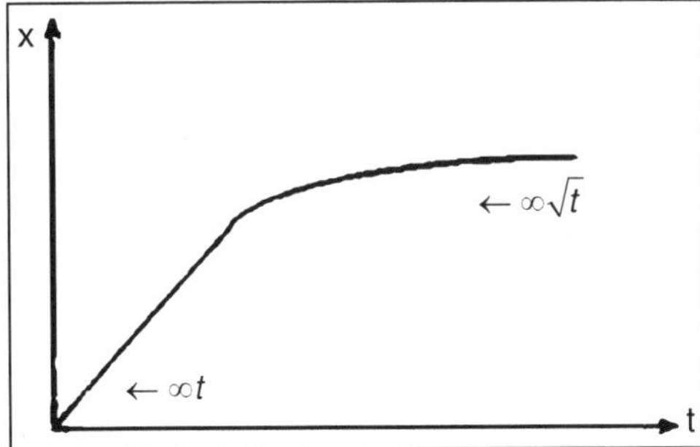

Fig. Oxide Thickness Verses Time

Typical oxide thickness verses time dependency is shown in Figure above. For small values of *t*, the oxide growth equation can be approximated as

$$x = \frac{B}{A}t$$

and thus

$$\frac{B}{A}$$

is known as linear rate constant. For large values of t, the above equation can be approximated as $x = \sqrt{Bt}$ and thus B is known as parabolic rate constant. That is oxide thickness growth slows down with increase in oxide thickness and can be readily seen from:

$$\frac{dx}{dt} = \frac{B}{A+2x}$$

Dry oxidation of silicon is typically used to grow a thin, high quality oxide that is used in transistor gates and capacitors. Oxide grown in dry oxygen ambient has excellent insulating properties and is denser, free of defects. Wet oxidation of silicon is typically used to grow thick oxides that are used as diffusion barriers. Silicon dioxide acts as an effective mask against many impurities, allowing dopants into silicon only in regions that are not covered with oxide.

The oxide thickness grown on silicon is dependent on the oxidation time and temperature. Wet oxidation method offers faster growth rate compared to dry oxidation. The linear and parabolic rate constants have larger values in wet oxidation case than in case of dry oxidation.

This is because equilibrium concentration of oxidising species in oxide (C_A) is approximately three orders of magnitude greater for water than in dry oxygen. Therefore, for growing thick oxide within a realistic time, wet oxidation is a better choice.

The linear and parabolic rate constants increase with temperature in both dry and wet oxidation methods. B increases in both cases through the diffusion coefficient (D) increase due to temperature. The reason for increase in

$$\frac{B}{A}$$

is through k_s.

$$A = 2D\left(\frac{1}{k_S} + \frac{1}{h}\right)$$

$$B = 2D\left(\frac{C_A}{N_1}\right)$$

$$\frac{B}{A} = \frac{C_A}{N_1\left(\frac{1}{k_S} + \frac{1}{h}\right)}$$

It can be seen that for small duration, the oxidation is a reaction controlled process and longer duration it is a diffusion controlled process.

From the Deal and grove model, we have seen that the oxide growth rate is affected by time, temperature, and pressure. Thickness of oxide is raised by an increase in oxidation time, oxidation temperature, or oxidation pressure. Other factors that affect thermal oxidation growth rate for SiO_2 include: the crystallographic orientation of the wafer; the wafer's doping level; the presence of halogen impurities in the gas phase and the presence of plasma during growth.

4

Digital Design in Binary System and Numbers

BINARY NUMBERS

Since digital circuits deal with binary values, we will begin with a quick introduction to binary numbers. A bit, having either the value of 0 or 1, can represent only two things or two pieces of information. It is, therefore, necessary to group many bits together to represent more pieces of information. A string of n bits can represent 2^n different pieces of information. For example, a string of two bits results in the four combinations: 00, 01, 10, and 11. By using different encoding techniques, a group of bits can be used to represent different information, such as a number, a letter of the alphabet, a character symbol, or a command for the microprocessor to execute.

The use of decimal numbers is quite familiar to us. However, since the binary digit is used to represent information within the computer, we also need to be familiar with binary numbers. Note that the use of binary numbers is just a form of representation for a string of bits. We can just as well use octal, decimal, or hexadecimal numbers to represent the string of bits. In fact, you will find that hexadecimal numbers are often used as a shorthand notation for binary numbers.

The decimal number system is a positional system. In other words, the value of the digit is dependent on the position of the digit within the number. For example, in the decimal number 48, the decimal digit 4 has a greater value than the decimal digit 8 because it is in the tenth position, whereas the digit 8 is in the unit position. The value of the number is calculated as $(4 \times 10^1) + (8 \times 10^0)$.

Like the decimal number system, the binary number system is also a positional system. The only difference between the two is that the binary system is a base-2 system, and so, it uses only two digits, 0 and 1, instead of ten. The binary numbers from 0 to 15 (decimal) are shown in Figure. The range from 0 to 15 has 16 different combinations. Since $2^4 = 16$, therefore, we need a 4-bit binary number (*i.e.*, a string of four bits) to represent this range.

When we count in decimal, we count from 0 to 9. After 9, we go back to 0, and have a carry of a 1 to the next digit. When we count in binary numbers, we do the same thing, except that we only count from 0 to 1. After 1, we go back to 0 and have a carry of a 1 to the next bit. The decimal value of a binary number can be found just like that for a decimal number, except that we raise the base number 2 to a power rather than the base number 10 to a power. For example, the value for the decimal number 658 is

$$65810 = (6 \times 10^2) + (5 \times 10^1) + (8 \times 10^0) = 600 + 50 + 8 = 65810$$

Similaly, the decimal value for the binary number 10110112 is

$$10110112 = (1 \times 2^6) + (0 \times 2^5) + (1 \times 2^4) + (1 \times 2^3) + (0 \times 2^2) + (1 \times 2^1) + (1 \times 2^0)$$
$$= 64 + 16 + 8 + 2 + 1 = 9110$$

To get the decimal value, the least significant bit (in this case, the rightmost 1) is multiplied with 2^0. The next bit to the left is multiplied with 2^1, and so on. Finally, they are all added together to give the value 9110.

Notice the subscript 10 in the decimal number 65810, and the 2 in the binary number 10110112. This subscript is used to denote the base of the number whenever there might be confusion as to what base the number is in.

Table. Numbers from 0 to 15 in Binary, Octal, and Hexadecimal Number Systems

Decimal	Binary	Octal	Hexadecimal
0	0000	0	0
1	0001	1	1
2	0010	2	2
3	0011	3	3
4	0100	4	4
5	0101	5	5
6	0110	6	6
7	0111	7	7
8	1000	10	8
9	1001	11	9
10	1010	12	A
11	1011	13	B
12	1100	14	C
13	1101	15	D
14	1110	16	E
15	1111	17	F

Converting a decimal number to its binary equivalent can be done by successively dividing the decimal number by 2 and keeping track of the remainder at each step. Combining the remainders together (starting with the last one) forms the equivalent binary number. For example, using the decimal number 91, we divide it by 2 to get 45 with a remainder of 1. Then we divide 45 by 2 to get 22 with a remainder of 1. We continue in this fashion until the end as shown here.

|91 1
|45 1
|22 0
|11 1
|5 1
|2 0
2 1 911

Least significant bit 245 1 222 0 211 1
= 1011011 2 51 2 20
1 Most significant bit

Concatenating the remainders together, starting with the last one (most significant bit) results in the binary number 1011011_2.

Binary numbers usually consist of a long string of bits. A shorthand notation for writing out this lengthy string of bits is to use either octal or hexadecimal numbers. Since the octal system is base-8 and the hexadecimal system is base-16 (both of which are a power of 2), a binary number can be converted easily to an octal or hexadecimal number, or vice versa.

Octal numbers only use the digits from 0 to 7 for the eight different combinations. When counting in octals, the number after 7 is 10 as shown in Figure. To convert a binary number to octal, we simply group the bits into groups of threes, starting from the right (least significant bit). The reason for this is because $8 = 2^3$. For each group of three bits, we write the equivalent octal digit for it. For example, the conversion of the binary number 11100112 to the octal number 1638 is shown here.

001 110 011 1 63

Since the original binary number has seven bits, we need to extend it with two leading 0's to get three bits for the leftmost group. Note that when we are dealing with negative numbers, we may require extending the number with leading 1's instead of 0's.

Converting an octal number to its binary equivalent is just as easy. For each octal number, we write down the equivalent three bits. These groups of three bits are concatenated together to form the final binary number. For example, the conversion of the octal number 5724_8 to the binary number 1011110101002 is shown here.

5 7 2 4101 111 010 100

The decimal value of an octal number can be found just like that for a binary or decimal number, except that we raise the base number 8 to a power instead of the base number 2 or 10 to a power. For example, the octal number 5724_8 has the value:

$$57248 = (5 \times 8^3) + (7 \times 8^2) + (2 \times 8^1) + (4 \times 8^0) = 2560 + 448 + 16 + 4 = 302810$$

Hexadecimal numbers are treated basically the same way as octal numbers except with the appropriate changes to the base. Hexadecimal (or hex for short) numbers use base-16, and thus require 16 different digit symbols, as shown in Figure. Converting binary numbers to hexadecimal numbers involves grouping the bits into groups of fours since $16 = 2^4$. For example, the conversion of the binary number 110110110112 to the hexadecimal number 6DB16 is shown below. Again, we need to extend it with a leading 0 to get four bits for the leftmost group.

0110 1101 1011 6 D B

To convert a hex number to a binary number, we write down the equivalent four bits for each hex digit, and then concatenate them together to form the final binary number. For example, the conversion of the hexadecimal number $5C4A_{16}$ to the binary number 01011100010010102 is shown here.

5 C 4 A0101 1100 0100 1010

The following example shows how the decimal value of the hexadecimal number C4A16 is evaluated.

$$C4A_{16} = (C \times 16^2) + (4 \times 16^1) + (A \times 16^0)$$
$$= (12 \times 16^2) + (4 \times 16^1) + (10 \times 16^0)$$
$$= 3072 + 64 + 10 = 314610$$

POSITIVE BINARY NUMBERS

Binary representations of positive can be understood in the same way as their decimal counterparts. For example

$$86_{10} = 1*64 + 0*32 + 1*16 + 0*8 + 1*4 + 1*2 + 0*1$$

or

$$86_{10} = 1*2^6 + 0*2^5 + 1*2^4 + 0*2^3 + 1*2^2 + 1*2^1 + 0*2^0$$

or

$$86_{10} = 1010110_2$$

The subscript 2 denotes a binary number. Each digit in a binary number is called a bit. The number 1010110 is represented by 7 bits. Any number can be broken down this way, by finding all of the powers of 2 that add up to the number in question (in this case 2^6, 2^4, 2^2 and 2^1). You can see this is exactly analagous to the decimal deconstruction of the number 125 that was done earlier. Likewise we can make a similar set of observations:

- To multiply a number by 2 you can simply shift it to the left by one digit, and fill in the rightmost digit with a 0. To divide a number by 2, simply shift the number to the right by one digit.
- To see how many digits a number needs, you can simply take the logarithm (base 2) of the number, and add 1 to it. The integer part of the result is the number of digits. For instance, $\log_2(86)+1=7.426$. The integer part of that is 7, so 7 digits are needed.
- With n digits, 2^n unique numbers (from 0 to 2^n–1) can be represented. If n=8, 256 (=2^8) numbers can be represented 0–255.

It is often convenient to handle groups of bits, rather than individually. The most common grouping is 8 bits, which forms a byte. A single byte can represent 256 (2^8) numbers. Memory capacity is usually referred to in bytes. Two bytes is usually called a word, or short word (though word-length depends on the application). A two-byte word is also the size that is usually used to represent integers in programming languages. A long word is usually twice as long as a word. A less common unit is the nibble which is 4 bits, or half of a byte. It is cumbersome for humans to deal with writing, reading and remembering individual bits, because it takes many of them to represent even fairly small numbers. A number of different ways have been developed to make the handling of binary data easier for us.

Decimal	Hexadecimal	Binary
0	0	0000
1	1	0001
2	2	0010
3	3	0011
4	4	0100
5	5	0101
6	6	0110
7	7	0111
8	8	1000
9	9	1001
10	A	1010
11	B	1011
12	C	1100
13	D	1101
14	E	1110
15	F	1111

The most common is hexadecimal. In hexadecimal notation, 4 bits (a nibble) are represented by a single digit. There is obviously a problem with this since 4 bits gives 16 possible combinations, and there are only 10 unique decimal digits, 0 to 9. This is solved by using the first 6 letters (A..F) of the alphabet as numbers. The table shows the relationship between decimal, hexadecimal and binary.

There are some significant advantages to using hexadecimal when dealing with electronic representations of numbers (if people had 16 fingers, we wouldn't be saddled with the awkward decimal system). Using hexadecimal makes it very easy to convert back and forth from binary because each hexadecimal digit corresponds to exactly 4 bits ($\log_2(16) = 4$) and each byte is two hexadecimal digit. In contrast, a decimal digit corresponds to $\log_2(10) = 3.322$ bits and a byte is 2.408 decimal digits. Clearly hexadecimal is better suited to the task of representing binary numbers than is decimal.

As an example, the number $CA3_{16} = 1100\ 1010\ 0011_2$ ($1100_2 = C_{16}$, $1010_2 = A_{16}$, $0011_2 = 3_{16}$). It is convenient to write the binary number with spaces after every fourth bit to make it easier to read. Converting back and forth to decimal is more difficult, but can be done in the same way as before.

$$3235_{10} = C_{16}*256 + A_{16}*16 + 3_{16}*1 = C_{16}*16^2 + A_{16}*16^1 + 3_{16}*16^0$$

or

$$3235_{10} = 12*256 + 10*16 + 3*1 = 12*16^2 + 10*16^1 + 3*16^0$$

Octal notation is yet another compact method for writing binary numbers. There are 8 octal characters, 0...7. Obviously this can be represented by exactly 3 bits. Two octal digits can represent numbers up to 64, and three octal digits up to 512. A byte requires 2.667 octal digits. Octal used to be quiete common, it was the primary way of doing low level I/O on some old DEC computers. It is much less common today but is still used occasionally (e.g., to set read, write and execute permissions on Unix systems)

In summary:

Bit: a single binary digit, either zero or one.

Byte: 8 bits, can represent positive numbers from 0 to 255.

Hexadecimal: A representation of 4 bits by a single digit 0..9,A..F. In this way a byte can be represented by two hexadecimal digits

long word: A long word is usually twice as long as a word.

Nibble: 4 bits, half of a byte.

Octal: A representation of 3 bits by a single digit 0..7. This is used much less commonly than it once was (early DEC computers used octal for much of their I/O)

Word: Usually 16 bits, or two bytes. But a word can be almost any size, depending on the application being considered—32 and 64 bits are common sizes

SIGNED BINARY INTEGERS

It was noted previously that we will not be using a minus sign (–) to represent negative numbers. We would like to represent our binary numbers with only two symbols, 0 and 1.

There are a few ways to represent negative binary numbers. The simplest of these methods is called ones complement, where the sign of a binary number is changed by simply toggling each bit (0's become 1's and vice-versa).

This has some difficulties, among them the fact that zero can be represented in two different ways (for an eight bit number these would be 0000 0000 and 1111 1111)., we will use a method called two's complement notation which avoids the pitfalls of one's complement, but which is a bit more complicated. To represent an n bit signed binary number the leftmost bit, has a special significance. The difference between a signed and an unsigned number is given in the table below for an 8 bit number.

The value of bits in signed and unsigned binary numbers

	Bit 7	Bit 6	Bit 5	Bit 4	Bit 3	Bit 2	Bit 1	Bit 0
Unsigned	$2^7 = 128$	$2^6 = 64$	$2^5 = 32$	$2^4 = 16$	$2^3 = 8$	$2^2 = 4$	$2^1 = 2$	$2^0 = 1$
Signed	$-(2^7)$ = h128	$2^6 = 64$	$2^5 = 32$	$2^4 = 16$	$2^3 = 8$	$2^2 = 4$	$2^1 = 2$	$2^0 = 1$

Let's look at how this changes the value of some binary numbers

Binary	Unsigned	Signed
0010 0011	35	35
1010 0011	163	–93
1111 1111	255	–1
1000 0000	128	–128

If Bit 7 is not set (as in the first example) the representation of signed and unsigned numbers is the same. However, when Bit 7 is set, the number is always negative. For this reason Bit 7 is sometimes called the sign bit. Signed numbers are added in the same way as unsigned numbers, the only difference is in the way they are interpreted. This is important for designers of arithmetic circuitry because it means that numbers can be added by the same circuitry regardless of whether or not they are signed.

To form a two's complement number that is negative you simply take the corresponding positive number, invert all the bits, and add 1. The example below illustrated this by forming the number negative 35 as a two's complement integer:

$35_{10} = 0010\ 0011_2$

invert → $1101\ 1100_2$

add 1 → $1101\ 1101_2$

So 1101 1101 is our two's complement representation of -35. We can check this by adding up the contributions from the individual bits

$1101\ 1101_2 = -128 + 64 + 0 + 16 + 8 + 4 + 0 + 1 = -35.$

The same procedure (invert and add 1) is used to convert the negative number to its positive equivalent. If we want to know what what number is represented by 1111 1101, we apply the procedure again

$? = 1111\ 1101_2$

invert → $0000\ 0010_2$

add 1 → $0000\ 0011_2$

Since 0000 0011 represents the number 3, we know that 1111 1101 represents the number –3.

POSITIVE BINARY FRACTIONS

The representation of unsigned binary fractions proceeds in exactly the same way as decimal fractions. For example

$0.625_{10} = 1*0.5 + 0*0.25 + 1*0.125 = 1*\ 2^{-1} + 0*\ 2^{-2} + 1*\ 2^{-3} = 0.101_2$

Each place to the right of the decimal point represents a negative power of 2, just as for decimals they represent a negative power of 10. Likewise, if there are m bits to the right of a decimal, the precision of the number is 2^{-m} (versus 10^{-m} for decimal). Though it is possible to represent numbers greater than one by having digits to the left of the decimal place we will restrict ourselves to numbers less than one. These are commonly used by Digital Signal Processors.

The largest number that can be represented by such a representation is $1\text{–}2^{-m}$, the smallest number is 2^{-m}. For a fraction with 15 bits of resolution this gives a range of approximately 0.99997 to 3.05E–5.

Note that this representationis easily extended to represent all positive numbers by having the digits to the left of the decimal point represent the integer part, and the digits to the right representing the fractional part. Thus

$6.625_{10} = 110.101_2$

SIGNED BINARY FRACTIONS

Signed binary fractions are formed much like signed integers. We will work with a single digit to the left of the decimal point, and this will represent the number –1 (= $-(2^0)$). The rest of the representation of the fraction remains unchanged. Therefore this leftmost bit represents a sign bit just as with two's complement integers. If this bit is set, the number is negative, otherwise the number is positive. The largest positive number that can be represented is still $1\text{-}2^{-m}$ but the largest negative number is -1. The resolution is still $1\text{–}2^{-m}$.

There is a terminology for naming the resolution of signed fractions. If there are m bits to the right of the decimal point, the number is said to be in Q*m*format. For a 16 bit number (15 bits to the right of the decimal point) this results in Q15 notation.

BINARY NUMBER SYSTEMS

For centuries, most civilizations have performed arithmetic by counting in groups of ten. True, there are societies that even today, count differently, but such groups are rare. When we speak of groups of ten, we mean for example, that one hundred (100) is really ten groups of ten; one thousand is really one hundred groups of ten or more precisely, ten groups of ten groups of ten; one million would actually be ten groups of ten groups of ten groups of ten groups of ten groups of ten. Sounds confusing, but numerically, we would write:

$$1 \text{ million} = 10 \times 10 \times 10 \times 10 \times 10 \times 10 \text{ or,}$$

$$1 \text{ million} = 10(10(10(10(10(10)))))$$

The latter representation illustrates the grouping by tens concept perhaps a bit more clearly.

As man began to build machines to perform computations, it seemed both natural and logical to emulate the human way of counting as well as doing arithmetic. When Sir Charles Babbage built his *analytic engine,* he devised a series of gears that would mesh with a series of wheels – each wheel represented a numeric position (units, tens, hundreds, thousands, and so forth), and each position was engineered to represent any of ten possible digits. This was truly a 10-state machine. The problem was that the machine's excessive complexity required a science of machined parts that was far beyond the capability of extant technology. Nevertheless, Babbage's ideas found their way into 20th century adding machines and electric calculators – definitely NOT the hand held kind though!

Electrically, it was possible to represent each digit by a different voltage level, and some analog computers successfully used this ploy though serious accuracy was a challenge never readily solvable. When John Atanasoff created the first electronic digital computer in 1933, he devised computational circuits around a different concept. His circuits would represent only *two possible states: on and off*. This suggested a bivalent system in which only two values could exist: a value of 1 was assigned to the on state and a value of 0 was used to represent the off state.

What Atanasoff did was make use of what had been previously an academic concept – performing arithmetic in a numeric base other than ten. In Atanasoff's system, arithmetic had to be performed using only two digits (0 and 1). He gave us a practical use for the *binary* number system also referred to as the base 2 system.

It turns out that there are an infinite number of number systems, each with its own base. But before we investigate specific number systems that are used in various aspects of computing, it will prove worthwhile to discuss the general attributes ascribed to all number systems.

Every number system consists of a group size for counting purposes. This group size is called the base of the number system. For example, the decimal number system with which we are most familiar, has a base of 10. The binary system described earlier has a base of two. Each number system consists of a set of digits. Digits are numbered consecutively with zero (0) always being the first digit in sequence. Each remaining digit is generated by adding one to the previous digit until the cardinality of the set of digits equals the size of the base. [For those unfamiliar with set theory terminology, the *cardinality of a set* is the number of elements contained in the set]. The base number itself is never a digit. With all number systems, when one is added to the largest digit in the system, the resulting number consists of a sum of zero and a carry of one. In the base ten system, we represent this phenomenon as follows:

$$\begin{array}{r} 9 \\ +1 \\ 10 \end{array}$$

In the binary system, we would have 1 + 1 = 10. Note that we have the same two digits for our summation in each case but in the decimal system, our sum represents the *value ten* whereas in our binary system, the sum has an equivalent "value" of two. This concept will become clearer as we progress so do not be alarmed if what you have just read seems a bit fuzzy.

One final notion about number systems concerns positional value. Consider the base 10 number 33,333. Each '3' represents a different quantity. The left most '3' – also known as the *most significant digit*, represents the value thirty thousand, whereas the '3' on the right – also known as the *least significant digit*, represents the value of three. Note that in base ten, it appears that the positional values are associated with successive powers of the base. This is, in fact, the case. In fact, every number system has a units position – the position located immediately to the left of the *fraction point*. The corresponding power of the base is zero for the units position and any number raised to the zero power, does yield one from whence the units position takes its name. As we proceed in the direction of increasing significance (moving left from the fraction point), the powers increase so that we have a tens position, hundreds position, thousands position and so forth. The following diagram illustrates this.

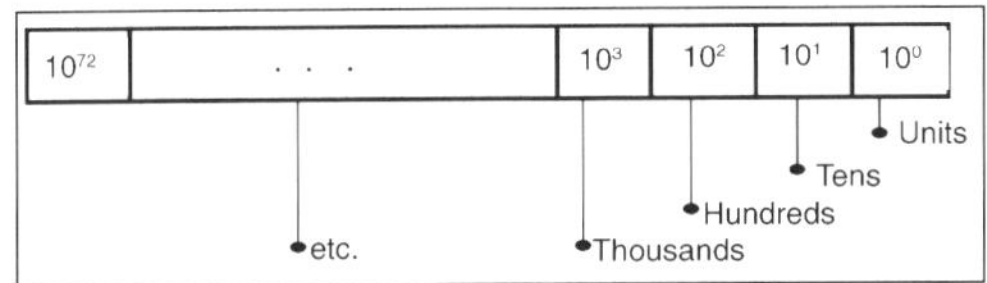

Strictly speaking, computers are modeled on the binary or base 2 number system. The only permissible digits are 0 and 1. We will also discuss the base 8 (octal) and base 16 (hexadecimal) number systems because they bear a close relationship to binary and are often used as shorthand representations of binary numbers. Since numbers in any base can be used to represent quantities, our first task will be to learn how to convert any base ten integer into its equivalent number in binary. We will follow this with similar explanations for converting base ten integers into both the octal and hexadecimal systems and finally, we will generalise the procedure so that you could (if you so wished) convert a base ten integer into its equivalent representation in any other numeric base.

Binary Number System

The Binary number system is called Base 2 because there are only two numbers that make up its set of digits, 0 and 1. Binary numbers are the language of electricity, where a 1 = voltage and a 0 = no voltage. [NOTE: more precisely, a value of '1' is represented by a "high" voltage, generally 3-5 volts; a value of '0' is represented by a "low" voltage, normally 0-2 volts. A voltage level not in either range would signify a circuit that is beginning to degrade or fail]. Actually, these *bi*nary digi*ts* (called *bits* for short) are the only numeric forms a computer can recognise and is the state by which all data is moved within a computer. The relative ordering of the digits correspond to successive powers of two,

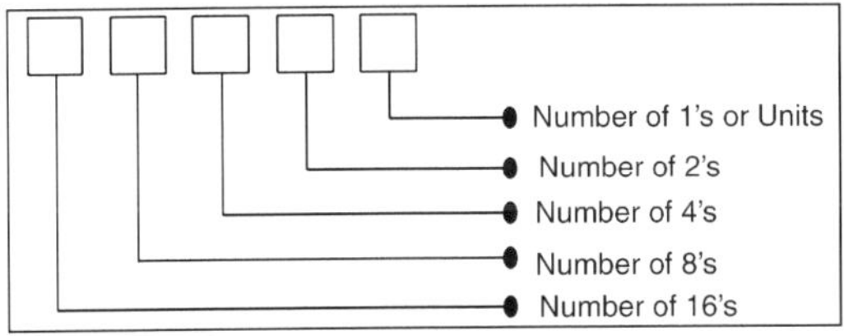

Decimal	Binary
1	1
2	10
3	11
4	100
5	101
6	110
7	111
8	1000
9	1001
10	1010
11	1011
12	1100
13	1101
14	1110
15	1111
16	10000

As the above table indicates, we can count in binary just as readily as we can count in decimal. We simply need more digits to represent the same quantities.

Binary and Octal Number Systems

Converting from binary (base 2) to octal (base 8) begins simply by arranging the binary digits into groups of three (*triplets*). If the total number of binary digits is not divisible by three, leading zeros may be added. It is important to remember to begin grouping *from* the least significant bit *towards* the most significant bit. Next, each triplet is converted to the octal number equivalent according to the coding system previously described. Recall that the range of equivalencies is from 0 (000) to 7 (111). This process is referred to as 3-bit substitution code. Our example of $(1100110)_2$ would yield $(146)_8$.

Method	Add Leading Zeros		
	← 001	100	110
	1	4	6

The conversion from octal to binary is a simple process reversal that begins by converting the most significant octal digit (left most digit) to its binary equivalent. Repeat the conversion with each octal number, working your way towards the least significant digit (right most digit). After converting each octal number, concatenate all binary numbers together. Our example of $(6132)_8$ will yield $(110001011010)_2$.

What happens when a binary point (base 2 equivalent of decimal point) is present? In this case, we group the binary numbers into triplets beginning at the binary point and working outward in both directions. If the number of binary digits is not divisible by three, leading or trailing zeros may be added at either end without altering the value of the number to be converted. Next, each triplet is replaced by its octal equivalent according to the chart above. Our example of $(1100110.10)_2$ would yield $(146.4)_8$.

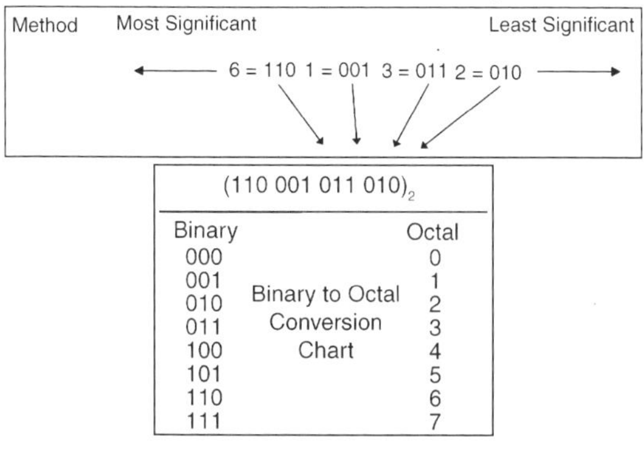

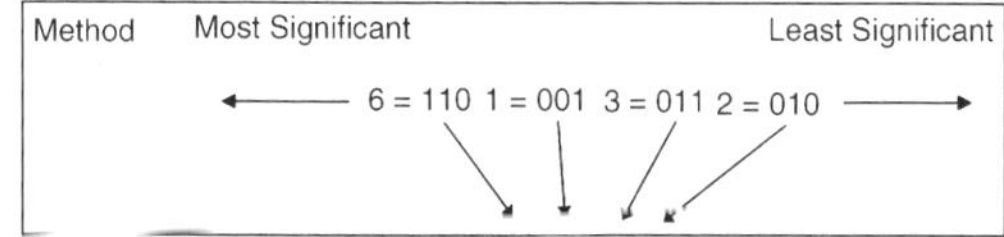

Although we have not yet discussed fractions, they do exist in all number systems. Just as the base 10 system denotes the fractional part of a number by inserting digits to the *right* of a decimal point, all other bases have their own fraction points – be they *binary points, octal points, hexadecimal points* or whatever.

Working with these fraction points in an octal to binary conversion is straightforward. Simply convert each number to the left and right of the decimal point. After converting each octal number, merge all binary numbers together separated by the fraction point. We simply begin at the fraction point, moving to the left for each successive binary triplet and to the right for each successive fractional binary triplet. See the example below for clarification. Our example of $(613.2)_8$ will yield $(110001011.010)_2$.

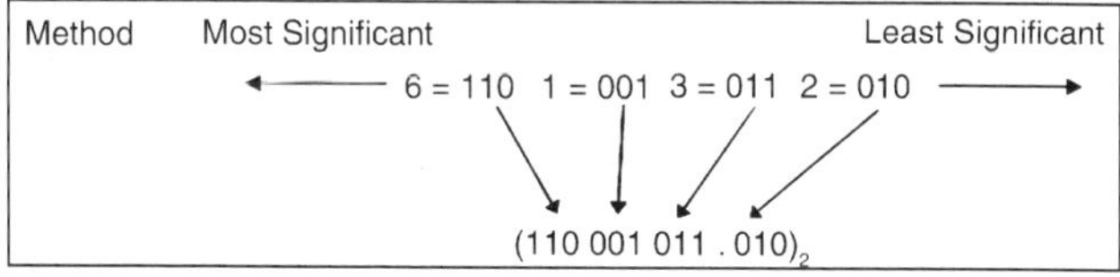

DECIMAL AND HEXADECIMALN UMBER SYSTEMS

Converting from base 10 to base 16 begins by dividing the decimal number by sixteen. As before, this modulo division will produce an integer quotient and most likely, an integer remainder as well. When the quotient divides evenly, the remainder will be zero. Otherwise, the remainder will be in the [1,15] interval. The next step will require us to divide the *new* quotient by sixteen again yielding another quotient and remainder. We continue with this process, until we produce a quotient of zero along with a non-zero remainder. Unlike previous conversions using the division algorithm in which we recorded the remainders in order, and wrote them in a right to the left sequence, we must first identify those remainders that were greater than nine. We must replace remainders 10 through 15 with the corresponding hexadecimal digits A through F.Then we write the sequence as before. The equivalent number in the new base is the string of remainders recorded. Our example below converts $(937)_{10}$ to $(3A9)_{16}$.

Method

$$16\overline{)937} = 58,\quad 928,\quad 9 \qquad 16\overline{)58} = 3,\quad 48,\quad 10 \qquad 16\overline{)3} = 0,\quad 0,\quad 3$$

Shorthand Method

937÷16=58 R 9 LSD
58÷16=3 R 10 (A)
3÷16=0 R 3 MSD

$(937)_{10} = (3A9)_{16}$

Converting from base 16 to base 10 proceeds with positional values and the multiplication algorithm explained previously. In this case, the positional values are functions of successive powers of sixteen. After determining the appropriate positional values (be careful as the numbers get large quickly), simply multiply the hexadecimal digits by their corresponding positional values, and then sum the products to obtain the equivalent decimal value. In our example, we calculate the base 10 value of $(F3B6)_{16}$ to be $(62,390)_{10}$.

Method	F	3	B	6	-Beginning Integers
	16^3	16^2	16^1	16^0	-Base to Positional
Powers					
	4096	256	16	1	-Multiples of Base
number					
	15(F) x 4096	3 x 256	11(B) x 16	6 x 1	-Multiply Position's Digit
by the					
					Multiple of the Base
number					
	61440 +	768 +	176 +	6	-Add Products to find
base 10					
					Equivalent = 62,390

The base ten to base sixteen digit conversion chart is depicted in the following chart:

Decimal to Hexadecidmal Conversion Chart

Decimal	Hex
0	0
1	1
2	2
3	3
4	4
5	5
6	6
7	7
8	8
9	9
10	A
11	B
12	C
13	D
14	E
15	F

IMAGINATION OF BINARY CODES

It would be stretching our imagination to suggest that Sir Francis had digital audio on his minde (sic) when he wrote the prophetic words

Nonetheless, this basic idea forms the basis of everything we do in digital computing, digital communications, and digital audio/video. In 1832, Samuel F. B.

Morse used the very same idea to propose that telegram *words* be coded into *binary addresses* or *binary codes* that could be transmitted over telegraph lines and decoded at the receiving end to unravel the telegram. Morse abandoned his scheme, illustrated in Figure, as too complicated and, in 1838, proposed his fabled Morse code for coding letters (instead of words) into objects (dots, dashes, spaces) capable of a threefold difference onely (sic).

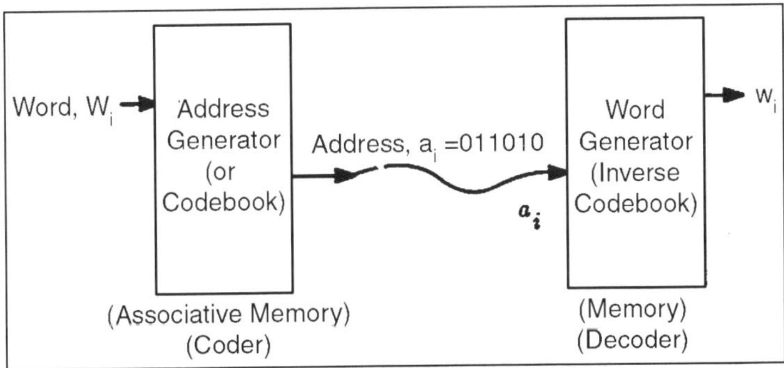

Fig. Generalized Coder-Decoder

The basic idea of Figure is used today in cryptographic systems, where the "address a_i" is an encyphered version of a messagew_i; in vector quantizers, where the "address a_i" is the address of a close approximation to data w_i; in coded satellite transmissions, where the "address a_i" is a data word w_i plus parity check bits for detecting and correcting errors; in digital audio systems, where the "address a_i" is a stretch of digitized and coded music; and in computer memories, where a_i is an address (a coded version of a word of memory) and w_i is a word in memory.

In this chapter we study three fundamental questions in the construction of binary addresses or binary codes. First, what are plausible schemes for mapping symbols (such as words, letters, computer instructions, voltages, pressures, etc.) into binary codes? Second, what are plausible schemes for coding likely symbols with short binary words and unlikely symbols with long words in order to minimize the number of binary digits (bits) required to represent a message?

Third, what are plausible schemes for "coding" binary words into longer binary words that contain "redundant bits" that may be used to detect and correct errors? These are not new questions.

They have occupied the minds of many great thinkers. Sir Francis recognized that arbitrary messages had binary representations. Alan Turing, Alonzo Church, and Kurt Goedel studied binary codes for computations in their study of computable numbers and algorithms.

BASIC EXPLANATION

Did you know that everything a computer does is based on ones and zeroes? It's hard to imagine, because you hear people talking about the absolutely gargantuan (huge) numbers that computers "crunch".

But all those huge numbers - they're just made up of ones and zeros. It's kind of like the computer is made up of a bunch of lightswitches, and each lightswitch controls just one lightbulb.

On or Off. One or Zero. But if you took *all* of those lightbulbs together, and said "Let's make each sequence of On-and-Off represent a different number!" Well then, you could get some pretty large numbers. What do I mean by sequence? Let's say you had two lightswitches.

There are four different ways we could flip those switches:

Both Off
First Off, Second On
First On, Second Off
Both On
Binary Code takes each of those combinations and assigns a num berto it, like this:
BothOff= 0
First Off, Second On = 1
First On, Second Off = 2
Both On = 3

Intermediate Explanation

Another way of thinking about it is this: let's give each lightbulb a point value. We'll say the first lightbulb is worth two points, and the second one is worth one point. Now take a look at the combinations:

Both Off = 0 + 0 = 0
First Off, Second On = 0 + 1 = 1
First On, Second Off = 2 + 0 = 2
Both On = 2 + 1 = 3

If we added in another lightbulb, we would make it worth twice as many points as the 2 pointer. Then, if all the bulbs were on, the point value would be 4 + 2 + 1 = 7. And if we added in *another* bulb, we would make it worth 8 points (twice as much as four). Now if they are all turned on, that's worth 8 + 4 + 2 + 1 = 15. As you can see, it's going to take a lot of lightbulbs to make a really big number!

Finally, even though we're giving point values to each of those lightbulbs, when we write them down we still only write them as ones and zeros. One means On, and Zero means Off.

So let's say we had 8 lightbulbs, and they were set up like this:

Off Off On On Off On Off Off.

The point values of those eight bulbs are: 0 + 0 + 32 + 16 + 0 + 4 + 0 + 0 (remember - we only give points if they're turned on!)

And that adds up to 52. So we would say the sequence of bulbs is worth 52. But how do we write it? We write it like this: 00110100

So now we can say $00110100_{binary} = 52$.

8421 BCD CODE

In the 8421 Binary Coded Decimal (BCD) representation each decimal digit is converted to its 4-bit pure binary equivalent.

For example: $57_{dec} = 0101\ 0111_{bcd}$

Addition is analogous to decimal addition with normal binary addition taking place from right to left. For example,

6	0110	BCD for 6	42	0100 0010	BCD for 42
+3	0011	BCD for 3	+27	0010 0111	BCD for 27
	1001	BCD for 9		0110 1001	BCD for 69

Where the result of any addition exceeds 9(1001) then six (0110) must be added to the sum to account for the six invalid BCD codes that are available with a 4-bit number. This is illustrated in the example below

8	1001	BCD for 8
+7	0111	BCD for 7
	1111	exceeds 9 (1001) so
	0110	add six (0110)
	0001 0101	BCD for 15

Note that in the last example the 1 that carried forward from the first group of 4 bits has made a new 4-bit number and so represents the "1" in "15". In the examples above the BCD numbers are split at every 4-bit boundary to make reading them easier. This is not necessary when writing a BCD number down. This coding is an example of a binary coded (each decimal number maps to four bits) weighted (each bit represents a number, 1, 2, 4, etc.) code.

4221 BCD CODE

Decimal	4221	1's complement
0	0000	1111
1	0001	1110
2	0010	1101
3	0011	1100
4	1000	0111
5	0111	1000
6	1100	0011
7	1101	0010
8	1110	0001
9	1111	0000

The 4221 BCD code is another binary coded decimal code where each bit is weighted by 4, 2, 2 and 1 respectively. Unlike BCD coding there are no invalid representations. The decimal numbers 0 to 9 have the following 4221 equivalents the 1's complement of a 4221 representation is important in decimal arithmetic. In forming the code remember the following rules.

- Below decimal 5 use the right-most bit representing 2 first.
- Above decimal 5 use the left-most bit representing 2 first.
- Decimal 5 = 2+2+1 and not 4+1.

GRAY CODE

Gray coding is an important code and is used for its speed, it is also relatively free from errors. In pure binary coding or 8421 BCD then counting from 7 (0111) to 8 (1000) requires 4 bits to be changed simultaneously. If this does not happen then various numbers could be momentarily generated during the transition so creating spurious numbers which could be read.

The first 16 Gray Coded Numbers are Indicated below.

Decimal	Gray Code
0	0000
1	0001
2	0011
3	0010
4	0110
5	0111
6	0101
7	0100
8	1100
9	1101
10	1111
11	1110
12	1010
13	1011
14	1001
15	1000

Gray coding avoids this since only one bit changes between subsequent numbers. To construct the code there are two simple rules. First start with all 0s and then proceed by changing the least significant bit (lsb) which will bring about a new state.

To convert a Gray-coded number to binary then follow this method:

- The binary number and the Gray-coded number will have the same number of bits
- The binary MSB (left-hand bit) and Gray code MSB will always be the same
- To get the binary next-to-MSB (i.e. next digit to the right) add the binary MSB and the gray code next-to-MSB. Record the sum, ignoring any carry.

- Continue in this manner right through to the end. Gray coding is a non-BCD, non-weighted reflected binary code.

CATEGORY OF CYCLIC CODES

Cyclic codes fall under the category of codes called linear codes. Definition 6 A linear code of dimension k and length n over a field $\mathbb{F}$ is a k-dimensional subspace of $\mathbb{F}^n$. Such a code is called an n,k code. If the minumum distance of the code is d, then the code is called an $[n,k,d]$ code.

The row-space of the following matrix forms a $[7,3,4]$ code.

$$\begin{pmatrix} 1 & 0 & 1 & 1 & 1 & 0 & 0 \\ 0 & 1 & 0 & 1 & 1 & 1 & 0 \\ 0 & 0 & 1 & 0 & 1 & 1 & 1 \end{pmatrix}$$

Notice that there are 4 ones in each row. Any binary n-vector can be thought of as the coeffiecients of an nth degree polynomial in $\mathbb{Z}_2$. Looking at the first row in the matrix we can relate it to the polynomial

$$g(x) = 1 + x^2 + x^3 + x^4$$

in the polynomial ring $\mathbb{Z}_2[x]$. Given a linear code the following proposition is clear.

Proposition 1 Let C be a linear code. Then $d(C)$ equals the smallest Hammming weight of all nonzero code vectors. There are many different varieties of linear codes, such as the well known Hamming codes. We now want to consider a brand of linear codes called cyclic codes. Then we will define BCH codes, which are a specific type of cyclic code.

Definition 7 A linear code C is called cyclic if $(c_1, c_2, \ldots, n_c) \in C$ implies $(c_n, c_1, c-2, \ldots, n_{n-1}) \in C$

For example, if $(1,1,1,0,1,0)$ is in a cyclic code, then $(0,1,1,1,0,1)$ is also in the code. The definition is recursive and so every cyclic shift of the elements is in the code. In order to formalize the discussion we can use the following isomorphism

$$(a_0, a_1, \ldots, a_{n-1}) \in \mathbb{F}_q^n \rightleftharpoons a_0 + a_1 x + \ldots + a_{n-1} x^{n-1} \in \mathbb{F}_q[x]$$

between the n-dimensional vector space $\mathbb{F}_q^n$ and the residue class of polynomials $\mathbb{F}_q[x]/(x^n - 1)$. Recall from algebra that

$$\mathbb{F}_q[x]/(x^n - 1) = \left\{a_0 + a_1 x + \ldots + a_{n-1} x^{n-1} \middle| a_i \in \mathbb{F}_q, 0 \le i \le n\right\}$$

We can now speak of codeword in C as the polynomials $c(x)|\mathbb{F}_q[x]/(x^n-1)$, and the following theorem follows

Theorem 2 A linear code C in $\mathbb{F}_q^n$ is a cyclic code iff C is an ideal in $\mathbb{F}_q[x]/(x^n-1)$.

Proof. $(\Rightarrow)$ If C is an ideal in $\mathbb{F}_q[x]/(x^n-1)$ and $c(x)=c_0+c_1x+c_{n-1}x^{n-1}$ is any codeword, then $xc(x)$ is also a codeword. Thus

$$(c_{n-1},c_0,c-1,...,c_{n-2})\in C \quad (\Leftarrow)$$

If C is cyclic, then for every codeword $c(x)$ the word $xc(x)$ is also a codeword. Therefore $x^i c(x)$ is in C for every i, and since C is linear $a(x)c(x)$ is in C for every polynomial $a(x)$. It follows that C is an ideal.

The next theorem states the general structure of cyclic codes.

Theorem 3 Let C be a cyclic code of length n over a finite field $\mathbb{F}$. To each codeword $(a_0,a_1,...,a_n-1)\in C$, associate the polynomial $a_0+a_1x+...+a_{n-1}x^{n-1}$ in $\mathbb{F}[x]$. Among all the nonzero polynomials obtained from C in this way, let $g(x)$ have the smallest degree. We may assume that $g(x)$ is monic. This polynomial is called the generating polynomial for C. Then

- $g(x)$ is uniquely determined by C.
- $g(x)$ is a divisor of x^{n-1}.
- C is exactly the set of coefficients of the polynomials of the form $g(x)f(x)$ with deg $(f)\leq n-1$.
- Write $x^n-1=g(x)h(x)$. Then $m(x)\in\mathbb{F}[x]/(x^n-1)$ corresponds to an element of C iff $h(x)m(x)\equiv 0 \bmod (x^n-1)$.

Proof:

- The codespace is spanned by the rows of a matrix. Because the row space of the matrix is closed under subtraction the difference between any two vectors, or polynomials, in the space is still in the space. So if there was another monic generating polynomial $g_1(x)$ with minimum degree, then $g(x)-g_1(x)$ is also in the space. But this is a contradiction since this new polynomial has smaller degree than both $g(x)$ and $g_1(x)$.

- Divide $g(x)$ into $x^n - 1$. By the division algorithm we get $x^n - 1 = g(x)h(x) + r(x) \Rightarrow -r(x) \equiv g(x)h(x) \bmod (x^n - 1)$ where deg $r(x) <$ deg $g(x)$.

But multiplying $g(x)$ by powers of x corresponds to a cyclic shift of the corresponding code vector and multiplication by a polynomial is thus a linear combination of the code vectors in the generating matrix and all of their cyclic shifts. So $r(x)$ is in the code space of C. Since deg $r(x) <$ deg $g(x) <$, $r(x)$ must be equal to zero. Thus $g(x) \mid x^n - 1$. Notice then that $\mathbb{F}[x]/(x^n - 1)$ is not a field since it has zero divisors.

- Let $m(x)$ correspond to an element in C. Divide $g(x)$ into $m(x)$ $m(x) = g(x)f(x) + r_1(x)$ with deg $r(x) <$ deg $g(x)$. From (2) we know that $g(x)f(x)$ corresponds to a codeword and by assumption so does $m(x)$. So, since C is closed under subtraction, $m(x) - g(x)f(x)$ mod $(x^n - 1)$ is also a codeword. But this is just $r_1(x)$ and has degree less than $g(x)$. It follows that $r_1(x) \equiv 0$ and $m(x)$ is the product of the generating polynomial with another polynomial. Because each codeword has length n, deg $m(x) \leq n - 1$ so deg $f(x) \leq n - 1 -$ deg $g(x)$. Conversely, we have shown that the product of any two polynomials $g(x)f(x)$ is in C. Thus, these polynomials yield all of the code words of C.
- By (2) we can write $x^n - 1 = g(x)h(x)$. Let $m(x)$ correspond to an element in C. Then by (3) $h(x)m(x) = h(x)g(x)f(x) = (x^n - 1)f(x) \equiv 0$ mod $(x^n - 1)$ Next suppose that $m(x)h(x) \equiv 0$ mod $(x^n - 1)$.

Then $m(x)h(x) = (x^n - 1)q(x) = h(x)g(x)q(x)$ for some polynomial $q(x)$. We can now divide through by $h(x)$ and get $m(x) = g(x)q(x)$ and so $m(x)$ coresponds to a codeword of C.

Remark: There will be a unique polynomial $h(x)$ such that $g(x)h(x) = x^n - 1$. When thinking of the codespace as a product of polynomials with $g(x)$ instead of the row space of a binary matrix, this polynomial $h(x)$ will take the place of the check matrix in the decoding process.

Let $c(x) = b(x)g(x)$ be a codeword then

$$c(x) = h(x)g(x)h(x) = b(x)(x^n - 1)$$

Then if $m = \deg b(x)$ the coeffiecients of the m highest powers match the coeffiecients of the m lowest powers of x in the polynomial $c(x)h(x)$.

Further, if the generating polynomial

$g(x) = g_0 + g_1 x + ... + g_{n-k} x^{n-k}$ has degree $n-k$, then the codewords $g(x), xg(x) + x^2 g(x), ... + g_{n-k} x^{n-k}$ form a basis for the codespae C with a generating matrix

$$\begin{pmatrix} g_0 & g_1 & \cdots & g_{n-k} & 0 & 0 & ... & 0 \\ 0 & g_1 & ... & g_{n-k-1} & g_{n-k} & 0 & ... & 0 \\ 0 & 0 & ... & & & & ... & 0 \\ 0 & 0 & ... & & g_0 & g_1 & ... & g_{n-k} \end{pmatrix}$$

and we can Realise a code word as an information sequence $(a_0, a-1, ... a_{k-1})$ times the matrix $aG = \left(a_o + a_1 x + ... + a_{k-1} x^{k-1}\right) g(x)$ mod $x^n - 1$ As stated before, we can treat the polynomial $h(x)$, having the condition that $g(x)h(x) \equiv 0$ mod $x^n - 1$, as the check polynomial of the code C. Knowing that we want $g_0 h_i + g_1 h_{i-1} + ... + g_{n-k} h_{i-n-k} = 0$ for $i = 0, 1, ..., n-1$, it follows that the check matrix for the code is

$$\begin{pmatrix} 0 & 0 & ... & 0 & h_k & ... & h_1 & h_0 \\ 0 & 0 & ... & h_k & ... & h_1 & h_0 & 0 \\ \vdots & & & & & & & \vdots \\ h_k & ... & h_1 & h_0 & 0 & ... & 0 & 0 \end{pmatrix}$$

Lets now consider an example using a 3×7 matrix of 0's and 1's.

$$\begin{pmatrix} 1 & 0 & 1 & 1 & 1 & 0 & 0 \\ 0 & 1 & 0 & 1 & 1 & 1 & 0 \\ 0 & 0 & 1 & 0 & 1 & 1 & 1 \end{pmatrix}$$

Because each of the three rows are linearly independent we know from linear algebra that the row space of the matrix is a 3-dimensional subspace of $\mathbb{F}^7$ where $\mathbb{F}$ is the field $\mathbb{Z}_2$. In fact, since the algorithm is based on left multiplication by a code vector, the row space is the range of the matrix. In essence, because each code vector corresponds to a polynomial, the code space corresponds to cyclic shifts of the coefficients of polynomials, which is just all polynomial products $a(x)g(x)$ where $g(x)$ is the generating polynomial for the code C.

It turns out that in this case all the cyclic shifts of the first row vector account for every vector in the row space. Thus we have a cyclic code and the generating polynomial is

$$g(x) = 1 + x^2 + x^3 + x^4$$

which corresponds to the first row of the generating matrix.

Also, $(x^4 + x^3 + x^2 + 1)(x^3 + x^2 + 1) = x^7 - 1$ and so the check polynomial is

$$h(x) = x^3 + x^2 + 1$$

and we get the following check matrix

$$H = \begin{pmatrix} 0 & 0 & 0 & 1 & 1 & 0 & 1 \\ 0 & 0 & 1 & 1 & 0 & 1 & 0 \\ 0 & 1 & 1 & 0 & 1 & 0 & 0 \end{pmatrix}$$

Now taking the codeword c_g corresponding to $g(x)$ we get

$$Hc_g = \begin{pmatrix} 0 & 0 & 0 & 1 & 1 & 0 & 1 \\ 0 & 0 & 1 & 1 & 0 & 1 & 0 \\ 0 & 1 & 1 & 0 & 1 & 0 & 0 \end{pmatrix} \begin{pmatrix} 1 \\ 0 \\ 1 \\ 1 \\ 1 \\ 0 \\ 0 \end{pmatrix} = \begin{pmatrix} 0 \\ 0 \\ 0 \\ 0 \\ 0 \\ 0 \\ 0 \end{pmatrix}$$

mod 2

In fact, for every codeword c in the space C, the product Hc will be the zero vector. This will be an extremely important detail when creating an algorithm for a single-error correcting BCH code

ERROR-CORRECTING CODES

The theory of error detecting and correcting codes is that branch of engineering and mathematics which deals with the reliable transmission and storage of data. Information media are not 100% reliable in practice, in the sense that noise (any form of interference) frequently causes data to be distorted. To deal with this undesirable but inevitable situation, some form of redundancy is incorporated in the original data.

With this redundancy, even if errors are introduced (up to some tolerance level), the original information can be recovered, or at least the presence of errors can be detected. We saw in class how adding to the original message the parity bit or the arithmetic sum allows the detection of a (certain type of) error. However, that kind of redundancy doesn't allow for the correction of the error. Error-correcting codes do exactly this: they add redundancy to the original message in such a way that it is possible for the receiver to detect the error and correct it, recovering the original message.

This is crucial for certain applications where the re-sending of the message is not possible (for example, for interplanetary communications and storage of data). The crucial problem to be resolved then is how to add this redundancy in order to detect and correct as many errors as possible in the most efficient way. This document starts with an overview of the applications of the error-correcting codes, followed by a session describing the fundamental concepts. It then addresses some of the issues involved in the design of the codes. The last session is a brief introduction of the very last discovery in the theory of error-correcting codes: the turbo codes. Most of us have used error-correcting codes, perhaps without knowing it.

The International Standard Book Number (ISBN) system identifies every book with a ten-digit number, such as 0-226-53420-0. The first nine digits are the actual number but the tenth is added according to a mathematical formula based on the first nine. If a single one of the digits is changed, as in a misprint when ordering a book, a simple check verifies that something is wrong.

Other common numbers with check digits include the Chemical Abstracts Service (CAS) method of identifying chemical compounds and the United States Postal Service (USPS) nine-digit "zip+four" code, which when it is bar-coded includes a tenth check digit hidden to the human eye but visible to the bar code reader. Some high-end computer memory chips, "ECC RAM," use extended nine-bit bytes. The ninth bit, or "check bit," is always set so that the total number of ones in the extended byte is even. (This rule is called a "checksum." Specifically, if the sum of the nine bits is not even, an error has been detected.) This ninth bit is not used or even seen by your software, but the computer hardware itself will signal an error if it ever reads an extended byte that contains an odd number of ones.

All these processes can *detect* a single error within a very short message: nine decimal digits for the ISBN and USPS codes, a varying number of decimal digits for the CAS numbers, eight binary digits for ECC RAM. However, they cannot *correct* any error that is detected. Furthermore, some combinations of two or more errors occurring within the message will not be detected.

CODING THEORY

Coding theory deals with ways of extending the check-bit and checksum ideas to enable detection and even correction of large numbers of errors within messages. It builds on some beautiful mathematical ideas that we can hint at here. Let's focus on binary messages, the stuff of which computer communications is built. The first thing is to break up the message into "words" of fixed size. We will focus on how to correct errors occurring within individual words.

If the words are n bits long, then there are 2^n possible messages each word can carry. The set of all such words has an interesting geometry based on the *Hamming distance*. The distance between two words is just the number of places in which they differ. For example, let's consider the case n=7. The distance between $\underline{0}10011\underline{1}$ and $\underline{1}10010\underline{1}$ is two, because these words differ in the first and sixth places.Words can also be added to each other. This is done by the usual odd-even rules of addition, performed bit by bit. Specifically, 0+0=0, 0+1 = 1, 1+0 = 1, and 1+1 = 0. Thus 0100111 + 1100101 = 1000010.

Suppose you want to send a message in the form of a binary string, i.e., a string of 0's and 1's. Due to noise in the medium over which you are transmitting the message, be it a telephone line, a radio link or the Internet, it is possible that some of the 0's will be received as 1's and vice versa. Let us denote the probability that any one bit is garbled in transmission by p. Typically, p will be small, say about one in a million, but that may still be too large for our purposes. If we are transferring a file that has 10 million bits, the chance that the file is incorrectly received is nearly 90%. How can we improve the reliability of the transmission? The simplest method is to repeat ourselves three times.

For example, if we want to send the string 010, we transmit 000111000. If a single error occurs in transmission, this may be received for instance as 001111000. At the receiver, we use what is called majority decoding. We look at each set of 3 bits and decode it as the bit which occurs most often. Thus, the received message above is correctly decoded as 010 even though an error occured in transmission. In the above example, it could have happened that two errors occured and we received the string 101111000. This would be decoded as 110, which is different from the original message. This shows that we cannot eliminate errors altogether, we can only make them less likely.

To see how much less likely, let us work out the probability that a single bit is incorrectly received. Since each bit is transmitted three times, it has to be incorrectly received at least twice in order for the decoder to get it wrong. The probability of this is $3p^2+p^3$, which, for p equal to one in a million works out to about one in 330 billion. Thus, using the repetition code, a 10 Mbit file can be sent correctly with probability 99.997%. While this coding scheme dramatically reduces the likelihood of errors, it is rather inefficient. Can we do better?

Here's one way, using the famous (7,4) Hamming code. This works with a block of 4 bits at a time rather than with a single bit. Suppose we want to transmit the block 0011. Look at the picture below. We write these bits in the intersections of the circles. Next, we fill in the part of each circle lying outside the intersection using the rule that the total number of 1's contained in each circle should be even.

The extra bits are called parity bits—they ensure even parity for the number of 1's in each circle. The message we transmit will consist of all seven bits, in a pre-arranged order so that the receiver can reconstruct the diagram. Suppose now that a single error occurs, so that the receiver gets a 1 in the intersection of the top two circles. The receiver proceeds to work out the parity for each of the circles, and finds that the top two circles both have odd parity (odd number of 1's), while the bottom circle has even parity. This enables her to conclude that an error has occurred, and that the error is in the bit position common to the top two circles. Therefore, the intended message can be decoded correctly. It is not hard to see that a similar approach works if the error is in some other bit position. However, if there are two errors, then the decoding procedure fails in that it yields a message different from the intended one.

APPLICATIONS OF ERROR CORRECTING CODES

The increasing reliance on digital communication and the emergence of the digital computer as an essential tool in a technological society have placed error-correcting codes in a most prominent position. We cite here a few specific applications, in an attempt to indicate their practicality and importance. Many computers now have error-correcting capabilities built into their random access memories; it is less expensive to compensate for errors through the use of error-correcting codes than to build integrated circuits that are 100% reliable. The single error-correcting Hamming codes, and linear codes in general, are of use here.

Disk storage is another area of computing where error-coding is employed. Storage capacity has been greatly increased through the use of disks of higher and higher density. With this increase in density, error probability also increases, and therefore information is now stored on many disks using error-correcting codes. In 1972, the *Mariner* space probe flew past Mars and transmitted pictures back to earth. The channel for such transmissions is space and the earth's atmosphere. Solar activity and atmospheric conditions can introduce errors into weak signals coming from the spacecraft. In order that most of the pictures sent could be correctly recovered here on earth, the following coding scheme was used. The source alphabet consisted of 64 shades of gray. The source encoder encoded each of these into binary 6-tuples and the channel encoder produced binary 32-tuples. The source decoder could correct up to 7 errors in any 32-tuple. This was done with the Reed-Muller codes.

In 1979, the *Voyager* probes began transmitting Colour pictures of Jupiter. For Colour pictures the source alphabet needed to be much larger and was chosen to have 4096 Colour shades. The source encoder produced binary 12-tuples for each Colour shade and the channel encoder produced 24-tuples (i.e. 24 dimensional vectors). This code would correct up to 3 errors in any 24-tuple, and is the Golay code. The increasing popularity of digital audio is due in part to the powerful error-correcting codes that the digitization process facilitates. Information is typically stored on a small aluminized disk as a series of microscopic pits and smooth areas, the pattern representing a sequence of 0's and 1's.

A laser beam is used as the playback mechanism to retrieve stored data. Because the data is digital, error correction schemes can be easily incorporated into such a system. Given an error-coded digital recording, a digital audio system can on playback, correct errors introduced by fingerprints, scratches, or even imperfections originally present in the storage medium. The compact disc system pioneered by Philips Corporation of the Netherlands in cooperation with Sony Corporation of Japan, allowing playback of pre-recorded digital audio disks, is an excellent example of the application of error-correcting codes to digital communication; *crossinterleaved Reed-Solomon* codes are used for error correction in this system. Digital audio tape (DAT) systems have also been developed, allowing digital recording as well as playback.

Error-correcting codes are particularly suited when the transmission channel is noisy. This is the case of wireless communication. Nowadays, all digital wireless communications use error-correcting codes.

FUNDAMENTAL CONCEPTS

We begin with a few definitions, in order to develop a working vocabulary. Let A be an alphabet of q symbols. For example, A = {a,b,c,...,z) is the standard lower case alphabet for the English language, and A = (0,1) is the binary alphabet.

Definition A block code of length n containing M codewords over the alphabet A is a set of M n-tuples where each n-tuple takes its components from A. We refer to such a block code as an [n,M]-code over A. In practice, we most frequently take A to be the binary alphabet.

Given a code C of block length n over an alphabet A, those specific n-tuples over A which are in C are referred to as codewords. Note that while the channel encoder transmits codewords, the n-tuples received by the channel decoder may or may not be codewords, due to the possible occurrence of errors during transmission.

Example: Suppose the information we are to transmit comes from the set of symbols {A, B, C, D}. For practical considerations we associate sequences of 0's and l's with each of these symbols.

A $\rightarrow$ 00

B $\rightarrow$ 01

C $\rightarrow$ 10

D $\rightarrow$ 11

This is the source encoding.

Now we want to add some redundancy (channel encoding).

A $\rightarrow$ 00 $\rightarrow$ 00000

B $\rightarrow$ 01 $\rightarrow$ 10110

C $\rightarrow$ 10 $\rightarrow$ 01011

D $\rightarrow$ 11 $\rightarrow$ 11101

We have just constructed a [5,4]-code over a binary alphabet. That is, we constructed a code with 4 codewords, each being a 5-tuple (block length 5), with each component of the 5-tuple being O or 1. The code is the set of n-tuples produced by the channel encoder (as opposed to the source encoder). The source encoder transforms messages into k-tuples (k=2 in the example above) over the code alphabet A, and the channel encoder assigns to each of these information k-tuples a codeword of length n (n=5 in the example). Since the channel encoder is adding redundancy, we have n > k and hence we have message expansion. While the added redundancy is desirable from the point of view of error control, it decreases the efficiency of the communication channel by reducing its effective capacity. The ratio k to n is a measure of the fraction of information in the channel which is non-redundant. Definition The rate of an [n,M]-code which encodes information k-tuples is

$$R = K/n$$

The rate of the simple code given in example 1 is 2/5.

The quantity r = n-k is sometimes called the redundancy of the code.

A fundamental parameter associated with an [n,M]-code C is the Hamming distance for C. Before we can define the Hamming distance for a code, we must define the Hamming distance between two codewords.

Definition The Hamming distance d(x,y) between two codewords x and y is the number of coordinate positions in which they differ.

Example: Over the alphabet A = (0,1}, the codewords x and y x=(10110) y=(11O11) have Hamming distance d(x,y) =3.

Example: The codewords u and v over the alphabet A = (0,1,2), given by u = (21002) v=(12001) have Hamming distance d(u,v) =3.

Definition Let C be an [n,M]-code.

The Hamming distance d of the code C is d= min {d(x,y): x,y belong to C, x != y}.

In other words, the Hamming distance of a code is the minimum distance between two distinct codewords, over all pairs of codewords.

Example: Consider C = $(c_0, c_1, c_2, c_3\}$ where c_0=(00000) c_1 =(10110) c_2= (01011) c_3= (11101) This code has distance d = 3. It is the [5,4] code constructed earlier.

Suppose we have an [n,M]-code C with distance d. We need to adopt a strategy for the channel decoder (or just decoder). When the decoder receives an n-tuple r it must make some decision. This decision may be one of

- *No errors have occurred*: Accept r as a codeword.
- *Errors have occurred*: Correct r to a codeword c.
- *Errors have occurred*: No correction is possible.

In general, the decoder will not always make the correct decision; for example, consider the possibility of an error pattern occurring which changes a transmitted codeword into another codeword. The goal is that the decoder take the course of action which has the greatest probability of being correct. One usually makes the assumption that errors are introduced by the channel at random, and that the probability of an error in one coordinate is independent of errors in adjacent coordinates.

The decoding strategy we shall adopt, called nearest neighbour decoding, can then be specified as follows. Nearest Neighbour Decoding If an n-tuple r is received, and there is a unique codeword c that belongs to C such that d(r,c) is a minimun, then correct r to the c. If no such c exists, report that errors have been detected, but no correction is possible. By nearest neighbour decoding, a received vector is decoded to the codewords"closest" to it, with respect to Hamming distance. The figure below illustrates the concept. The red dots are the codewords, the green ones are the received words that can be corrected to the closest codewords and the blue dot is a received word that is equidistant from 2 codewords and therefore cannot be corrected.

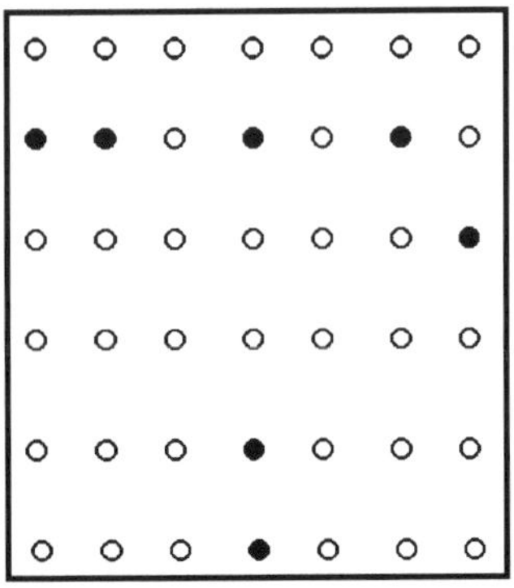

A code is said to correct e errors if a decoder using the above scheme is capable of correcting any pattern of e or fewer errors introduced by the channel. In this case, the decoder can correct any transmitted codeword which has been altered in e or fewer coordinate positions.

Theorem Let C be an [n,M]-code having distance d = 2e+1. Then C can correct e errors. If used for error detection only, C can detect 2e errors.

Let c be one codeword of C and S be the set of all n-tuples over the alphabet of C and define S(c) = {x belongs to S: d(x,c) $\Leftarrow$ e).

S(c) is called the sphere of radius e about the codeword c. It consists of all n-tuples within distance e of the codeword c, which we think of as being at the Centre of the sphere. In the figure below, the"spheres" arount the codewords of the previous figure. The radius is in this case is 1.

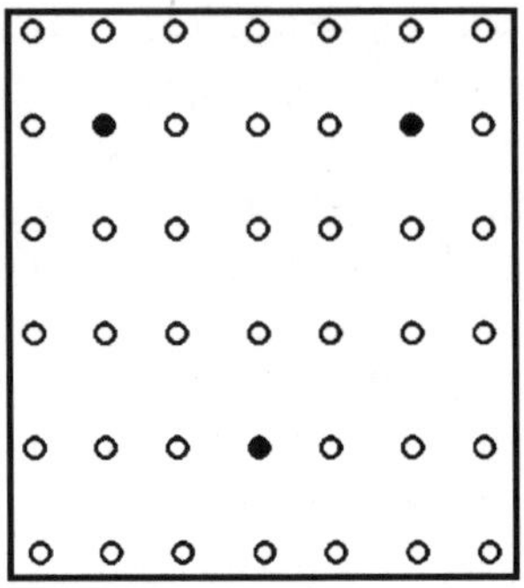

Given 2 codewords, their spheres don't intersect, hence if codeword c is transmitted and t⇐ e errors are introduced, the received word r is an n-tuple in the sphere S(c) and thus c is the unique codeword closest to r.

The decoder can always correct any error pattern of this type. If we use the code only for error detection, then at least 2e+1 errors must occur in a codeword to carry it into another codeword.

If at least 1 and at most 2e errors are introduced, the received word will never be a codeword and error detection is always possible.

The spheres have to be disjoint in order to be able to correct errors. But if a received word happens to be in the space between the spheres, correction is not possible: that space doesn't belong to any sphere, i.e. to any codeword: the error is detected but not corrected.

Therefore, we want to find codes such that the the spheres around the codewords are disjoint (to correctly correct the errors) but as close as possible to each other, to correct ALL the errors.

If all the space is taken by the sphere, all the words received will fall into the sphere of a codeword and therefore can be corrected to that codeword. These are the perfect codes. Lets give the formal definition of perfect codes.

Definition A perfect code is an e-error-correcting [n, M]-code over an alphabet A such that every n-tuple over A is in the sphere of radius e about some codework.

In a perfect code of block length n, the sphere of radiuse about codeworks are not only disjoint (since the code is e-error-correcting) but exhaust the entire space of n-tuples.

ISSUES IN THE DESIGN OF ERROR-CORRECTING CODES

Designing a good code is a very hard problem. Also, lots of factors need to be taken in account. For instance, in practice the code should be designed appropriately depending on the expected rate of errors for the particular channel being employed. If it is known that the probability that the channel will introduce 2 errors into any transmitted codeword is extremely small, then it is likely not necessary to construct a code that will correct all 2-bit error patterns.

A single error correcting code will likely suffice. Conversely, if double errors are frequent and a single error correcting code is being used, then decoding errors will be frequent. From this brief introduction, a number of questions arise.

- Given n, M and d, can we determine if an [n Al] code with distance d exists?
- Assuming such a code does exist, how would one be constructed in practice?
- How should information k-tuples be associated with codeword n-tuples to facilitate efficient channel encoding?
- How should channel decoding be performed?

Lets examine the issue of channel decoding in greater detail. When a word is received and it's not a codeword, finding the closest codeword involves computing the distance from the received word to each codeword, requiring M comparisons. While this might be acceptable for small M, in practice we usually find M quite large. Suppose a code C is used with M = 2^50 codewords, which is not unrealistic. If we could carry out 1 million distance computations per second, it would take around 20 years to make a single correction.

Clearly, this is not tolerable and more efficient techniques are required. Wishing to retain the nearest neighbour decoding strategy, we seek more efficient techniques for its implementation. The theory of error-correcting code is concerned with constructing codes for various values of n, M and d, and the consideration of appropriate encoding and decoding techniques.

Most of the best known codes are algebraic in nature like the linear codes. Linear codes are an important and general class of error-correcting codes. For example, theBCH codes used for CDs are linear codes. They are heavily mathematical and an appropriate description here is not possible. The next session describes instead the last"discovery" in the theory of error correcting codes, the Turbo codes.

Turbo Codes

We introduced the Nearest Neighbour Decoding, we didn't mention the fact that what the decoding actually does is maximizing the probability P(r|c) that r is received, given that c is sent. i.e. choosing the nearest codeword is equivalent to choosing the most likely input message c given the received tuple r. This decoding strategy is known as maximum likelihood decoding. We can then state the decoding problem in the following manner. Let U be the original message, X the encoded message and Y the message received after the noisy channel. This is schematized in the following figure.

The receiver knows Y and the model of the encoder, i.e. is has a knowledge of how the original message was originally encoded.

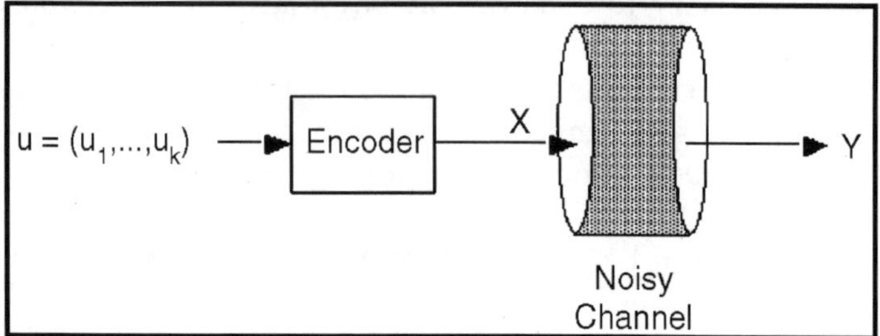

The problem now is finding the U that maximizes the probability that U was sent given that Y was received, i.e. the probability P(U | Y).

Having stated the decoding problem in probabilistic terms, we can take advantage of various methods that deal with probability estimations. A class of such methods is the class of graphical models.

The simplest graphical model:

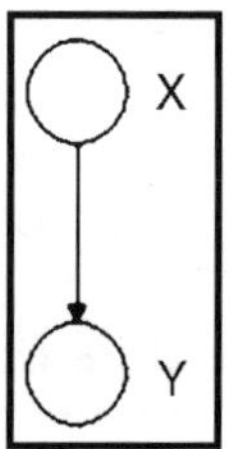

A classical problem of probability estimation is finding the probability distribution P(X | Y) where X and Y are 2 random variables.

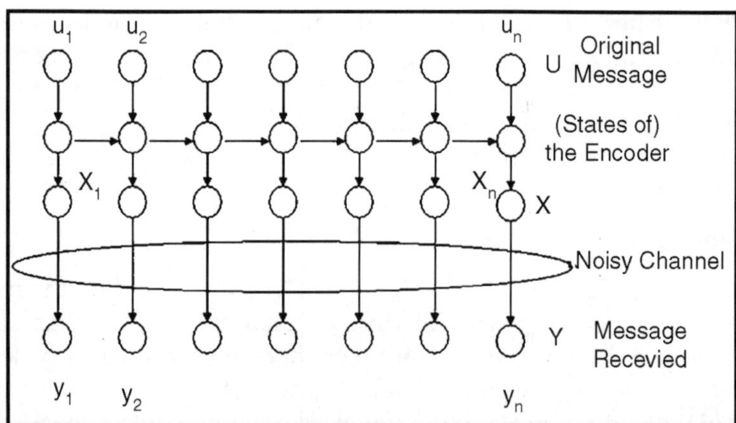

The graphical model framework allows for such probability estimations to be calculated in an efficient way. Note that the calculation of P(X | Y) from P(Y | X) and P(X) is just Bayes rule – P(X | Y) = P(Y | X)P(X)/P(Y) –

and that graphical model inference algorithms can be viewed as generalizations of Bayes rule to arbitrary graphs.

Lets now see how we can view error correcting codes as graphical models. In the figure below, a convolution code as a graphical model.Let U be the original message, X the encoded message and Y the message received after the noisy channel.

Now, the decoding problem of finding U that maximizes p(U I Y) can be tackled within the framework of graphical models.

We have introduced the convolutional codes mainly because the Turbo codes are two convolutional codes put together. The following figure illustrate the graphical representation of a turbo code.

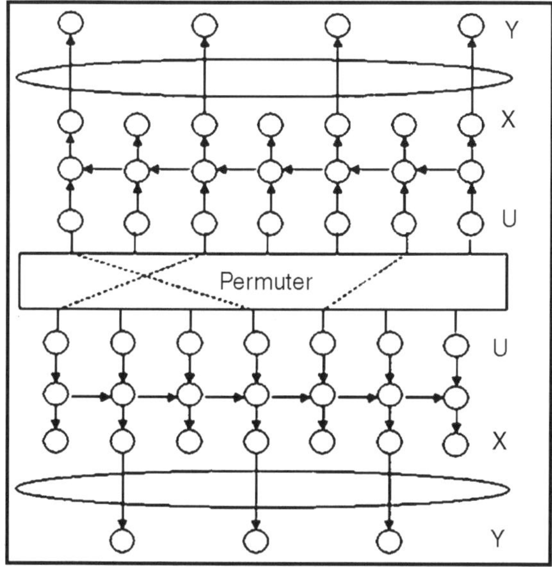

It should be mentioned that general inference in graphical models is NP-hard, and, in particular, exact inference, calculating P(U I Y), for turbo codes is infeasible and therefore an approximate inference algorithm is used.

Turbo codes were discovered very recently. There is much excitement about them. They outperform all other kinds of error-correcting codes in the case of long block lengths, although the reason why they do so is not yet clear.

NUMBER SYSTEMS AND CODES

Computers are information processing tools that give meaning to raw data. Data should be represented in a form that permits efficient storage and ease of manipulation (although these two goals might be conflicting at times). In data communications, we are also concerned with issues such as security, error detection and error correction.

Data are usually represented in some form of code for compactness and uniformity. For instance, a 'yes' response may be represented by '*Y*", and a 'no' by '*N*'; the four seasons by '*S*' (spring), '*M*' (summer), '*A*' (autumn) and '*W*' (winter). We can see that the choices are rather arbitrary. In real-life, codes are used for identification purposes, such as the Singapore NRIC number and the NUS student's matriculation number.

Computers are also number crunching devices. Numeric values are represented in a form that eases arithmetic computations and preserves accuracy as much as possible. The atomic unit of data is the *bit* (*b*inary dig*it*), stemming from the fact that the elementary storage units in a computer are electronic switches where each switch holds one of the two states: *on* (usually represented by 1), or *off* (0), as shown below.

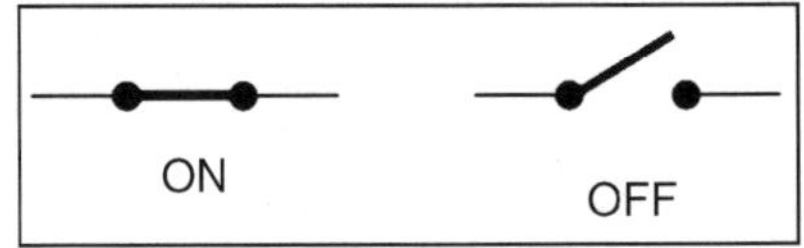

Fig. Two States of a Switch

Ultimately, all data, numeric or non-numeric, are represented in sequence of bits of zeroes and ones. Eight bits constitute a *byte* (a *nibble* is half a byte, or four bits, but this is rarely used these days), and a *word*, which is a unit for data storage and transfer, is usually in multiples of byte, depending on the width of the system bus. Computers nowadays typically use 32-bit or 64-bit words.

A sequence of bits allows for a range of representations. Storage units can be grouped together to provide larger range of representations. For convenience, we shall refer to these representations as values, though they may represent non-numeric data. For example, the four seasons could be represented by two switches, to cover a range of four values 00, 01, 10 and 11, as shown below.

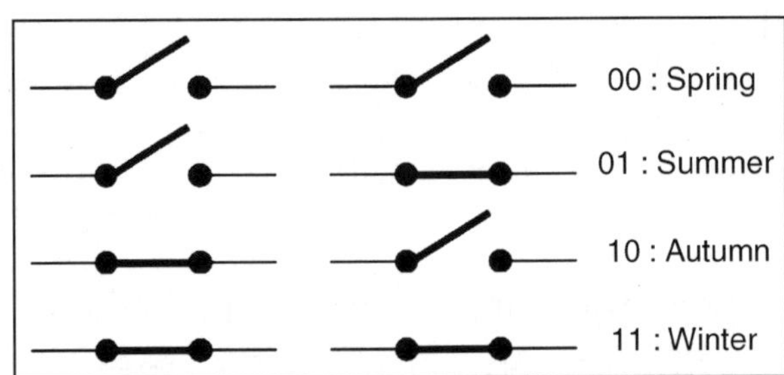

Fig. Representing Four States Using Two Switches (Two Bits)

In general, N bits can represent up to 2^N distinct values. Conversely, to represent a range of M values, the number of bits required is $\lceil \log_2 M \rceil$.

1 bit → represents up to 2 values (0, 1)
2 bits → represents up to 4 values (00, 01, 10, 11)
3 bits → represents up to 8 values (000, 001, 010, 011, 100, 101, 110, 111)

4 bits → represents up to 16 values (0000, 0001, 0010, ..., 1110, 1111)
32 values → requires 5 bits
40 values → requires 6 bits
64 values → requires 6 bits
100 values → requires 7 bits
1024 values → requires 10 bits

POSITIONAL NUMERALS

Different numeral systems have emerged in civilizations over the history. They can be roughly categorised as follows:

- Non-positional numeral systems
- Relative-position numeral systems
- Positional numeral systems

In the decimal numeral system, the positional weights are powers of ten. In general, a decimal number $(a_n\, a_{n-1} \ldots a_0 . f_1 f_2 \ldots f_m)$ has the value,

$(a_n \times 10^n) + (a_{n-1} \times 10^{n-1}) + \ldots + (a_0 \times 10^0) + (f_1 \times 10^{-1}) + (f_2\ 10^{-2}) + \ldots + (f_m \times 10^{-m})$

For example, $28.75 = (2 \times 10^1) + (8 \times 10^0) + (7 \times 10^{-1}) + (5 \times 10^{-2})$

BASES OF NUMBER SYSTEMS

The *base* or *radix* of a number system is the number of digits present. The decimal numeral system has a base or radix of 10, where the set of 10 symbols (digits) is {0, 1, 2, 3, 4, 5, 6, 7, 8, 9}. The weights are in powers of ten.

Number systems on other bases are defined likewise. For instance, a base-four number system consists of the set of four symbols {0, 1, 2, 3} with weights in powers of four. Hence, a base-four number 123.1, also written as $(123.1)_4$ for clarity, has the value,

$(1 \times 4^2) + (2 \times 4^1) + (3 \times 4^0) + (1 \times 4^{-1})$ or 16 + 8 + 3 + 0.25 or 27.25 in decimal, or $(27.25)_{10}$. Listing base-four integers in increasing value yields

0, 1, 2, 3, 10, 11, 12, 13, 20, 21, 22, 23, 30, 31, 32, 33, 100, 101, 102, 103, 110, ...

In general, a base-b number $(a_n\, a_{n-1} \ldots a_0 . f_1 f_2 \ldots f_m)_b$ has the value,

$(a_n ´ b^n) + (a_{n-1} ´ b^{n-1}) + \ldots + (a_0 ´ b^0) + (f_1 ´ b^{-1}) + (f_2 ´ b^{-2}) + \ldots + (f_m ´ b^{-m})$

The point that separates the integer part and fraction part is known as the *radix point*. The weights are in powers of b. The above forms the basis for conversion of a base-b number to decimal. We may also adopt the computationally more efficient *Horner's rule* to convert a base-b integer to decimal. This method simply factors out the powers of b, by rewriting,

$(a_n ´ b^n) + (a_{n-1} ´ b^{n-1}) + \ldots + (a_0 ´ b^0)$ as $(((a_n ´ b) + a_{n-1}) ´ b + \ldots) ´ b + a_0$.

For example,

$$(123)_4 = ((1 ´ 4) + 2) ´ 4 + 3 = 27.$$

For fractions, we may adopt the same method by replacing multiplication with division. Special names are given to number systems on certain bases.If you think that the base is a small value, you are wrong. The Babylonians used a sexagesimal system with base 60!

Table. Bases of Positional Numeral Systems

Base	Name	Base	Name
2	Binary	9	Nonary
3	Ternary	10	Decimal (or Denary)
4	Quaternary	12	Duodecimal
5	Quinary	16	Hexadecimal
6	Senary	20	Vigesimal
7	Septimal	60	Sexagesimal
8	Octal (or Octonary)		

For a base b that is less than ten, the symbols used are the first b Arabic symbols: 0, 1, 2, …, b–1. For bases that exceed ten, more symbols beyond the ten Arabic symbols must be introduced.

The hexadecimal system uses the additional symbols A, B, C, D, E and F to represent 10, 11, 12, 13, 14 and 15 respectively. The binary number system is of particular interest to us as the two symbols 0 and 1 correspond to the two bits that represent the states of a switch. The octal and hexadecimal number systems are also frequently encountered in computing.

Examples: Convert the following numbers into their decimal equivalent.

$(1101.101)_2 = (1 \times 2^3) + (1 \times 2^2) + (1 \times 2^0) + (1 \times 2^{-1}) + (1 \times 2^{-3})$

$= 8 + 4 + 1 + 0.5 + 0.125 = 13.625$

$(572.6)_8 = (5 \times 8^2) + (7 \times 8^1) + (2 \times 8^0) + (6 \times 8^{-1})$

$= 320 + 56 + 2 + 0.75 = 378.75$

$(2A.8)_{16} = (2 \times 16^1) + (10 \times 16^0) + (8 \times 16^{-1})$

$= 32 + 10 + 0.5 = 42.5$

$(341.24)_5 = (3 \times 5^2) + (4 \times 5^1) + (1 \times 5^0) + (2 \times 5^{-1}) + (4 \times 5^{-2})$

$= 75 + 20 + 1 + 0.4 + 0.16 = 96.56$

DECIMAL TO BINARY CONVERSION

One way to con vert a decimal number to its binary equivalent is the *sum-of-weights* method. Here, we determine the set of weights (which are in powers of two) whose sum is the number in question.

Examples: Convert decimal numbers to their binary equivalent using sum-of-weights.

$(9)_{10} = 8 + 1 = 2^3 + 2^0 = (1001)_2$

$(18)_{10} = 16 + 2 = 2^4 + 2^1 = (10010)_2$

$(58)_{10} = 32 + 16 + 8 + 2 = 2^5 + 2^4 + 2^3 + 2^1 = (111010)_2$

$(0.625)_{10} = 0.5 + 0.125 = 2^{-1} + 2^{-3} = (0.101)_2$

We observe that multiplying a number x by 2^n is equivalent to shifting left the binary representation of x by n positions, with zeroes appended on the right. For example, the value $(5)_{10}$ is represented as $(101)_2$, and $(40)_{10}$ (which is 5 ×2^3) is $(101000)_2$. Similarly, dividing a number x by 2^n is equivalent to shifting right its binary representation by n positions. This may help in quicker conversion.

The second method deals with the integral portion and the fractional portion separately, as follows:

- Integral part: *Repeated division-by-two*
- Fractional part: *Repeated multiplication-by-two*

CONVERSION BETWEEN BASES

Decimal to Base-*R* Conversion

We may extend the repeated-division and repeated-multiplication techniques to convert a decimal value into its equivalent in base R:

- Integral part: *Repeated division-by-R*
- Fractional part: *Repeated multiplication-by-R*

Truncation and Rounding

In the course of converting values – especially fractional values – between bases, there might be instances when the values are to be corrected within a specific number of places. This may be done by *truncation* or *rounding*.

In truncation, we simply chop a portion off from the fraction. Table shows the truncation of the value $(0.12593)_{10}$ with respect to the number of decimal places desired.

Table. Truncation Versus Rounding

Numberof Decimal Places	Truncated Value	Rounded Value
4	$(0.1259)_{10}$	$(0.1259)_{10}$
3	$(0.125)_{10}$	$(0.126)_{10}$
2	$(0.12)_{10}$	$(0.13)_{10}$
1	$(0.1)_{10}$	$(0.1)_{10}$

In rounding, we need to examine the leading digit of the portion we intend to remove. We divide the digits of the numeral system into two equal-sized sets, where the first set contains the 'lighter' digits (Example: 0, 1, 2, 3, 4 in the decimal system), and the second set the 'heavier' digits (5, 6, 7, 8, 9).

If that digit being examined belongs to the former set, we remove the unwanted portion; if it belongs to the latter set, we 'promote' (round up) the last digit in the retained portion to its next higher value (propagating the promotion if necessary). Table 2-2 above shows the rounding of the value $(0.12593)_{10}$ with respect to the specified number of decimal places.

For bases that are odd numbers, there exists a 'middle' digit that may be classified either as a 'light' digit or a 'heavy' digit. For example, $(6.143)_7$ may be rounded to two places either as $(6.14)_7$ or $(6.15)_7$.

Conversion between Binary and Octal/Hexadecimal

There exist simple techniques for conversion between binary numbers and octal (or hexadecimal) numbers.

They are given below:

- *Binary → Octal:* Partition (from the radix point outwards) in groups of 3; each group of 3 bits corresponds to an equivalent octal digit.
- *Octal → Binary:* Expand each octal digit into an equivalent group of 3 bits.
- *Binary → Hexadecimal:* Partition (from the radix point outwards) in groups of 4; each group of 4 bits corresponds to an equivalent hexadecimal digit.
- *Hexadecimal → Binary:* Expand each hexadecimal digit into an equivalent group of 4 bits.

Examples: Conversion between binary and octal/hexadecimal.

$(10\ 111\ 011\ 001.\ 101\ 11)_2 = (2731.56)_8$

$(2731.56)_8 = (010\ 111\ 011\ 001.\ 101\ 110)_2$

$(101\ 1101\ 1001.\ 1011\ 1)_2 = (5D9.B8)_{16}$

$(5D9.B8)_{16} = (0101\ 1101\ 1001.\ 1011\ 1000)_2$

Octal and hexadecimal notations are commonly encountered in computing literature as they provide more compact writing compared to binary notation, and they can be easily converted into binary form.

On closer observation, it is not difficult to uncover the underlying principle behind these techniques. (Hint: 8 is 2^3, and 16 is 2^4.) With this, we can design similar techniques to convert between related bases, such as between binary and quaternary (base 4), and between ternary (base 3) and nonary (base 9).

General Conversion

Apart from the techniques for conversion between related bases such as those discussed above, for general conversion between two bases, the approach is to convert the given value first into decimal, followed by converting the decimal value into the target base.

ARITHMETIC OPERATIONS ON BINARY NUMBERS

						1	1	0	0	1
					×	1	0	1	0	1
						1	1	0	0	1
				1	1	0	0	1		
+		1	1	0	0	1				
	1	0	0	0	0	0	1	1	0	1

Arithmetic operations on binary numbers are similar to those on decimal. The example below shows the multiplication of two binary numbers. The multiplicand (11001) is multiplied with every bit of the multiplier (10101), and the partial products are then added.

Binary addition is performed in the same manner as in decimal. The examples below compare addition in the two systems.

Binary								Decimal				
		1	1	0	1	1				6	4	8
+		1	0	0	1	1		+		5	9	7
	1	0	1	1	1	0			1	2	4	5

The addition is performed column by column, from right to left. At each column three bits are added: one bit from each of the two numbers, and a *carry-in* bit. The column addition generates the *sum* bit, and a *carry-out* bit that is propagated to the next column to its left as the carry-in. Table shows the eight possible outcomes of a column addition.

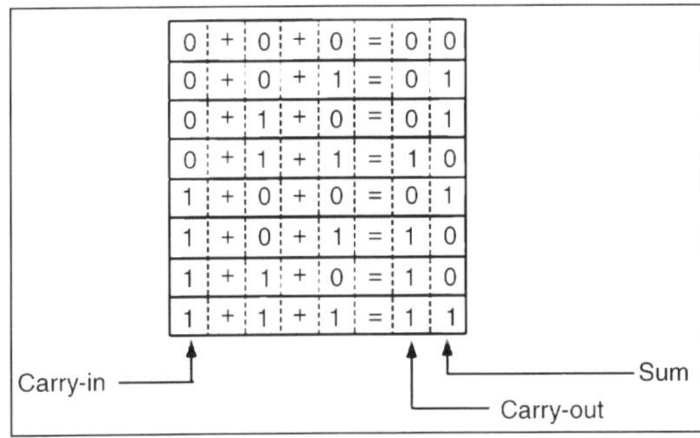

Fig. Bit Addition Figure

Subtraction is performed in a similar fashion.

Binary								Decimal				
		1	0	0	1	0				8	2	3
–		0	1	1	0	1		–		3	9	7
		0	0	1	0	1				4	2	6

Like addition, the subtraction is performed column by column. However, unlike traditional method where we might progressively 'borrow' from a few columns before returning to the column we are currently working on, we adopt the 'borrow-in' and 'borrow-out' concept that allows the information to be propagated one column at a time. Table 2-4 shows the eight possible outcomes of a column subtraction. A column subtraction generates a *difference* bit and a *borrow-out* bit that becomes the *borrow-in* of the next column to its left.

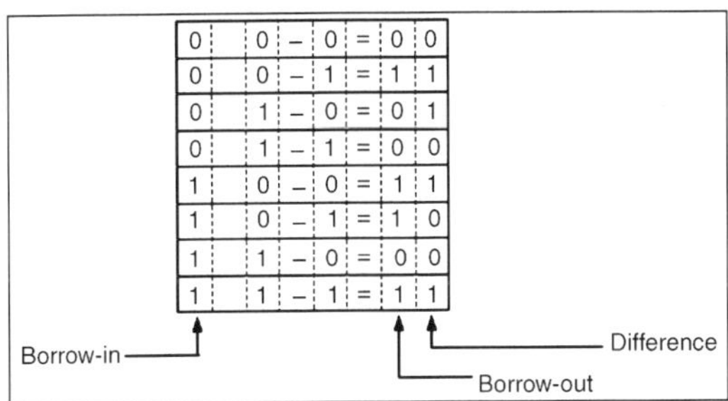

Fig. Bit Subtraction Figure

NEGATIVE NUMBERS

Till now, we have only considered *unsigned* numbers, which are non-negative values. We shall now consider a few schemes to represent *signed* integers (positive and negative values). The four common representations for signed binary numbers are:

Sign-and-Magnitude

The *sign-and-magnitude* representation employs a prefix bit to indicate the *sign* of the number, followed by the *magnitude* field. A positive value has a sign bit of 0 while a negative value a sign bit of 1. Figure shows how the value –75 is represented in an 8-bit sign-and-magnitude scheme. We may write the value as $(11001011)_{sm}$.

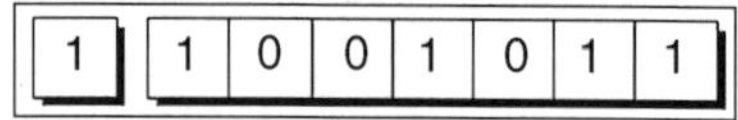

Fig. An 8-bit Sign-and-Magnitude representation of –75

An 8-bit sign-and-magnitude representation allows values between –127 (represented as $(11111111)_{sm}$) and +127 (represented as $(01111111)_{sm}$) to be represented. There are two representations for zero: $(00000000)_{sm}$ and $(10000000)_{sm}$. In general, an n-bit sign-and-magnitude representation can cover this range of values $[-(2^{n-1} - 1), 2^{n-1} - 1]$.

To negate a value, we invert the sign bit.

Excess

The *excess* system (also known as *biased* system) is another popular system in use for negative numbers. To represent a value, a fixed bias is added to this value, and the binary representation of the result is the desired representation. For example, assuming a 5-bit excess system, if the bias chosen is 16, then it is called the excess-16 system. In the excess-16 system, the value –16 is represented as $(00000)_{ex16}$ (since –16 + 16 = 0), and the value 3 would be represented as $(10011)_{ex16}$ (since 3 + 16 = 19). Since the excess system is just a simple translation of the binary system, $(00000)_{ex16}$ would represent the most negative value (which is –16), and $(11111)_{ex16}$ would represent the largest positive value (which is 15).

Table. The 4-bit Excess-8 system

Value	Excess-8	Value	Excess-8
–8	0000	0	1000
–7	0001	1	1001
–6	0010	2	1010
–5	0011	3	1011
–4	0100	4	1100
–3	0101	5	1101
–2	0110	6	1110
–1	0111	7	1111

The bias is usually chosen so that the range of values represented has a balanced number of positive and negative values. Hence we pick 16 as the bias for a 5-bit excess system, resulting in the range of [–16, 15]. (Sometimes we pick 15 as the bias, which results in the range [–15, 16].) For a 4-bit system, a reasonable bias would be 8 (or sometimes 7). In general, for n bits, the bias is usually 2^{n-1}. Table below shows the 4-bit excess-8 system.

1's Complement

For positive values, the *ones' complement* representation is identical to the binary representation. For instance, the value 75 is represented as $(01001011)_2$ in an 8-bit binary system, as well as $(01001011)_{1s}$ in an 8-bit 1's complement system. What about negative values? Given a number X which can be expressed as an n-bit binary number, its negated value, $-X$, can be obtained in 1's complement form by this formula:

$$-X = 2^n - X - 1$$

For example, 75 is represented as $(01001011)_2$ in an 8-bit binary system, and hence –75 is represented as $(10110100)_{1s}$ in the 8-bit 1's complement system. (Note that $2^8 - 75 - 1 = 180$ whose binary form is 10110100.) We observe that we can easily derive $-X$ from X in 1's complement by inverting all the bits in the binary representation of X. Note also that the first bit serves very much like a sign bit, indicating that the value is positive (negative) if the first bit is 0 (1). However, the remaining bits do not constitute the magnitude of the number, particularly for negative numbers.

An 8-bit 1's complement representation allows values between –127 (represented as $(10000000)_{1s}$) and +127 (represented as $(01111111)_{1s}$) to be represented. There are two representations for zero: $(00000000)_{1s}$ and $(11111111)_{1s}$. In general, an n-bit 1's complement representation has a range $[-(2^{n-1} - 1), 2^{n-1} - 1]$. To negate a value, we invert all the bits. For example, in an 8-bit 1's complement scheme, the value 14 is represented as $(00001110)_{1s}$, therefore –14 is represented as $(11110001)_{1s}$.

2's Complement

The *two's complement* system share some similarities with the ones' complement system. For positive values, it is the same as the binary representation. The first bit also indicates the sign of the value (0 for positive, 1 for negative). Given a number X which can be expressed as an n-bit binary number, its negated value, $-X$, can be obtained in 2's complement form by this formula:

$$-X = 2^n - X$$

For example, 75 is represented as $(01001011)_2$ in an 8-bit binary system, and hence –75 is represented as$(10110101)_{2s}$ in the 8-bit 2's complement system. (Note that $2^8 - 75 = 181$ whose binary form is 10110101.) Again, we observe that we can easily derive $-X$ from X in 2's complement by inverting all the bits in the binary representation of X, and then adding one to it.

An 8-bit 2's complement representation allows values between –128 (represented as $(10000000)_{2s}$) and +127 (represented as $(01111111)_{2s}$) to be represented, and hence it has a range that is one larger than that of the 1's complement representation.

This is due to the fact that there is only one unique representation for zero: $(00000000)_{2s}$. In general, an n-bit 2's complement representation has a range $[-2^{n-1}, 2^{n-1} - 1]$.

To negate a value, we invert all the bits and plus 1. For example, in an 8-bit 2's complement scheme, the value 14 is represented as $(00001110)_{2s}$, therefore –14 is represented as $(11110010)_{2s}$.

ERROR DETECTING

R. W. Hamming wrote the paper that both opened and closed this field in 1950. His interest was in providing a means of self-checking in computers, which were just being developed at the time he wrote this. the paper appeared in the Bell System Technical Journal, April, 1950. Definitely worth tracking down in the library and reading.

BIT STRINGS AS ADDRESSES IN BINARY HYPERCUBES

The best starting point for understanding ECC codes is to consider bit strings as addresses in a binary hypercube. A hypercube is a generalization of a cube to various dimensions; we're probably most familiar with the notion of a four-dimensional hypercube. Here's a picture of binary hypercubes for several different dimensionalities:

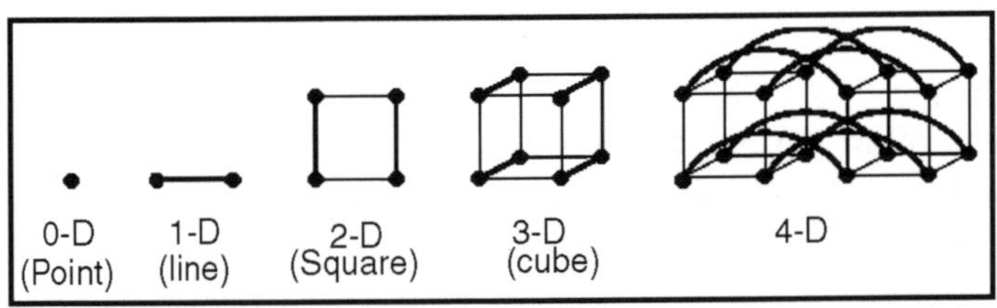

Each of them was created by copying the one to the left twice, and connecting corresponding vertices.

We can assign each vertex in a hypercube a location in a coordinate space determined by the dimensionality of the hypercube.

- A zero-dimensional hypercube requires no coordinates to know where you are
- A one-dimensional hypercube can use one bit to tell whether you're at the bottom or the top of the line segment.
- A two-dimensional hypercube can use two bits: first bit is left vs. right, second is inherited from the line.
- A three-dimensional hypercube can use a bit to tell front square from back, and inherit two bits from the square.

- A four-dimensional hypercube can use a bit to tell left cube from right, and inherits three bits from cube.

This can continue through as many dimensions as you want. The Hamming distance between two bit strings is the number of bits you have to change to convert one to the other: this is the same as the number of edges you have to traverse in a binary hypercube to get from one of the vertices to the other.

The basic idea of an error correcting code is to use extra bits to increase the dimensionality of the hypercube, and make sure the Hamming distance between any two valid points is greater than one.

- If the Hamming distance between valid strings is only one, a single-bit error results in another valid string. This means we can't detect an error.
- If it's two, then changing one bit results in an invalid string, and can be detected as an error. Unfortunately, changing just one more bit can result in another valid string, which means we can't know which bit was wrong: so we can detect an error but not correct it.
- If the Hamming distance between valid strings is three, then changing one bit leaves us only one bit away from the original error, but two bits away from any other valid string. This means if we have a one-bit error, we can figure out which bit is the error; but if we have a two-bit error, it looks like one bit from the other direction. So we can have single bit correction, but that's all.
- Finally, if the Hamming distance is four, then we can correct a single-bit error and detect a double-bit error. This is frequently referred to as a SECDED (Single Error Correct, Double Error Detect) scheme.

Parity

The simplest case is by adding a parity bit. Suppose we have a three-bit word (so the bit strings define points in a cube). If we add a fourth bit, we can decree that any time we want to switch a bit in the original three-bit string, we also have to switch the parity bit.

If we start with 000 in the left cube, so the full string is 0000, changing any one of the original three bits requires us to change to the other cube: 1001, 1010, and 1100. Now if we change a second bit, we have to move back to the left cube: 0011, 0101, 0110. And if we change the third bit, we move back to the right cube: 0111.

So, there is a Hamming distance of two between any two valid strings. If we get a one-bit error, we know it is an error because it's on one of the invalid vertices.

This can be computed by counting the number of 1's, and making sure it's always even (so this is called even parity). We could have selected exactly the opposite set of vertices as the valid ones, which would have given us odd parity. We picked even parity because we'll be using it in the next step.

INFORMATION PROTECTED IN HAMMING CODES

Several Teletext packets contain information protected by hamming. This is a method for encoding data such that errors in reception can be detected, and if the error is sufficiently small, corrected. Teletext uses two hamming codes. The simpler and more robust version encodes 4 bits of data in one 8-bit byte. This code is used extensively to protect the fields that have meaning to the Teletext system itself.

The more efficient code encodes 18 bits of data in three 8-bit bytes. It is used for to protect some of the data fields inside Teletext packets. Both codes are designed such that any one-bit error can be corrected, and any 2-bit error can be detected.There's an error correction code that separates the bits holding the original value (data bits) from the error correction bits (check bits), and the difference between the calculated and actual error correction bits is the position of the bit that's wrong. It's called a Hamming Code. Error correction codes are a way to represent a set of symbols so that if any 1 bit of the representation is accidentally flipped, you can still tell which symbol it was. For example, you can represent two symbols x and y in 3 bits with the values x=111 and y=000.

If you flip any one of the bits of these values, you can still tell which symbol was intended. If more than 1 bit changes, you can't tell, and you probably get the wrong answer. So it goes; 1-bit error correction codes can only correct 1-bit changes. If b bits are used to represent the symbols, then each symbol will own 1+b values: the value representing the symbol, and the values differing from it by 1 bit.

In the 3-bit example above, y owned 1+3 values: 000, 001, 010, and 100. Representing n symbols in b bits will consume n*(1+b) values. Hamming code is a set of error-correction code s that can be used to detect and correct bit errors that can occur when computer data is moved or stored. Hamming code is named for R. W. Hamming of Bell Labs. Like other error-correction code, Hamming code makes use of the concept of parity and parity bit s, which are bits that are added to data so that the validity of the data can be checked when it is read or after it has been received in a data transmission. Using more than one parity bit, an error-correction code can not only identify a single bit error in the data unit, but also its location in the data unit.

In data transmission, the ability of a receiving station to correct errors in the received data is called forward error correction (FEC) and can increase throughput on a data link when there is a lot of noise present.

To enable this, a transmitting station must add extra data (called *error correction bits*) to the transmission. However, the correction may not always represent a cost saving over that of simply resending the information. Hamming codes make FEC less expensive to implement through the use of a *block parity* mechanism. Computing parity involves counting the number of ones in a unit of data, and adding either a zero or a one (called a *parity bit*) to make the count odd (for odd parity) or even (for even parity). For example, 1001 is a 4-bit data unit containing two one bits; since that is an even number, a zero would be added to maintain even parity, or, if odd parity was being maintained, another one would be added.

To calculate even parity, the XOR operator is used; to calculate odd parity, the XNOR operator is used. Single bit errors are detected when the parity count indicates that the number of ones is incorrect, indicating that a data bit has been flipped by noise in the line. Hamming codes detect two bit errors by using more than one parity bit, each of which is computed on different combinations of bits in the data.

8/4 HAMMING

The 8/4 code is quite simple; if the input nibble is the 4-bit value:

```
b3, b2, b1, b0
```

with b3 being the most significant bit, then the output hammed byte is the 8-bit value:

```
b3, b3^b2^b1, b2, !b2^b1^b0, b1,
!b3^b1^b0, b0, !b3^b2^b0
```

where ^ represents bitwise exclusive-or and ! is bitwise not.

This code has the property that every value is four bits different from all other such values. A one-bit error is therefore unambiguously correctable, being only one bit away from a valid code. A two-bit error is detectable but not correctable, being equidistant between two valid codes.

Here is the table of all 16 encoded nibbles:

Data nibble	Hammed byte
0 = 0 0 0 0	15 = 0 0 0 1 0 1 0 1
1 = 0 0 0 1	02 = 0 0 0 0 0 0 1 0
2 = 0 0 1 0	49 = 0 1 0 0 1 0 0 1
3 = 0 0 1 1	5E = 0 1 0 1 1 1 1 0
4 = 0 1 0 0	64 = 0 1 1 0 0 1 0 0
5 = 0 1 0 1	73 = 0 1 1 1 0 0 1 1
6 = 0 1 1 0	38 = 0 0 1 1 1 0 0 0
7 = 0 1 1 1	2F = 0 0 1 0 1 1 1 1
8 = 1 0 0 0	D0 = 1 1 0 1 0 0 0 0
9 = 1 0 0 1	C7 = 1 1 0 0 0 1 1 1
A = 1 0 1 0	8C = 1 0 0 0 1 1 0 0
B = 1 0 1 1	9B = 1 0 0 1 1 0 1 1
C = 1 1 0 0	A1 = 1 0 1 0 0 0 0 1
D = 1 1 0 1	B6 = 1 0 1 1 0 1 1 0
E = 1 1 1 0	FD = 1 1 1 1 1 1 0 1
F = 1 1 1 1	EA = 1 1 1 0 1 0 1 0
||||	||||||||

b3b2b1b0 b3 | b2 |b1 |b0 |

```
|     |     |     |
321 !210 !310 !320
```

Decoding a hammed byte back in to a 4-bit nibble can be done using the table below, or by a bit-twiddling algorithm. In the table, the hammed byte value is used to index the row (most significant 4 bits) and column (least significant four bits), and the decoded nibble is read out of the table. If a single bit error has been detected and corrected, the nibble is followed by ``!". If an uncorrecteable error has occurred, the table cell is ``.".

```
LSB
    | 0 1 2 3 4 5 6 7 8 9 A B C D E F |
MSB+----------------------------------+---
0 | 1!. 1 1!. 0! 1!.. 2! 1!. A!.. 7!| 0
1 |. 0! 1!. 0! 0. 0! 6!.. B!. 0! 3!. | 1
2 |. C! 1!. 4!.. 7! 6!.. 7!. 7! 7! 7 | 2
3 | 6!.. 5!. 0! D!. 6 6! 6!. 6!.. 7!| 3
4 |. 2! 1!. 4!.. 9! 2! 2. 2!. 2! 3!. | 4
5 | 8!.. 5!. 0! 3!.. 2! 3!. 3!. 3 3!| 5
6 | 4!.. 5! 4 4! 4!.. 2! F!. 4!.. 7!| 6
7 |. 5! 5! 5 4!.. 5! 6!.. 5!. E! 3!. | 7
8 |. C! 1!. A!.. 9! A!.. B! A A! A!. | 8
9 | 8!.. B!. 0! D!.. B! B! B A!.. B!| 9
A | C! C. C!. C! D!.. C! F!. A!.. 7!| A
B |. C! D!. D!. D D! 6!.. B!. E! D!. | B
C | 8!.. 9!. 9! 9! 9. 2! F!. A!.. 9!| C
D | 8 8! 8!. 8!.. 9! 8!.. B!. E! 3!. | D
E |. C! F!. 4!.. 9! F!. F F!. E! F!. | E
F | 8!.. 5!. E! D!.. E! F!. E! E. E!| F
---+----------------------------------+---
   | 0 1 2 3 4 5 6 7 8 9 A B C D E F |
```

The algorithmic approach shows what is going on more clearly. Assuming the hammed byte is

```
h7, h6, h5, h4, h3, h2, h1, h0
```

with h7 being the most significant bit, we compute

```
p = h7 ^ h6 ^ h5 ^ h4 ^ h3 ^ h2 ^ h1 ^ h0
c0 = h7 ^ h5 ^ h1 ^ h0
c1 = h7 ^ h3 ^ h2 ^ h1
c2 = h5 ^ h4 ^ h3 ^ h1
```

If the parity, p, is correct (equal to 1) then either 0 or 2 errors occurred. If all the check bits, c0, c1, c2 are correct (equal to 1) then the byte was received intact, (no errors) otherwise it was damaged beyond repair (two errors).

If p is 0, then there was a single bit error which can be recovered:

```
c0 c1 c2 meaning
1 1 1 error in bit h6
1 1 0 error in bit h4
```

```
1 0 1 error in bit h2
0 1 1 error in bit h0
0 0 1 error in bit h7
0 1 0 error in bit h5
1 0 0 error in bit h3
0 0 0 error in bit h1
```

The erroneous bit should be flipped. Note that there is actually no need to fix errors in bits h6, h4, h2 and h0 since they are not used in the decoded byte.

After flipping bits if necessary, the decoded byte is then:

```
h7, h5, h3, h1
```

24/18 HAMMING

The more efficient 24/18 code is based on the same principle of interleaved check bits and an overall parity bit as the simpler 8/4 code.

Assuming the input bits are:

```
b17, b16, b15, b14, b13, b12, b11, b10, b9, b8,
b7, b6, b5, b4, b3, b2, b1, b0
```

with b17 being the most significant bit, the six hamming check bits are:

```
c0 = ! b17 ^ b15 ^ b13 ^ b11 ^ b10 ^ b8 ^ b6 ^ b4 ^ b3 ^ b1 ^ b0
c1 = ! b17 ^ b16 ^ b13 ^ b12 ^ b10 ^ b9 ^ b6 ^ b5 ^ b3 ^ b2 ^ b0
c2 = ! b17 ^ b16 ^ b15 ^ b14 ^ b10 ^ b9 ^ b8 ^ b7 ^ b3 ^ b2 ^ b1
c3 = ! b10 ^ b9 ^ b8 ^ b7 ^ b6 ^ b5 ^ b4
c4 = ! b17 ^ b16 ^ b15 ^ b14 ^ b13 ^ b12 ^ b11
```

The bias is usually chosen so that the range of values represented has a balanced number of positive and negative values. Hence we pick 16 as the bias for a 5-bit excess system, resulting in the range of [-16, 15]. (Sometimes we pick 15 as the bias, which results in the range [-15, 16].) For a 4-bit system, a reasonable bias would be 8 (or sometimes 7). In general, for *n* bits, the bias is usually 2^{n-1}. Table below shows the 4-bit excess-8 system. c5 = b17 ^ b14 ^ b12 ^ b11 ^ b10 ^ b7 ^ b5 ^ b4 ^ b2 ^ b1 ^ b0

where ^ represents bitwise exclusive-or and ! is bitwise not.

c5 can alternatively and equivalently be computed as the odd parity of all the other data and check bits.

The output bytes are then:

```
c3, b3, b2, b1, c2, b0, c1, c0
c4, b10, b9, b8, b7, b6, b5, b4
c5, b17, b16, b15, b14, b13, b12, b11
```

in byte transmission order, with the most significant bit of each byte on the left.

To decode a hammed byte triplet:

```
h7, h6, h5, h4, h3, h2, h1, h0
h15, h14, h13, h12, h11, h10, h9, h8
h23, h22, h21, h20, h19, h18, h17, h16
```

with h0 being the least significant bit of the first byte and h23 being the most significant bit of the third byte, we compute

```
p = h23 ^ h22 ^ h21 ^ h20 ^... ^ h1 ^ h0
c0 = h0 ^ h2 ^ h4 ^ h6 ^ h8 ^ h10 ^ h12 ^ h14 ^ h16 ^ h18 ^ h20 ^ h22
c1 = h1 ^ h2 ^ h5 ^ h6 ^ h9 ^ h10 ^ h13 ^ h14 ^ h17 ^ h18 ^ h21 ^ h22
c2 = h3 ^ h4 ^ h5 ^ h6 ^ h11 ^ h12 ^ h13 ^ h14 ^ h19 ^ h20 ^ h21 ^ h22
c3 = h7 ^ h8 ^ h9 ^ h10 ^ h11 ^ h12 ^ h13 ^ h14
c4 = h15 ^ h16 ^ h17 ^ h18 ^ h19 ^ h20 ^ h21 ^ h22
```

If the parity, p, is correct (equal to 1) then either 0 or 2 errors occurred. If all the check bits, c0, c1, c2, c3, c4, c5 are correct (equal to 1), then the byte was received intact, (no errors) otherwise it was damaged beyond repair (two errors).

If p is 0, then there was a single bit error which can be recovered. For the check bits which are incorrect (equal to 0), add the following powers of two together:

```
if c0 = 0 add 1
if c1 = 0 add 2
if c2 = 0 add 4
if c3 = 0 add 8
if c4 = 0 add 16
```

The sum gives the bit position 1--24 corresponding to h0--h23 which is in error. The erroneous bit should be flipped. Note that there is actually no need to fix errors in bits h23, h15, h7, h3, h1 and h0 since they are not used in the decoded byte.

The output data bits, d17--d0 are:

```
h22, h21, h20, h19, h18, h17, h16, h14, h13, h12,
h11, h10, h9, h8, h6, h5, h4, h2
```

REPRESENT REAL NUMBERS ON COMPUTERS

FLOATING POINT REPRESENTATION

There are several ways to represent real numbers on computers. Fixed point places a radix point somewhere in the middle of the digits, and is equivalent to using integers that represent portions of some unit. For example, one might represent 1/100ths of a unit; if you have four decimal digits, you could represent 10.82, or 00.01. Another approach is to use rationals, and represent every number as the ratio of two integers. Floating-point representation - the most common solution - basically represents reals in scientific notation. Scientific notation represents numbers as a base number and an exponent. For example, 123.456 could be represented as 1.23456×10^2. In hexadecimal, the number 123.abc might be represented as $1.23\text{abc} \times 16^2$.

Floating-point solves a number of representation problems. Fixed-point has a fixed window of representation, which limits it from representing very large or very small numbers. Also, fixed-point is prone to a loss of precision when two large numbers are divided. Floating-point, on the other hand, employs a sort of "sliding window" of precision appropriate to the scale of the number. This allows it to represent numbers from 1,000,000,000,000 to 0.0000000000000001 with ease. Some of the greatest achievements of the 20th century would not have been possible without the floating point capabilities of digital computers. Nevertheless, this subject is not well understood by most programmers and is a regular source of confusion. In a February keynote address entitled *Extensions to Java for Numerical Computing* James Gosling asserted "95% of folks out there are completely clueless about floating-point." However, the main ideas behind floating point are not difficult, and we will demystify the confusion that plagues most novices.

Precision vs. Accuracy

Precision = tightness of specification. Accuracy = correctness. Do not confuse precision with accuracy. 3.133333333 is an estimate of the mathematical constant À which is specified with 10 decimal digits of precision, but it only has two decimal digits of accuracy.

As John von Neumann once said "There's no sense in being precise when you don't even know what you're talking about." Java typically prints out floating point numbers with 16 or 17 decimal digits of precision, but do not blindly believe that this means there that many digits of accuracy!

Calculators typically display 10 digits, but compute with 13 digits of precision. Kahan: the mirror for the Hubble space telescope was ground with great precision, but to the wrong specification. Hence, it was initially a great failure since it couldn't produce high resolution images as expected. However, it's precision enabled an astronaut to install a corrective lens to counter-balance the error. Currency calculations are often defined in terms of a give precision, e.g., Euro exchange rates must be quoted to 6 digits.

Also, in some cases, the floating point formats of different sizes for some machines belonged to different groups. When this happened, all the formats for any one architecture were placed within the discussion of one of the groups of formats included.

Particularly unsusual formats discussed below include the single precision floating-point format for the PDP-4, 7, 9 and 15, which can be thought of as a rearranged Group II or Group III format, and the double precision floating-point format of the ICL 1900, which applies the principle of making a double-precision float out of two single-precision floats, usually used with hardware Group I formats, to a base floating-point format which belongs to Group III; this appears to be the result of either successive implementations of the architecture evolving from hardware multiply to full hardware floating-point or the availability of hardware floating-point as an option.

Another case where the same machine had floating point formats belonging to different groups is the Harris 800, which added a Group II quad-precision floating-point format to an existing architecture whose single and double precision floating-point formats belonged to Group III: here, the hardware level seems to have remained constant, and what happened was that the larger size of the quad-precision format made it reasonable to use a full word instead of a partial word for the exponent, and once that was done, a Group II format appeared more reasonable than a Group III format.

DECIMAL CASES

- 3.141592653589...
- 2.71828...
- 6.023×10^{23} (N_A)
- 6.626×10^{-32} ($\hbar$)

In programming, a floating point number—123.45×10^{-6} is expressed as—123.45E—6. In general, a floating-point number can be written as

$\pm M \times B^E$

where

- M is the fraction mantissa or significand.
- E is the exponent.
- B is the base, in decimal case B = 10.

BINARY CASES

As an example, a 32-bit word is used in MIPS computer to represent a floating-point number:

S	E	M

1 bit..... 8 bits............. 23 bits

representing: $(-1)^S \times M \times 2^E$

- The implied base is 2 (not explicitly shown in the representation).
- The exponent can be represented in signed 2's complement (but also see biased notation later).
- The implied decimal point is between the exponent field E and the significand field M.
- More bits in field E mean larger range of values representable.
- More bits in field M mean higher precision.
- Zero is represented by all bits equal to 0: $0000...000_2$

NORMALIZATION

To efficiently use the bits available for the significand, it is shifted to the left until all leading 0's disappear (as they make no contribution to the precision).

The value can be kept unchanged by adjusting the exponent accordingly.

Moreover, as the MSB of the significand is always 1, it does not need to be shown explicitly. The significand could be further shifted to the left by 1 bit to gain one more bit for precision. The first bit 1 before the decimal point is implicit. The actual value represented is

$$(-1)^S \times (1.+M) \times 2^E$$

However, to avoid possible confusion, in the following the default normalization does not assume this implicit 1 unless otherwise specified.

Zero is represented by all 0's and is not (and cannot be) normalized.

Example

A binary number X = 0.0001101001101 can be represented in 14-bit floating-point form in the following ways (1 sign bit, a 4-bit exponent field and a 9-bit significand field):

- $x = 0.0001101001101 \times 2^0$ $\boxed{0}\boxed{000}\boxed{000110100}$
- $x = 0.001101001101 \times 2^{-1}$ $\boxed{0}\boxed{111}\boxed{001101001}$
- $x = 0.001101001101 \times 2^{-2}$ $\boxed{0}\boxed{1111}\boxed{001101001}$
- $x = 0.1101001101 \times 2^{-3}$ $\boxed{0}\boxed{1101}\boxed{110100110}$
- $x = 1.101001101 \times 2^{-4}$ $\boxed{0}\boxed{1100}\boxed{101001101}$ with an implied 1.0:

By normalization, highest precision can be achieved.

BIASED NOTATION FOR EXPONENT

To simplify the hardware for comparing two exponents (to use simpler integer sorting rather than subtraction), we may want to avoid 2's complement representation for the exponent.

This can be done by simply adding 1 (a bias) at the MSB of the exponent field and the resulting representation is called biased notation.

Consider a 5-bit exponent field (range of exponents: $-2^4 \sim 2^4 - 1$):

Decimal Exponent	Signed-2's Complement (Excess-16)	Biased Noation of Biased Notation	Decimal Value
15	01111	11111	31
14	01110	11110	30
...	...	...	...
1	00001	10001	17
0	00000	10000	16(bias)
–1	11111	01111	15
...	...	...	...
–15	10001	00001	1
–16	10000	00000	0

The bias depends on number of bits in the exponent field. If there are e bits in this field, the bias is Bias = 2^{c-1}, which lifts the representation (not the actual exponent) by half of the range to get rid of the negative parts represented by 2's complement. The range of actual exponents represented is still the same. With the biased exponent, the value represented by the notation is:

$$(-1)^S \times (1.+M) \times 2^{E-Bias}$$

FLOATING-POINT NOTATION OF IEEE 754

The IEEE 754 floating-point standard uses 32 bits to represent a floating-point number, including 1 sign bit, 8 exponent bits and 23 bits for the significand. As the implied base is 2, an implied 1 is used, i.e., the significand has effectively 24 bits including 1 implied bit to the left of the decimal point not explicitly represented in the notation.

Note in particular that in IEEE 754 notation, the bias for the 8-bit exponent is $127_{10} = 01111111_2$ (instead of $2^{c-1} = 2^7 = 128$).

The 8-bit exponent field:

Decimal Exponent Complement	Signed 2's	Biased Notation of Biased Noation	Decimal Value
For Infinities			
127	01111111	11111111	255
...	...	...	...
2	00000010	10000001	129
1	00000001	10000000	128
0	00000000	01111111	127
–1	11111111	01111110	126
–2	11111110	01111101	125
...	...	...	...
–126	10000010	00000001	1
For Denorms	10000001	00000000	0

Note:

- Zero exponent is represented by $011111111_2 = 127_{10}$the bias of the notation;
- The range of exponents representable is from—126 to 127;
- The exponent 11111111 (with all zero significand) is reserved to represent infinities or not-a-number (NaN) which may occur when, e.g., a number is divided by zero;
- The smallest exponent 00000000 is reserved to represent denormalized numbers (smaller than 2^{-126} which cannot be normalized) and zero, e.g.,0.001×2^{-126} is represented by:

0	00000000	001000...0

OTHER IMPLIED BASES

Given e bits for the exponent field, the range of exponent values representable is $-2^{e-1} \sim 2^{e-1} - 1$ and the range of magnitudes representable is about

$$-2^{-2e-1} \sim 2^{2e-1} - 1$$

For example, if e = 4, the range of exponent values representable is

$$-2^3 = -8 \sim 2^3 - 1 = 7$$

and the range of magnitudes representable is

$$2^{-8} = -1/256 \sim 2^7 = 128$$

This range can be extended by (a) increasing number of bits for exponent, or (b) increasing the implied base from 2 to 4, 8, 16, etc. (or in general, 2^q). For example, when the implied base is $2^q = 2^2 = 4$, the range of magnitudes representable is

$$4^{-8} = 1/65536 \sim 4^7 = 16384$$

Normalization

If the implied base is $B = 2^q$, the significand must be shifted multiple of q bits at a time so that the exponent can be correspondingly adjusted to keep the value unchanged. If at least one of the first q bits of the significand is 1, the representation is normalized. Obviously, the implied 1 can no longer be used.

Examples

- Normalize 0.000000101×4^3. Note that the base is 4 (instead of 2)

$$0.00000101 \times 4^3 = 0.000101 \times 4^2 = 0.0101 \times 4^1$$

Note that the significand has to be shifted to the left two bits at a time during normalization, because the smallest reduction of the exponent necessary to keep the value represented unchanged is 1, corresponding to dividing the value by 4.

Similarly, if the implied base is $B = 2^3 = 8$, the significand has to be shifted 3 bits at a time. In general, if $B = 2^q$, normalization means to left shift the significand q bits at a time until there is at least one 1 in the highest q bits of the significand. Obviously the implied 1 can not be used.

- Represent –0.75 in biased notation with e = 5 bits for exponent field. The bias is $2^{5-1} = 2^4 = 16$ and implied base is 2.

$$-0.75_{10} = -0.11_2 = (-1.1 \times 2^{-1})_2$$

The biased exponent is –1 + 16 = 15, and the notation is (without implied 1):

1	10000	1100···

or (with implied 1):

1	01111	1000···

- Find the value represented in this biased notation:

1	10001	1100···

The biased exponent is 17, the actual exponent is 17 – 16 = 1, the value is (without implied 1):

$-0.11_2 \times 2^1 = 1.1_2 = -1.1_2 = -1.5_{10}$
or (with implied 1):
$-1.11 \times 2^1 = -11.1_2 = -3.5_{10}$
Examples of IEEE 754:

- -0.3125

$-0.3125_{10} = -(0.25 + 0.0625)_{10} = -0.0101_2 = -1.01_2 \times 2^{-2} = 1.01_2 \times 2^{-2} = -1.01_2 \times 2^{125-127}$

The biased exponent is $-2 + 127 = 125$,

1	01111101	01000000000000000000000

- 1.0

$1.0 = 1.0 \times 2^0$

The biased exponent is 127,

0	01111111	00000000000000000000000

- 37.5

$$37.5_{10} = (100101.1)_2 = 1.001011 \times 2^5$$

The based exponent: $127 + 5 = 132 = (10000100)_2$,

0	00000100	00101100000000000000000

.

- -78.25

$$-78.25_{10} = -(1001110.01)_2 = -1.00111001 \times 2^6$$

The biased exponent: $127 + 6 = 133 = (10000101)_2$,

1	10000101	00111000000000000000000

- $0.01 \times 2^{-126} = 1.0 \times 2^{-128}$

As the most negative exponent representable is -126, this value is a denorm which cannot be normalized:

0	00000000	01000000000000000000000

Can you answer the following questions regarding 32–bit IEEE 754 floating-point representation and explain why?:

- What is the largest magnitude (absolute value) representable?

$$1.11\cdots1 \times 2^{127} = (2 - 2^{-23}) \times 2^{127} \approx 2^{128}$$

- What is the smallest magnitude (absolute value) representable?

$$0.00\cdots01 \times 2^{126} = 2^{-23} \times 2^{126} = 2^{-149}$$

- What is the largest gap between two consecutive numbers?

$$2^{127} \times 2^{-23} = 2^{104}$$

- What is the smallest gap between two consecutive numbers?

$$2^{-126} \times 2^{-23} = 2^{-149}$$

COMPUTER STORAGE AND BINARY NUMBER SYSTEMS

Any data stored inside a computer must be converted into machine language. Because the computer is build mainly form electronic circuits whereby they operate on an ON/OFF basis, the best numbering system to use is binary which supports two states – on/off and is given in the digits 0 and 1.

Other numbering systems used to represent data in the computer are octal and hexadecimal (both of which could be factored into powers of 2) but they are eventually converted to binary before storage. Because a computer memory is finite, only a certain amount of information can be stored in a given computer at any given time. Furthermore, the size of the numbers are also limited, for example, if the computer is an 8-bit system, the largest integer value it can hold is $2^7 = 255$. If an attempt is made to calculate a number greater than 255 it will result in an error – arithmetic overflow. *However, the computer is generally designed to allow more than 8 bit size numbers to be manipulated even if it is an 8-bit system.*

Binary used in ASCII (American Standard Code for Information Interchange). This is a table used to represent all characters in the English alphabet, including the digits 0-9. It currently use an 8-bit system, thus is able to represent 256 different characters.

Unicode – a new coding scheme that will be out soon. Uses a 16-bit system hence will represent 2^{16} characters - 65,536. This will be adequate to incorporate all of the characters of all of the languages in the world – hence the word UNICODE.

NUMBER SYSTEMS

Binary (base 2), decimal (base 10), Octal (base 8) and Hexadecimal (base 16) and their conversions from one system to the other. Storing floating point number – sign of number, mantissa, sign of exponent and exponent.

Binary addition, subtraction (2's complement), multiplication and division. Also addition in Hexadecimal.

Binary Representation of Sound and Images

Earlier, most data in a computer were textual and numbers, but now data is not only textual and numbers, it also includes lots of information in sound and image format. Both sounds and images are represented in the computer by the same binary digits as shown earlier. However, sound in their initial format is analog information, *i.e.*, when represented in a sound wave, it is sinusoidal unlike digital data that is discrete and thus in a graph represented as histograms.

In the case of sound (represented by a sound wave), the amplitude (height) is a measure of the loudness, the period measures the time of a wave (from the crest of one wave to that of another), the frequency measures the total number of cycles per unit of time (usually second) and given as cycles/

second also called hertz. This measure is also known as the pitch – giving the highness or lowness of the sound. To store a waveform, the analog signal must first be digitised. To do so, we take a sampling at fixed time intervals during the sound wave.

That is, the amplitude of the sound is measured which virtually converts a sinusoidal wave to that of a histogram. These histogram values are all discrete numbers (whole numbers) and can now be easily stored in the computer as binary. For each sound wave converted, the more sampling per wave (>= 40,000 samples per wave recommended), the higher the quality of the sound reproduced. Also, the bit depth helps to decide the accuracy of the reproduced sound. The bit depth is the amount of bits used to represent each sample, the greater the amount, the better the representation. Most audio encoding scheme today uses 16 or 24 bits per sample allowing for 65,000 to 16,000,000 distinct amplitude levels.

The various audio-encoding formats in use today includes WAV, AU, Quicktime, RealAudio and perhaps the most popular is MP-3 (MPEG – Motion Picture Experts Group). These take samples at 41,000 samples/second using 16 bits per sample. An image such as a photograph is also analog data which need to be converted to be stored as digital.

The photograph is a continuous set of intensity of colour values which if sampled correctly can be converted to digital. This process is often called scanning – which is measuring the intensity values of distinct points located at regular intervals. These points are called pixels and the more the amount of pixels the more accurate the picture is reproduced. However, there is a "supposedly" cutoff point – the naked human eyes cannot distinguish anything closer than.01mm. Anything closer than this is seen as a continuous image.

A high quality digital camera stores about 3-5 million pixels per picture. Thus a 3 x 5 picture will use about 250,000 pixels/sq. in. or 500 pixels per inch (assuming we are using about 4 million pixels for the picture so (square root of (4000000/15) = 250,000 pixels per inch). And each pixel is separated by 1/500 of an inch (1/500 * 24 → .05mm – assuming that 1 inch = 24 mm), thus we get very good quality pictures. Note each pixel can use different amount of binary digits to represent it. For true colour, it uses 24 bits, per colour – and each colour is made up of a combination of RGB.

In this case therefore, it uses 8 bits per colour. To get a black and white image, a gray scale may be used to reduce some of the total amount of bits used top represent the picture – that is rather than using 24 bits, we could use 16 bits with different distribution. After the pixels are encoded (from left to right, row by row) known as raster graphics, it is stored. This is the method used by JPEG (Joint Photographer Experts Group), GIF (Graphics Interchange formality and BMP (bit map image). *Note:* Even though we can have colour represented by 24 bits giving a total of over 16 million colours, we really don't use all of these colours at any one time – it is more like 256 colours. This allows the space requirements for storing an image to be reduced drastically. Both sound and image require a huge amount of storage against text or numbers.

For example, if we want to store a300-page book containing about 100,000 words and each word has an average of 5 characters, we would use:

100,000 x 5 x 8 = 4 million bits

Now to store 1 minute of sound at MP3 which is 41,000 samples per second at 16-bit depth:

44,100 x 16 x 60 = 42 million bits

And to store a photograph taken with a digital camera of 3 mega pixels using 24 bits per pixel:

3,000,000 x 24 = 72 million bits.

So to store information economically, we need to do data compression. One compression technique is the run-length encoding. This method replaces a sequence of identical values by a pair of values – $v_1, v_2, v_3, \ldots, v_n$ replaced with (V, n) meaning take V and replicate it n times. This produces a compression ratio which is obtained from the size of the uncompressed data divided by the size of the compressed data. There are many other codes that exist which allow us to compress data both for storage and transportation.

Another common compression method is the Lempel-Ziv (LZ77) code where we start by quoting the initial part of the message and the rest of the message is represented as a sequence of triples, consisting of two integers followed by a symbol from the message, each of which indicate how the next part of the message is to be constructed from the previous part. Example the compressed message: *xyxxyzy*(5, 4, *x*) and to decompress the message – the first number in the triple tells how far to count backwards in the first part of the message (in this case 5 symbols which leads us to the second *x*).

We now count the amount indicated by the second value in the triple going right – in this case 4 (which form the string is *xxyz*. We now append this (*xxyz*) to the end of the string producing the new string *xyxxyzyxxyz* and finally, we append the third value of the triple "*x*" to the end of this new string producing another new and finally decompressed string which is *xyxxyzyxxyzx*.

Another Example:

xyxxyzy (5,4,*x*) (0,0,*w*) (8,6,*y*) so:

- Xyxxyzy (5,4,x) gives xyxxyzyxxyzx.
- And now xyxxyzyxxyzx.(0,0,w) gives *xyxxyzyxxyzx w*
- And now *xyxxyzyxxyzxw* (8,6,*y*) gives xyxxyzyxxyzxwzyxxyzy.

Lempel-Ziv-Welsh came up with an easier scheme to encode data. Allocate umbers to the characters staring from 1, and group patterns when encountered Example:

Encode:

xyx z xyx m xyx z → 17 characters (including space)

121343 5 363 5 34 → 13 digits

Compression ratio of 17/13 → 1.33:1 or 1:1.33

This code had recently been updated but the base idea remains the same. Another scheme is the 4 bit encoding (and similar scheme can be developed) used to encode words in Hawaiian. The Hawaiian alphabet only has 11 characters 5 vowels and 6 consonants – so we could use 4 bits per character (rather than using the ASCII table) and hence shorten the amount of bits required to store a text.

We could even go a step further and decide which of these 11 characters are used most often (as in this alphabet A and H so use just 2 bits for these, then decide which are the next most commonly used, and use 3 bits for those, and so no. This code is known as a Variable Length Code.

This would further reduce the amount of bits necessary Example:

	H	A	W	A	I	I	TOTAL
ASCII ➔	01001000	01000001	01010111	01000001	01001001	01001001	48 bits
4 - bits ➔	0010	0000	0011	0000	0001	0001	24 bits
Var.len ➔	010	00	110	00	10	10	14 bits

So, we have a compression ratio of (48/24 → 2:1) for using the 4-bits encoding scheme over ASCII and we have a compression ratio of (24/14 → 1:1.7or 1.7:1) for using the Variable length over the 4-bit and a ratio of (48/14 → 1:3.4 or 3.4:1) when using the Variable length over the ASCII.

Encoding text, we have a Lossless Scheme, *i.e.*, when the data is decompressed, we do not lose data. However, upon the compression of images and/or sound, such methods as jpeg(jpg) and/or MP3 use a Lossy scheme, *i.e.*, when the sound or image is decompressed, some amount of data may be lost. But it is a tiny amount which our eyes nor ear cannot recognise. By using the Lossy scheme, compression ratio of ranging from 10:1 to 20:1 are possible.

Memories

Magnetic cores were used to construct computer memories from about 1955 to 1975. When the core become magnetised in a counterclockwise direction, the binary value of 0 is represented. The opposite would represent a 1. So the magnetic core was manipulated with the use of electrical current so as to store information. But it took too much of magnetic core to store a small amount of data. The invention of the transistor really revolutionises the manufacturing of computer. Typically, a transistor can switch states from 0 to 1 in about 2 billionths of a second. And millions of transistors can fit on a space of about 1cm^2. Thus, the rapid development of PC's.

BUILDING CIRCUITS

Truth Tables

Truth tables are derived from applying boolean algebra to binary inputs.

Example, if there are only two inputs – A and B, then there exist 4 possible combination and 16 possible binary functions. (Example given in handouts)

These circuits are built using gates. *A gate is a circuit that produces a digital output for one or more digital inputs. The output is a function of the input(s), and the type of function determines the name of the gate.*

The gate is the standard building block of all digital electronic system. Two of the most important gates are the NAND and NOR gates. But there are other gates such as the AND, OR, NOT, XOR (Exclusive OR). Some examples of these gates are:

OR gate:

A	B	F = A + B
0	0	0
0	1	1
1	0	1

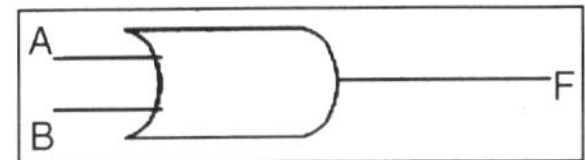

AND Gate

A	B	F = A. B
0	0	0
0	1	0

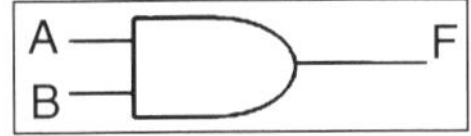

NOT Gate

A	Not A
0	1
1	0

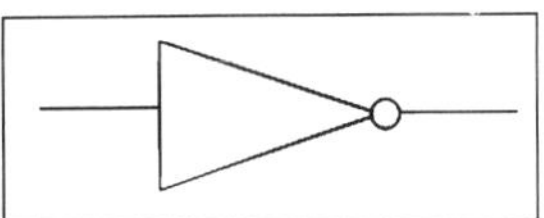

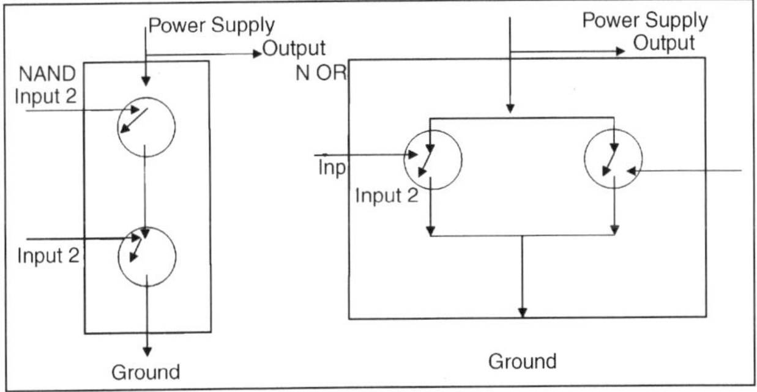

The internal construction of the AND gate is to place two transistors in serial connection. And to construct an OR gate the two transistors are placed in parallel. These are shown in the following diagrams:

CONSTRUCTION OF TRUTH TABLE AND CIRCUITS

To construct a truth table, as shown in the previous handout, we use 0 and 1 for the input and elaborate the amount of output depending upon the amount of input. If there are two inputs, then the truth table will consist of 4 different combinations. And if there are 3 inputs, the truth table will consist of 8 combinations.

In fact, if there are N inputs, then the truth table will consist of 2^n combinations. By ANDing or Oring these inputs, we can achieve the outputs, as demonstrated in the previous handout.

But to create circuits, we use the following algorithm:

- Create the truth table.
- Decide what you want for the output.
- In the output column, identify all 1 bits.
- Look at the inputs that correspond to the 1 bit output. Change all of the 0's to 1.
- Now create an expression combining the inputs (taking into consideration the changed bits.)
- AND these inputs to get the above mentioned expression.
- Now combine each of these expression by ORing them.
- Use this final result to build the circuit.

The following example will demonstrate:

A	B	Output Needed	Cases
0	0	0	
1	0	1	1
0	1	1	2
1	1	0	

Note in the output column, we want two 1 bits, where A = 1 and B = 0 for case 1 and A = 0 and B = 1 case 2. Let's look at case 1 first. We want to create an expression by using AND to get a 1 as the output, and we know that to get a 1 output from ANDing, both inputs must be 1, so we must get 1 for both inputs. To do so we take the NOT of B so we get A • (not)B.

Now for case 2: We do the same as above, only now we take the NOT of A so we get (not)A • B.

If we sh ould apply either of these expressions to the above table (individually of course) only the representative case will be true.

Now we OR these two expressions:

A • (not)B + (not)A • B

To build the circuit.

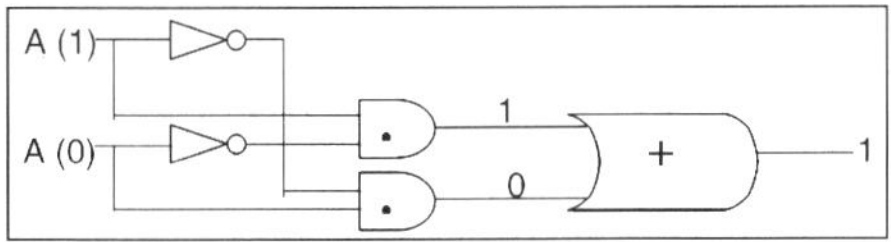

Building an Equality Tester

Again we need to follow the same procedure as above. *The truth table is as follows:*

For two inputs.

A	B	Output Needed	Cases
0	0	0	
1	0	1	1
0	1	1	2
1	1	0	

Note in the output column, we want two 1 bits, where A = 0 and B = 0 for case 1 and A = 1 and B = 1 case 2. Let's look at case 1 first. We want to create an expression by using AND to get a 1 as the output, and we know that to get a 1 output from ANDing, both inputs must be 1, so we must get 1 for both inputs. To do so we take the NOT of B and B so we get (not)A • (not)B.

Now for case 2: We do the same as above, only now we take A and B so we get A • B. If we should apply either of these expressions to the above table (individually of course) only the representative case will be true.

Now we OR these two expressions:

$$(\text{not})A \bullet (\text{not})B + A \bullet B$$

To build the circuit.

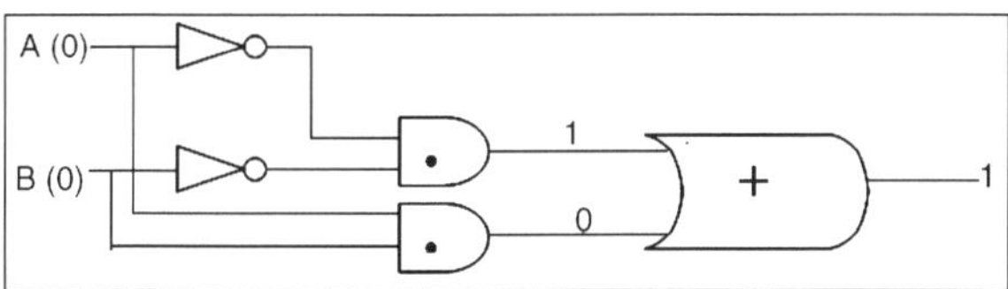

The One –Bit Adder

Because in this case when you are adding two bits, you would need a carry bit which can be 0 or 1, we must therefore have 3 inputs and two outputs (one being the sum and the other being the carry bit) as the following shows:

A	B	C	Sum	Carry
0	0	0	0	0
0	0	1	1	0
0	1	0	1	0
0	1	1	0	1
1	0	0	1	0
1	0	1	0	1

1	1	0	0	1
1	1	1	1	1

As with above, we need to create expressions for the four 1-bits in the sum column:

Case 1: (not)A • (not)B • C
Case 2: (not)A • B • (not)C
Case 3: A • (not)B • (not)C
Case 4: A • B • C

And Sum = ((*not*)*A* • (*not*)*B* • C) + ((*not*)*A* • *B* • (*not*)C) + (*A* • (*not*)*B* • (*not*)C) + (*A* • *B* • C)

Now we need to create expressions for the four 1-bits in the Carry Column:

Case 1: (not)A • B • C
Case 2: A • (not)B • C
Case 3: A • B • (not)C
Case 4: A • B • C

And Carry = ((*not*)*A* • *B* • C) + (*A* • (*not*)*B* • C) + (*A* • *B* • (*not*)C) + (*A* • *B* • C)

Control Circuits

According to the definition of an algorithm, the steps must be well ordered. The control circuits help in this ordering. The computer can only do one thing at a time. The control circuit helps to dictate what the computer will do at any one time. There are two types of control circuits: multiplexors and decoders (encoders). Multiplexors: These are in fact circuits that require 2^n inputs, n selector lines and will produce 1 output. The input lines are numbered 0, 1, 2, 3, ..., 2n. Shows a selector line multiplexor with 2 input lines. Decoders: This is the opposite to a multiplexor – it accepts n input lines and has 2n output lines. The difference is that decoders do not have selector lines.

Together, the decoder and multiplexor helps us to build a computer that will execute the correct instructions using the correct data values. Figure shows a typical decoder circuit – used to select a correct mathematical operation. If our computer is capable of only 4 operations – add, subtract, multiply and divide, we could use this decoder to select which operation needs to be selected. According to the diagram, n = 2 input lines and thus give 4 output lines (1 for add, 1 for subtract, 1 for multiply and 1 for divide). After selecting the correct operation to be carried out, the multiplexor will now make sure that the correct data is used to carry out the operation. Example, if the decoder had decoded that an addition operation needed to be done, then two multiplexors could be used to get the left operand and the right operand to carry out the addition. The decoder would have allowed the addition circuit to be selected, and now the multiplexors would send the left and right operand to the addition circuit to be processed.

5

Sequential Circuits Analysis and Design

DIGITAL ELECTRONICS IN SEQUENTIAL CIRCUITS

Digital electronics is classified into combinational logic and sequential logic. Combinational logic output depends on the inputs levels, whereas sequential logic output depends on stored levels and also the input levels.

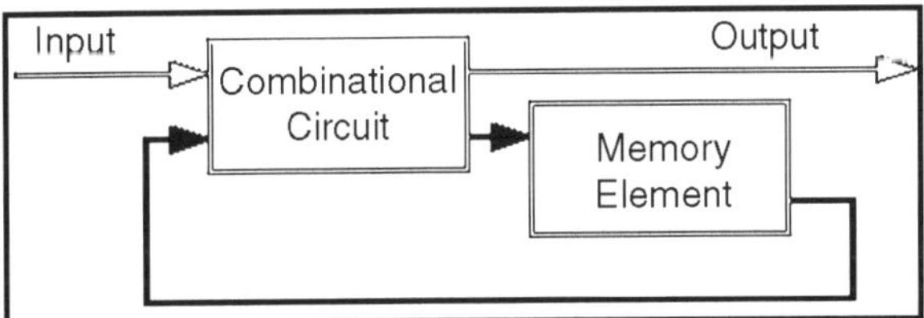

The following figure shows a way to consider sequential circuits.

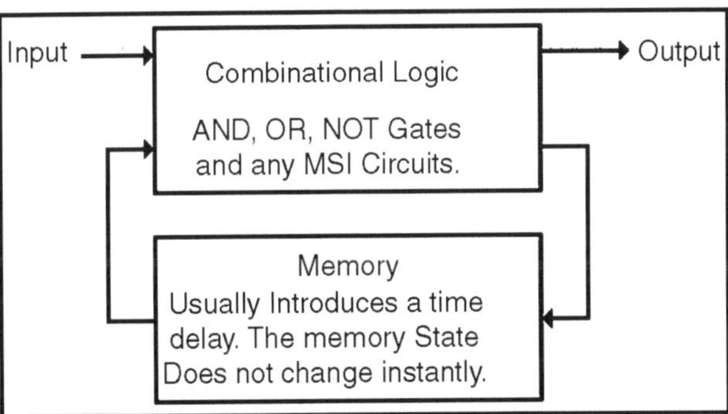

Fig: Sequential Logic Includes Combinational Logic and Memory

Sequential circuits can be characterized into two broad classes - synchronous and asynchronous. As a general rule, asynchronous circuits are faster, but much harder to design. We shall focus totally on synchronous circuits. By Q(T) we denote the state of a sequential circuit at time T - this is basically its memory. We watch the state of the circuit change from Q(T) to Q(T + 1) as the clock ticks. The constraint on synchronous circuits is that the state of the circuit changes after the input, thus we have a typical sequence as follows:

- At time T, we have input (denoted by X) and state Q(T).
- As a result of the input X and state Q(T), a new state is computed,

This becomes available to the input only at time (T + 1) and so is called Q(T + 1). The fact that the new state, computed as a result of X and Q(T), is not available to the input of the sequential circuit until the next time step greatly facilitates the design and analysis of the specific circuit.

DIGITAL CIRCUIT ANALYSIS AND DESIGN

Circuit analysis begins with a circuit diagram or a black box and ends with an identification of the sequential circuit implemented by the device - normally a truth table. The steps are:

- Identify the inputs and the outputs
- Express each output as a Boolean function of the inputs and the present state Q(T)
- Identify the circuit if possible.

CIRCUIT FOR ANALYSIS

We first study the analysis of digital circuits, then we study their design. There are a number of steps in the analysis of a circuit. Where to begin depends on what one has. When given a circuit diagram, the following steps are used to begin the analysis.

- Determine the inputs and outputs of the circuit. Assign variables to represent these.
- Determine the inputs and outputs of the flip-flops.
- Construct the Next State and Output Tables.
- Construct the State Diagram.
- If possible, identify the circuit. There are no good rules for this step.

Consider the following circuit. We want to discover what it does.

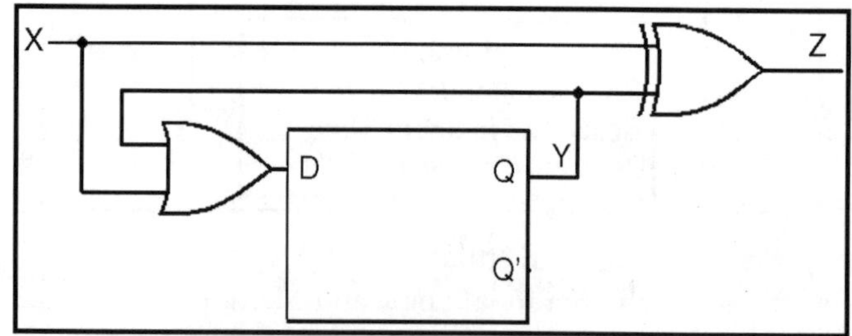

Fig: Circuit to Be Analysed

Step 1: Identify and Label the Inputs, Outputs, and Internal States

We use the following variables in the analysis of this circuit with a single flip-flop. X denotes the input, Y denotes the output of the flip-flop (Y' also), and Z the output of the circuit. If we had more than one flip-flop, we would label the flip-flop with a number beginning at 0 and use that as a subscript, so flip-flop 0 would have output Y0, etc.

Step 2: Determine the Inputs and Outputs of the Flip-flops

The next step is to determine the equations for Z, the output, and D, the input to the flip-flop. By inspection, we determine the following for the equations:

$$Z = X \oplus Y$$
$$D = X + Y$$

Step 3: Construct the Next State and Output Tables.

We begin this state by recalling the characteristic table of each flip-flop that is used in the design. Here we have only one flip-flip, a D with a very simple characteristic table that is better represented as an equation: Q(t+1) = D - the next state is what you put in now.

Noting that Q(t) = Y (the state of a flip-flop is also its output) we construct the following Next-State diagram for the flip-flop, based on the characteristic table of a D flip-flop and the equation we derived for the D input: D = X + Y.

One simple caution here is that the input to a flip-flop is a function of the present state only, having nothing to do with the next state (as we have no crystal balls). Thus Y = Q(t).

Here is the present state (PS)/ next state (NS) diagram for the circuit.

X	Q(t) = Y	D = X + Y	Q(t+1)
0	0	0	0
0	1	1	1
1	0	1	1
1	1	1	1

The output table is similarly constructed, using the equation Z = X Y = Z = X Q(t). Again, note that the output is not a function of the next state.

X	Q(t)	Z
0	0	0
0	1	1
1	0	1
1	1	0

These two tables are combined to form the transition/ output table.

X	Q(t) = Y	D = X + Y	Q(t+1)/ Z
0	0	0	0/ 0
0	1	1	1/ 1
1	0	1	1/ 1
1	1	1	1/ 0

At this point, we should produce a state table in the standard format. This involves assigning labels to each of the two states, currently identified only as Q(t) = 0 and Q(t) = 1. For lack of anything more imaginative, we label the states 0 and 1.

Present State	Next State/Output	
	X = 0	X = 1
0	0/ 0	1/ 1
1	1/ 1	1/ 0

Step 4: Construct the State Diagram

The final step in the process may be the creation of the state diagram.

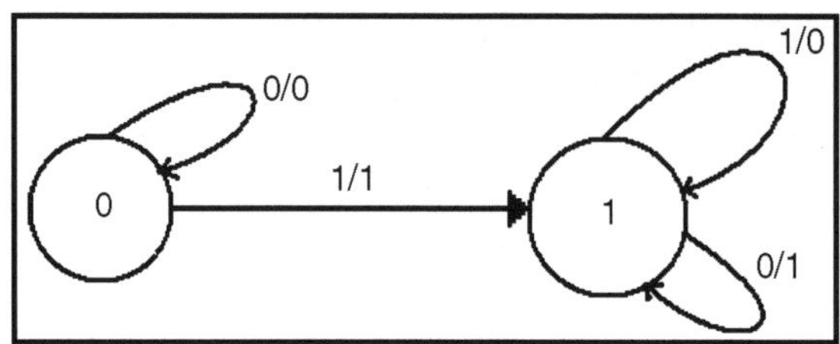

At this point, we have a complete description of the circuit. It may be possible to proceed from this diagram to obtain an understanding of what the circuit does.

Step 5: If possible, identify the circuit.

This circuit is a 2-state device, with memory represented by one bit. The circuit stays in state 0 from the start until a 1 is input at which time it transitions to state 1 and remains there.

Note the relation of the output to the input, depending on the state of the machine.

Input	Q(t)	Output	
0	0	0	For Q(t) = 0, the output is X
1	0	1	
0	1	1	For Q(t) = 1, the output is X'.
1	1	0	

A verbal description of the circuit is then that it copies its input to the output until the first 1 is encountered in the input stream. After that event, all input is output as complemented. What this circuit does is take the two's-complement of a binary integer, presented to the circuit least-significant bit first. It is easy to prove that such a strategy produces the two's complement of a number. First, consider a number ending in 1, say X_nX_n–1.... X_2X_{11}.

The one's complement of this number is Xn'Xn-1' .... X_2'X_1'0. Adding 1 to this produces the number Xn'X_{n-1}' .... X_2'X_1'1, in which the least significant 1 is copied and the remaining bits are complemented. The proof is completed by supposing that the number terminates in a one or more zeroes; i.e., its least significant bits are 10 .... 0, where the count of zeroes is not important.

The one's-complement of this number will end with 01 .... 1, where each zero in the original has turned to a 1. But 01 .... 1 + 1 = 10 .... 0, and up to the least significant 1 the two's-complement is a copy of the bits in the integer itself. Note that there is no carry out of the addition that produced the right-most 1, so the remainder of the integer is formed by the one's-complement. Thus we have a complete description of the circuit.

It is a serial two's-complementer.

CIRCUIT ELEMENTS OF LATCHES AND FLIP FLOPS

When showing either latches or flip-flops as circuit elements, it is undesirable to show the"internals" of the device.

Here are the symbols for SR and D latches.

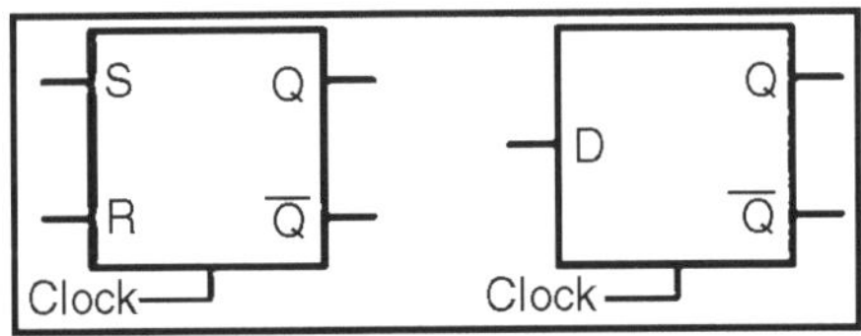

FLIP FLOPS

A flip-flop is a"bit box"; it stores a single binary bit. By Q(t), we denote the state of the flip-flop at the present time, or present tick of the clock; either Q(t) = 0 or Q(t) = 1. The student will note that throughout this textbook we make the assumption that all circuit elements function correctly, so that any binary device is assumed to have only two states.

A flip-flop must have an output; this is called either Q or Q(t). This output indicates the current state of the flip-flop, and as such is either a binary 0 or a binary 1. We shall see that, as a result of the way in which they are constructed, all flip-flops also output $\overline{Q(t)}$, the complement of the current state. Each flip-flop also has, as input, signals that specify how the next state, Q(t + 1), is to relate to the present state, Q(t).

Every flip-flop also has an input derived from the system clock, which allows it to function as a synchronous circuit. It also has connections to power and ground. Here are the symbols for SR and D flip-flops, which we have not yet defined.

Note the small triangles on the clock input; this says"edge triggered".

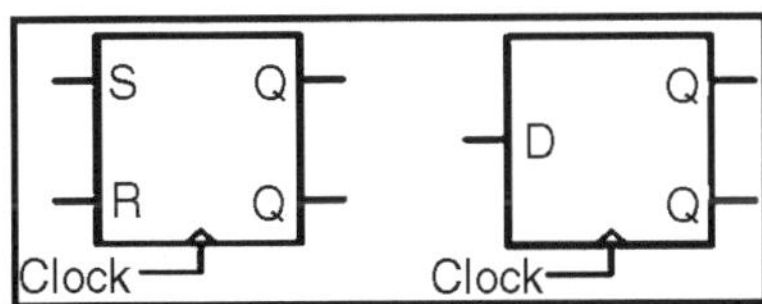

One should note that the more common symbols for flip-flops show the clock input coming in"from the side" as shown in the next figure.

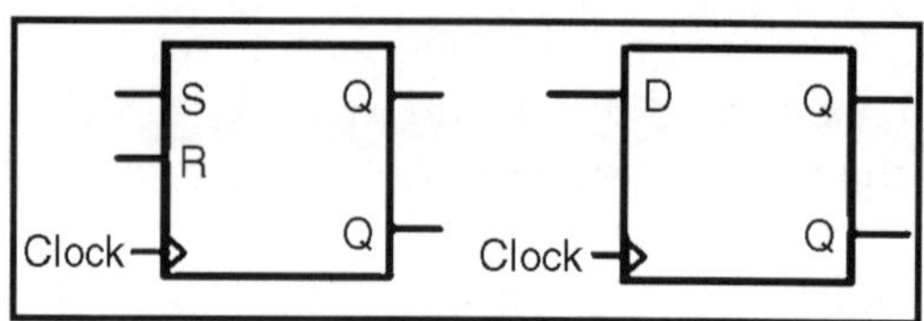

THE CLOCK

The most fundamental characteristic of synchronous sequential circuits is a system clock. This is an electronic circuit that produces a repetitive train of logic 1 and logic 0 at a regular rate, called the clock frequency. Most computer systems have a number of clocks, usually operating at related frequencies; for example - 2 GHz, 1GHz, 500MHz, and 125MHz. The inverse of the clock frequency is the clock cycle time. As an example, we consider a clock with a frequency of 2 GHz (2"109 Hertz). The cycle time is 1.0/ (2"109) seconds, or

0.5"10-9 seconds = 0.500 nanoseconds = 500 picoseconds.

Synchronous sequential circuits are sequential circuits that use a clock input to order events. Asynchronous sequential circuits do not use a common clock and, as hinted at above, are much harder to design and test. As we shall focus only on synchronous circuits, we immediately launch a discussion of the clock. The following figure illustrates some of the terms commonly used for a clock.

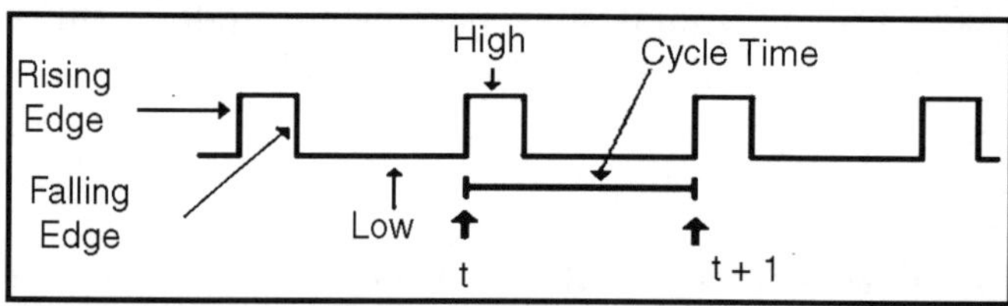

The clock input is very important to the concept of a sequential circuit. At each"tick" of the clock the output of a sequential circuit is determined by its input and by its state. We now provide a common definition of a"clock tick" - it occurs at the rising edge of each pulse. We use t to represent the time at a clock tick and (t + 1) to denote the time at the next clock tick - the difference between the two is the clock cycle time. Suppose a 2 GHz clock, which corresponds to a clock cycle time of 0.5 nanosecond. Strictly speaking, we should label our timings in nanoseconds: 1.0, 1.5. 2.0. 2.5, etc. The convention is just to count the ticks, referring to the present clock pulse as occurring at time t and the next one at time (t + 1).

DESCRIPTION

By definition, a flip-flop is an edge-triggered latch. We must now show how achieve edge triggering. In essence, what we shall do is take a level triggered latch and give it a clock pulse with a very short positive (logical 1)

phase. The key component of an edge-triggered flip-flop is a pulse generator that we studied in a previous chapter. This could be based on the gate delay of a NOT gate.

More modern devices likely use a different circuit so as to generate a shorter pulse.

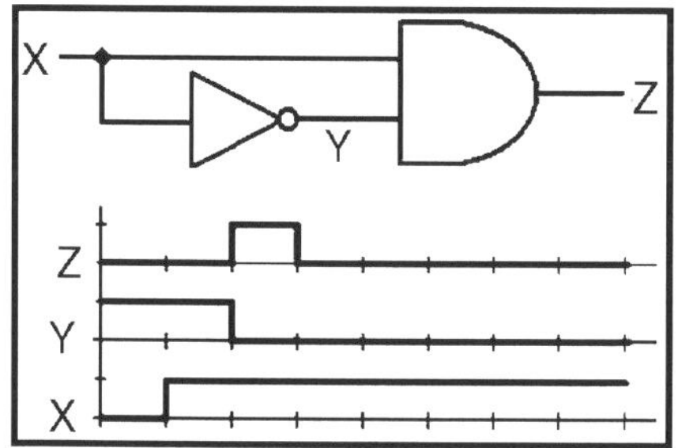

Now that we have a method to generate a short pulse, we can build an edge-triggered device. The following is a diagram of the components of two typical edge-triggered flip-flops. The top one is a SR flip-flop and the bottom one is a D flip-flop.

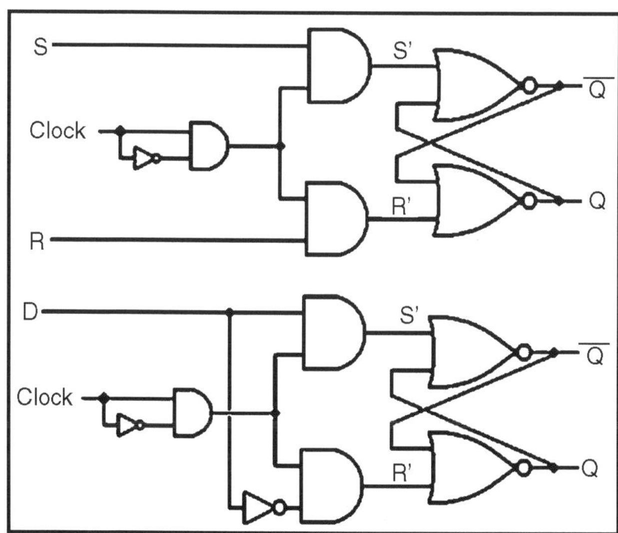

TYPES OF FLIP-FLOPS

In order to avoid the simple example, we shall examine the general case. Here is the circuit for consideration. We postulate a total of gate delays (including that of the D flip-flop) to be the time interval signified by Δ. Thus at time Δ after the D latch is first sensitive to its input, the circuitry has output a new value to become D. But the flip-flop is activated by a short pulse, of time length δ.

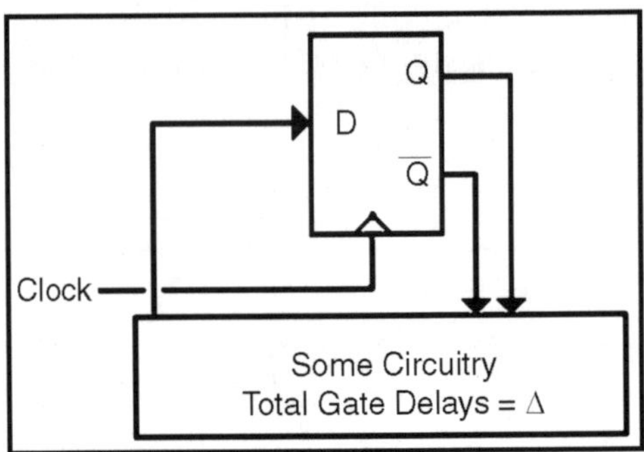

The timing diagram below shows the interruption of the uncontrolled feedback loop. By the time that the output of the circuitry has changed, the D flip-flop is no longer sensitive to input. The input will not become effective until the beginning of the next clock cycle.

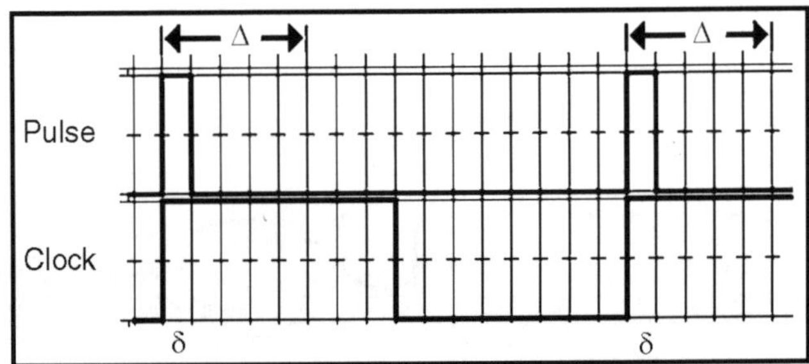

How can one insure that the relative timings of two circuits, the D flip-flop and the rest of the CPU circuitry, operate with the correct timings? This is one of the issues of central importance in the design of a CPU; since we have stumbled into it, let's talk about it.

To be more precise, define two total gate delays: Δ_{MIN} and Δ_{MAX}. Δ_{MIN} is the total time delay for the fastest CPU operation and Δ_{MAX}. the delay for the slowest. We must haveΔ_{MIN} >, or the circuit would occasionally display uncontrolled feedback. Conservatively, we might say Δ_{MIN} 1.5•. The next criterion is a bit more difficult to state precisely, but it might be stated something like T 1.5•δ Δ_{MAX}, where T denotes the clock period. What we say here is that the clock period must be long enough for the CPU output to"settle". Note, however, that a value of T much larger than MAX is just wasted time.

Put another way, the value of MAX determines the fastest clock that can reasonably be applied to the CPU. A good part of the art of CPU design is based on this issue. When we study the RISC (Reduced Instruction Set Computer) movement, we shall see that one of the issues was to remove the more complex instructions, thus reducing MAX and allowing for a faster clock. There is a trade-off here that we shall explore in later chapters. To be honest, the sum of CPU gate delays is not always the limiting factor in the clock speed.

Some recent CPUs have been designed with a"hot clock" that is hot in both ways. It is very fast, being of the order of 4 to 5 GHz. It is also hot in the literal sense, in that the CPU, operating at that clock rate, emits so much heat that it overheats itself. The art of CPU design is always bumping up against those messy laws of physics.

The SR Flip-Flop

We have completely defined the SR flip-flop, based on the idea of an SR latch. While we could just leave the topic and proceed, it is appropriate at this time to review the subject and state, in one place, what the student is expected to know. We begin with a depiction of the SR flip-flop that will be used in all future discussions.

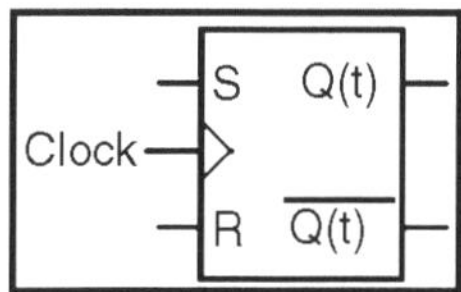

At this point we are no longer interested in the internal construction of the flip flop, but on its operational characteristics. These are given in the characteristic table.

S	R	Q(t + 1)
0	0	Q(t)
0	1	0
1	0	1
1	1	ERROR

We next address an issue that commonly arises in the use of flip-flops in circuit design. We have a number of scenarios. For each, we know Q(t) and what Q(t + 1) should be. The question is how to achieve that change. For example, if Q(t) = 0 and we want Q(t + 1) = 0, we have two choices: either S = 0 and R = 0, or S = 0 and R = 1.

The first option, keeps the state unchanged at 0; the second forces it to 0. As S = 0 is sufficient to do this without regard to the value of R, we say that the input is S = 0 and R = d; the d standing for"don't care". On the other hand, if Q(t) = 0 and Q(t + 1) = 1, only S = 1 and R = 0 will do. This is the only combination that will give a next state of 1 when the present state is 0.

If we have Q(t) = 1 and want Q(t + 1) = 0, then our choice is simple: S = 0 and R = 1. If we have Q(t) = 1 and want Q(t + 1) = 1, then we can choose either S = 0 and R = 0, or S = 1 and R = 0. This is denoted as S = d and R = 0.

The above discussions lead to the excitation table for the SR flip-flop.

Q(t)	Q(t + 1)	S	R
0	0	0	d

0	1	1	0
1	0	0	1
1	1	d	0

THE JK FLIP-FLOP: ENHANCING THE SR FLIP-FLOP

Recall the characteristic table of the SR flip-flop. We repeat the table here for emphasis.

S	R	Q(t + 1)
0	0	Q(t)
0	1	0
1	0	1
1	1	ERROR

The theoretician examining this table would note two facts immediately.

- The input S = 1 and R = 1 is disallowed; we would like to do something with it.
- The values for Q(t + 1) are three of the possible four Boolean functions of 1 variable.

Considering Q as a Boolean variable, we now show that there are exactly four Boolean functions of this Boolean variable.

These are f(Q) = 0, f(Q) = 1, f(Q) = Q, and f(Q) = $\overline{Q}$. We do this by showing the truth table for each of these functions and noting that there are only four different ways to put 0's and 1's into the two row entries for a function..

Table. The Four Boolean Functions of Boolean Variable Q

Q	0	Q	$\overline{Q}$	1
0	0	0	1	1
1	0	1	0	1

Given this, our enhanced SR flip-flop would have Q(t + 1) = $\overline{Q}$ (t) as one possible output. For maximal compatibility with the existing SR, we would want to leave the existing valid inputs alone and just make good use of the invalid one. What we get is a JK flip-flop.

Recalling that an SR flip-flop is so called because it is a Set-Reset device, we may ask for the meaning of JK. One is tempted to make up some story related to names in German, but the basic answer is that"I don't know".

In any case, we present the characteristic table for the JK flip-flop and then ask how one might modify an SR flip to achieve that goal.

Here is the desired characteristic table.

J	K	Q(t + 1)
0	0	Q(t)
0	1	0

1	0	1
1	1	$\overline{Q}$ (t)

Viewing this as a modification of the SR flip-flop, we ask how to generate each of S and R from the inputs J and K under the following constraints;

- Except when J = 1 and K = 1, we want to have S = J and R = K. Under these circumstances, the behaviour is identical.
- When J = 1, K = 1, and Q = 0, we want S = 1 and R = 0. This makes Q(t + 1) = 1.
- When J = 1, K = 1, and Q = 1, we want S = 0 and R = 1. This makes Q(t + 1) = 0.

When in doubt about how to create a circuit, we make a truth table. The inputs to the truth table are J, K, and Q (the present state). The outputs are S and R.

Row	**Q**	**J**	**K**	**S**	**R**
0	0	0	0	0	0
1	0	0	1	0	1
2	0	1	0	1	0
3	0	1	1	1	0
4	1	0	0	0	0
5	1	0	1	0	1
6	1	1	0	1	0
7	1	1	1	0	1

We have two patterns that almost work: S = J• $\overline{Q}$ and R = K"Q. Let's examine each case.

S = J• $\overline{Q}$ produces the expected result for all rows except row 6. In that row we have

Q = 1, J = 1, and K = 0. We want Q(t + 1) to be 1. But S = 0 and R = 0 will cause Q(t + 1) = Q(t) = 1, exactly what we want. So this simple formula causes no trouble.

R = K•Q produces the expected result for all rows except row 1. In that row we have Q = 0, J = 0, and K = 1. We want Q(t + 1) to be 0. But S = 0 and R = 0 will cause Q(t + 1) = Q(t) = 0, exactly what we want. So again we have no trouble.

Here is the basic detailed circuit for the JK flip-flop.

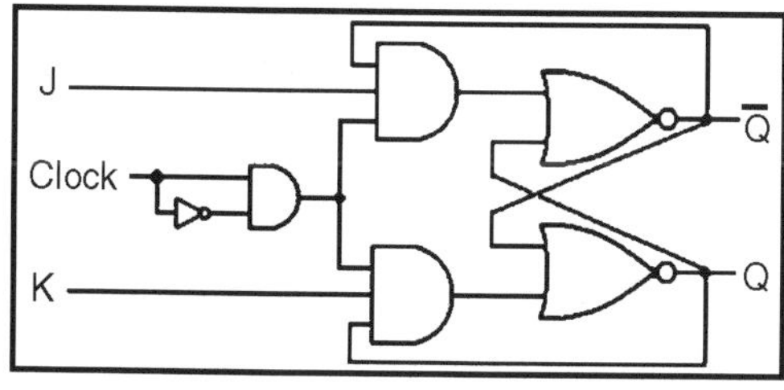

We now give the standard representation of the JK flip-flop as will be used in future discussions. Again, note the triangle symbol on the clock input, indicating edge triggering.

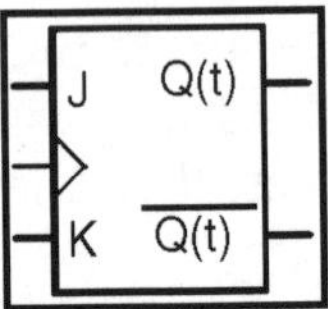

We have already presented the characteristic table for the JK flip-flop. Here it is again.

J	K	Q(t + 1)
0	0	Q(t)
0	1	0
1	0	1
1	1	$\overline{Q}$ (t)

We now create the excitation table for the JK.

If Q(t) = 0 and Q(t + 1) is to be 0, we can use either J = 0 and K = 0, or J = 0 and K = 1. If Q(t) = 0 and Q(t + 1) is to be 1, we can use either J = 1 and K = 0, or J = 1 and K = 1. Note that this is a new option, not available for an SR flip-flop. If Q(t) = 1 and Q(t + 1) is to be 0, we can use either J = 0 and K = 1, or J = 1 and K = 1.

Note that this also is a new option, not available for an SR flip-flop.

If Q(t) = 1 and Q(t + 1) is to be 1, we can use either J = 0 and K = 0, or J = 1 and K = 0. This gives the excitation table for the JK flip-flop.

Q(t)	Q(t + 1)	J	K
0	0	0	d
0	1	1	d
1	0	d	1
1	1	d	0

THE D FLIP-FLOP

We have already examined the D latch. The input is labeled"D" for"Data". The D flip-flop has a characteristic table identical to that of a D latch.

D	Q(t + 1)
0	0
1	1

The device is so simple that it does not require an excitation equation. We just use an excitation equation, which simply states"Give it what you want".

D = Q(t + 1)

The T Flip-Flop

This is the fourth and last of the major types of flip-flops. The input to this flip-flop is labeled"T" for toggle. When T = 0, the state remains the same. When T = 1, the flip-flop changes state. This gives rise to the following characteristic table.

T	Q(t + 1)
0	Q(t)
1	$\overline{Q}$ (t)

The excitation table for this flip-flop is almost obvious.

Q(t)	Q(t + 1)	T
0	0	0
0	1	1
1	0	1
1	1	0

This gives rise to an excitation equation.

$T = Q(t) \oplus Q(t+1)$

JK As a General Flip-Flop

We now take notice that the JK can be configured to function as any of the other 3 flip-flop types. In this, it is the most general type of flip-flop.

JK As a D Flip-Flop

To convert a JK to a D flip-flop, connect the inputs as follows.

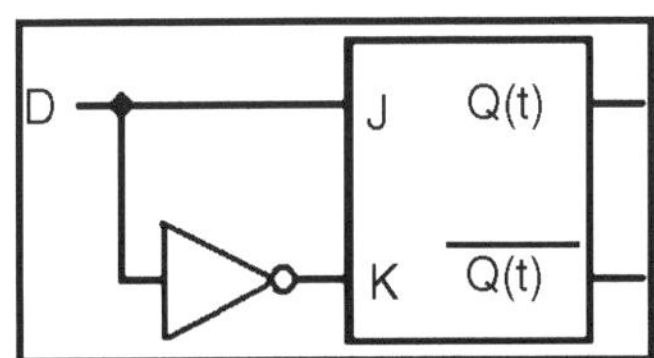

If D = 0, then J = 0, K = 1, and Q(t + 1) = 0, If D = 1, then J = 1, K = 0, and Q(t + 1) = 1.

The JK As A T Flip-Flop

To convert a JK to a T flip-flop, connect the inputs as follows.

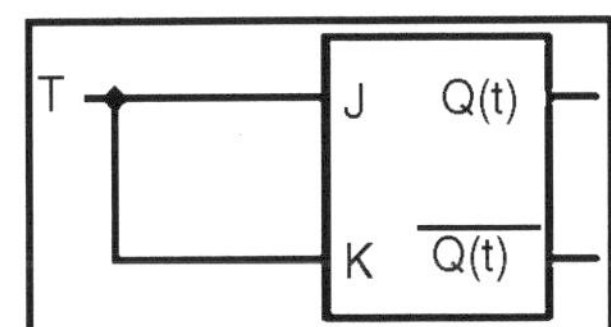

If T = 0, then J = 0, K = 0, and Q(t + 1) = Q(t).

If T = 1, then J = 1, K = 1, and Q(t + 1) = $\overline{Q}$ (t)

DESIGN OF SEQUENTIAL CIRCUITS

Having seen how to Analyse digital circuits, we now investigate how to design digital circuits. We assume that we are given a complete and unambiguous description of the circuit to be designed as a starting point. At this level, most design problems focus on one of two topics: modulo-N counters and sequence detectors.

Here is an overview of the design procedure for a sequential circuit:

- Derive the state diagram and state table for the circuit.
- Count the number of states in the state diagram (call it N) and calculate the number of flip-flops needed (call it P) by solving the equation 2P-1 < N 2P. This is best solved by guessing the value of P.
- Assign a unique P-bit binary number (state vector) to each state. Often, the first state = 0, the next state = 1, etc.
- Derive the state transition table and the output table.
- Separate the state transition table into P tables, one for each flip-flop.
- Warning: Things can get messy here; neatness counts.
- Decide on the types of flip-flops to use. When in doubt, use all JK's.
- Derive the input table for each flip-flop using the excitation tables for the type.
- Derive the input equations for each flip-flop based as functions of the input and current state of all flip-flops.
- Summarize the equations by writing them in one place.
- Draw the circuit diagram. Most homework assignments will not go this far, as the circuit diagrams are hard to draw neatly.

DESIGN PROBLEM: A MODULO-4 COUNTER

As our first design problem, let's consider a modulo-four counter. When the direction is not specified, we usually intend to build a modulo-four up-counter: 0, 1, 2, 3, 0, 1, 2, 3, etc. We solve these design problems by using the step-wise procedure listed above.

Step 1: Derive the state diagram and state table for the circuit.

Here is the state diagram. Note that it is quite simple and involves no input.

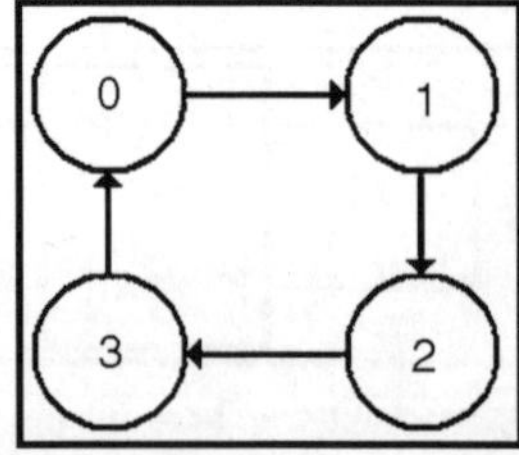

Fig. The State Diagram for a Modulo-4 Up Counter

The state table is simply a rearrangement of the state diagram into a tabular form.

Present State	Next State
0	1
1	2
2	3
3	0

Step 2: Count the number of states in the state diagram (call it N) and calculate the number of flip-flops needed (call it P) by solving the equation $2^{P-1} < N \leq 2^P$. The number of states on a modulo-N counter is simply N; these are labeled 0 through (N - 1). Specifically, a modulo-4 counter has four states: labeled 0, 1, 2, and 3.

We solve the equation $2^{P-1} < 4 \leq 2^P$ by noting that $2^1 = 2$ and $2^2 = 4$, so we have determined that $2^1 < 4 \leq 22$, hence P = 2. We shall see later that there are valid solutions with more than two flip-flops, but there are none with fewer.

Step 3: Assign a unique P-bit binary number (state vector) to each state.

Often, the first state = 0, the next state = 1, etc.

Some sequential circuits suggest an innovative numbering system, but modulo-N counters never do. We go with the obvious labeling, generated by assigning each decimal number its two-bit binary equivalent as an unsigned integer in the range from 0 to 3.

State	2-bit Vector
0	0 0
1	0 1
2	1 0
3	1 1

Step 4: Derive the state transition table and the output table.

There is no output table for any modulo-N counter, as the output associated with this type of table is the output on a transition, as is seen in a sequence detector. The transition table is a direct translation of the state table, using the assignments from the previous step.

Present	State	Next State
0	00	01
1	01	10
2	10	11
3	11	00

Step 5: Separate the state transition table into P tables, one for each flip-flop.

Here we separate the state transition table into 2 tables, one for each flip-flop. Note that the flip-flops will be numbered 0 and 1, with flip-flop 0 storing the least significant bit of the state information. Thus, we shall refer to the state information as Y1Y0.

Flip-Flop 1		Flip-Flop 0	
Present State	Next State	Present State	Next State
Y_1 Y_0	Y_1(T+1)	Y_1 Y_0	Y_0(T+1)
0 0	0	0 0	1
0 1	1	0 1	0
1 0	1	1 0	1
1 1	0	1 1	0

Step 6: Decide on the types of flip-flops to use. When in doubt, use all JK's.

Up to this point, we have made no assumptions about the type of flip-flop to use. In order to proceed any farther with the design, we must now commit to a specific type. In line with this author's preferences, he chooses to use two JK flip-flops.

Q(t)	Q(t+1)	J	K
0	0	0	d
0	1	1	d
1	0	d	1
1	1	d	0

The excitation table for a JK flip-flop is shown at right. Recall that the"d" stands for"Don't Care". For example, if we have Q(t) = 0 and want Q(t+1) = 0, we can use either J = 0, K = 0 or J = 0, K = 1. Similarly either J = 1 and K = 0 or J = 1 and K = 1 will take Q(t) = 0 to Q(t + 1) = 1.

Step 7: Derive the input table for each flip-flop using the excitation tables for the type.

First look at flip-flop 1, representing the high-order bit. Note that we compare the present state of Y1 to its next state in order to determine J1 and K1.

PS	NS	Input	
Y1 Y0	Y1	J1	K1
0 0	0	0	d
0 1	1	1	d
1 0	1	d	0
1 1	0	d	1

Note that in deciding on the input, we must match only the 0's and 1's. We ignore the don't-cares. Note that the"d" for"don't-care" is not a variable to be assigned a value. It is a value that does not need to be matched. At the moment, Y0 is included in the table for future use only. It plays no part in determining the values of J1 and K1.

Here is the table for Y0

PS	NS	Input	
Y1 Y0	Y0	J0	K0

0 0	1	1	d
0 1	0	d	1
1 0	1	1	d
1 1	0	d	1

Again, Y1 is included in the table for future use only. It plays no part in determining the values of J0 and K0.

Step 8 Derive the input equations for each flip-flop based as functions of the input and current state of all flip-flops.

At this point, we try to derive an expression that matches each column. Formal methods can be used, but generally are more trouble than they are worth. Here is this author's set of rules to match an expression to a given column.

- If a column does not have a 0 in it, match it to the constant value 1. If a column does not have a 1 in it, match it to the constant value 0.
- If the column has both 0's and 1's in it, try to match it to a single variable, which must be part of the present state. Only the 0's and 1's in a column must match the suggested function.
- If every 0 and 1 in the column is a mismatch, match to the complement of a function.
- If all the above fails, try for simple combinations of the present state.

Let's look at the input table for Y1.

PS	**NS**	**Input**	
Y_1 Y_0	Y_1	J_1	K_1
0 0	0	0	d
0 1	1	1	d
1 0	1	d	0
1 1	0	d	1

Note that the column for J1 has a 0 and a 1 in it as does the column for K1. Each column has two"don't cares" in it, but we ignore these. Because each column has both a 0 and a 1 in it, neither is a match for a constant function. We now try to match J1.

J1 does not match Y1, because Y1 is 0 in the same row (0 1) as J1 is 1. J1 matches Y0. In row 0 0, both Y0 and J1 are 1. In row 0 1, both Y0 and J1 are 1.

In rows 1 0 and 1 1, J1 is a"don't care", so we do not need to match it.

Similar logic shows that K1 matches Y0 also.

So now we have the following matches for J1 and K1.

PS	**NS**	**Input**	
Y_1 Y_0	Y_1	J_1	K_1
0 0	0	0	d
0 1	1	1	d
1 0	1	d	0
1 1	0	d	1

$J_1 = Y_0 \quad K_1 = Y_0$

We now examine Y_0

PS	NS	Input	
$Y_1\ Y_0$	Y_0	J_0	K_0
0 0	1	1	d
0 1	0	d	1
1 0	1	1	d
1 1	0	d	1

Note that there are no 0's in either the J0 or K0 column. The simplest (and best) match is

J0 = 1 and K0 = 1.

Step 9: Summarize the equations by writing them in one place.

Here they are.

J1 = Y0 K1 = Y0

J0 = 1 K0 = 1

This is a counter, so there is no Z output.

Step 10 Draw the circuit diagram.

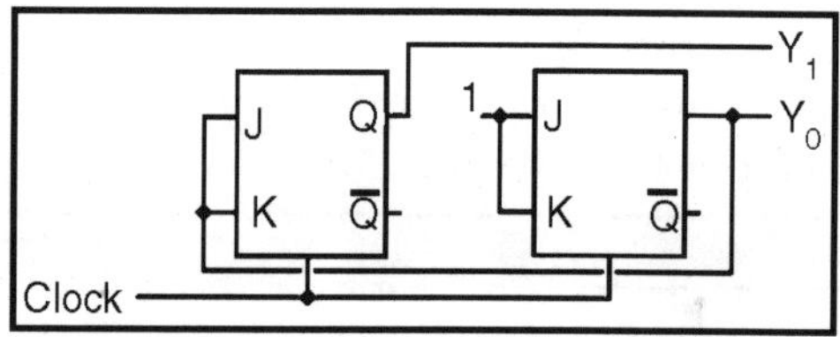

But wait - there is another solution hidden here. Recall that a JK flip-flop can be used to emulate a T flip-flop by setting the J input equal to the K input. Note that the design has the following interesting property.

J1 = K1 = Y0

J0 = K0 = 1

Given this, we can replace each JK flip-flop with a T flip-flop, arriving at this design.

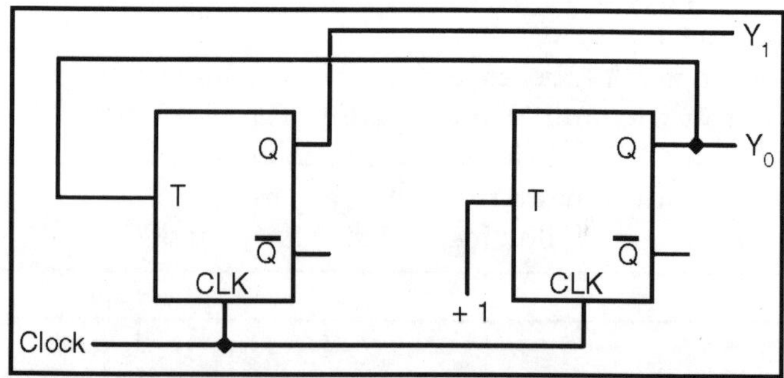

The modulo-4 counter just designed outputs binary codes for the time pulses. Specifically, we assume that it is initialized to Y1Y0 = 00 and then outputs 01, 10, 11, 00, 01, 10, 11, etc. A more realistic circuit would output

discrete pulses corresponding to the decoded output, so that first T0 = 1 and all others are 0, then T1 = 1 and all others are 0, etc. In order to produce the discrete signals T0, T1, T2, and T3, we need to add a decoding phase to the counter.

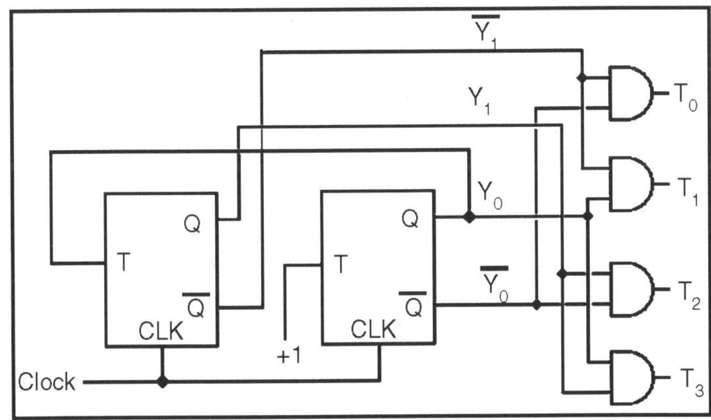

Note that the above design is simplified by the fact that the outputs Y1' and Y0' are available directly from the flip-flops and do not need to be synthesized using NOT gates. One can achieve a simpler design at the cost of additional flip-flops. The following design is called a one-hot design, in that it uses a shift register in which exactly one flip-flop at a time is storing a 1. This design also works as a modulo-4 counter and skips the decoder delays.

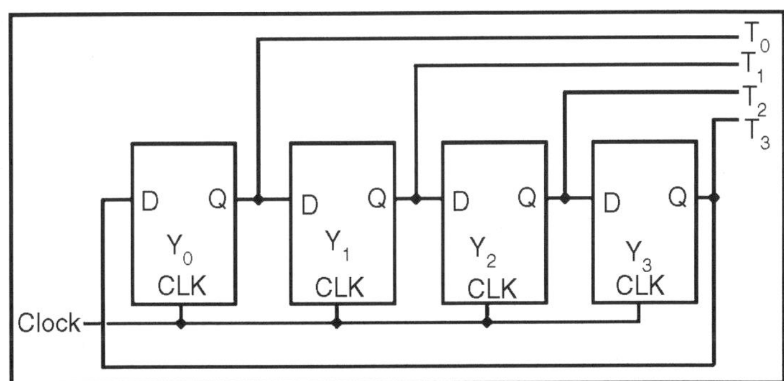

When the counter is initialized, we set Y0 = 1, and Y1 = Y2 = Y3 = 0. As the clock ticks, the single 1 is shifted by the shift register, so that the discrete signals become high in sequence.

DESIGN PROBLEM

The Modulo-4 Up-Down Counter. For the next design, we introduce a problem that uses input. This is a modulo-4 up-down counter. The input X is used to control the direction of counting.

If X = 0, the device counts up: 0, 1, 2, 3, 0, 1, 2, 3, etc.

If X = 1, the device counts down: 0, 3, 2, 1, 0, 3, 2, 1, etc.

Step 1: Derive the state diagram and state table for the circuit.

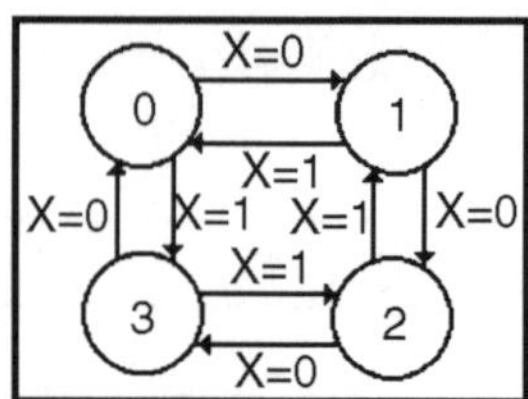

The state diagram for the modulo-4 up-down counter is shown at right. Notice that the X input is used to determine the counting direction. Again, this type of circuit does not have any output associated with the transitions; the output just reflects which of the four states the machine finds itself in at the moment. We now produce the sate table by translating the state diagram. As an aside, some students might prefer to begin the design process with the state table and omit the state diagram. That is certainly acceptable practice; whatever works should be used. Here, the state table depends on X - the input used to specify the counting direction.

Present State	Next State	
X = 0	X = 1	
0	1	3
1	2	0
2	3	1
3	0	2

Step 2: Count the number of states in the state diagram (call it N) and calculate the number of flip-flops needed (call it P) by solving the equation 2P-1 < N 2P. The number of states on a modulo-N counter is simply N; these are labeled 0 through (N - 1). Specifically, a modulo-4 counter has four states: labeled 0, 1, 2, and 3.

We solve the equation 2P-1 < 4 2P by noting that 21 = 2 and 22 = 4, so we have determined that 21 < 4 22, hence P = 2. We shall see later that there are valid solutions with more than two flip-flops, but there are none with fewer.

Step 3: Assign a unique P-bit binary number (state vector) to each state. Often, the first state = 0, the next state = 1, etc.

Some sequential circuits suggest an innovative numbering system, but modulo-N counters never do. We go with the obvious labeling, generated by assigning each decimal number its two-bit binary equivalent as an unsigned integer in the range from 0 to 3.

State	2-bit Vector
0	0 0
1	0 1
2	1 0
3	1 1

Step 4: Derive the state transition table and the output table.

There is no output table for any modulo-N counter, as the output associated with this type of table is the output on a transition, as is seen in a sequence detector. The transition table is a direct translation of the state table, using the assignments from the previous step.

The transition table for the modulo-4 up-down counter is as follows.

Present State		**Next State**	
X = 0		X = 1	
0	00	01	11
1	01	10	00
2	10	11	01
3	11	00	10

Step 5: Separate the state transition table into P tables, one for each flip-flop.

Here we separate the state transition table into 2 tables, one for each flip-flop. Note that the flip-flops will be numbered 0 and 1, with flip-flop 0 storing the least significant bit of the state information. Thus, we shall refer to the state information as Y1Y0.

Flip-Flop 1			Flip-Flop 0		
PS	Next State		PS	Next State	
Y_1Y_0	Y_1, X = 0	Y_1, X = 1	Y_1Y_0	Y_0, X = 0	Y_0, X = 1
0 0	0	1	0 0	1	1
0 1	1	0	0 1	0	0
1 0	1	0	1 0	1	1
1 1	0	1	1 1	0	0

The student who is paying attention at this point will notice an interesting feature concerning flip-flop 0; specifically that its next state does not depend on X. This is due to the fact that in considering a modulo-N counter, one moves from odd numbers to even numbers and from even numbers to odd numbers in both counting up and counting down.

Step 6: Decide on the types of flip-flops to use. When in doubt, use all JK's.

Up to this point, we have made no assumptions about the type of flip-flop to use. In order to proceed any farther with the design, we must now commit to a specific type. In line with this author's preferences, he chooses to use two JK flip-flops.

Q(t)	**Q(t+1)**	**J**	**K**
0	0	0	d
0	1	1	d
1	0	d	1
1	1	d	0

The excitation table for a JK flip-flop is shown at right. Recall that the"d" stands for"Don't Care". For example, if we have Q(t) = 0 and want Q(t+1) = 0, we can use either J = 0, K = 0 or J = 0, K = 1. Similarly either J = 1 and K = 0 or J = 1 and K = 1 will take Q(t) = 0 to Q(t + 1) = 1.

Step 7: Derive the input table for each flip-flop using the excitation tables for the type. Here is the input table for flip-flop 1. Note that the arrangement of the table has been altered to reflect the fact that we now have a binary input.

	X = 0				**X = 1**	
Y_1Y_0	Y_1	J_1	K_1	Y_1	J_1	K_1
0 0	0	0	d	1	1	d
0 1	1	1	d	0	0	d
1 0	1	d	0	0	d	1
1 1	0	d	1	1	d	0

Here is the input table for flip-flop 0.

	X = 0	**X = 1**				
Y_1Y_0	Y_0	J_0	K_0	Y_1	J_0	K_0
0 0	1	1	d	1	1	d
0 1	0	d	1	0	d	1
1 0	1	1	d	1	1	d
1 1	0	d	1	0	d	1

Step 8: Derive the input equations for each flip-flop based as functions of the input and current state of all flip-flops. At this point, we try to derive an expression that matches each column. Formal methods can be used, but generally are more trouble than they are worth. Here is this author's set of rules to match an expression to a given column.

- If a column does not have a 0 in it, match it to the constant value 1. If a column does not have a 1 in it, match it to the constant value 0.
- If the column has both 0's and 1's in it, try to match it to a single variable,
  which must be part of the present state. Only the 0's and 1's in a column must match the suggested function.
- If every 0 and 1 in the column is a mismatch, match to the complement of a function.
- If all the above fails, try for simple combinations of the present state.

The reader will note that there are two columns for each variable for which an equation is desired; one column for X = 0 and one column for X = 1.

For example, consider the table for flip-flop 0, just above. If we work on a column-by column basis, we shall arrive at four equations.

One for J0 when X = 0,

one for K_0 when X = 0,

one for J_0 when X = 1, and
one for K_0 when X = 1.
However, we need a single equation for J0 and a single equation for K0.

CONSIST OF REGISTERS

Registers are simply a set of flip-flops used to store a number of binary bits. A set of n flip-flops can store bits. 1-bit is a simple form of register.

This register consists of a 1-bit input (I), and enable signal (enb), and a 1-bit output (Q).

When enb=1, the register loads the value of I on the rising clock edge. When enb=0, the register holds the current value. We create a module by name of Reg1. The register consists of an interface of 4 ports, clk, I, enb, Q. // 1-bit Register

```
module Reg1(clk, I enb, Q);
endmodule
```

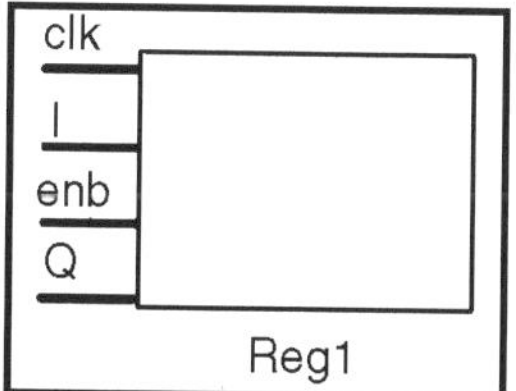

We next specify what type the ports are. In this example, clk, I, and enb are input ports. Port Q is an output port.

Notice in addition to declaring Q as an output, we must also declare Q as a register so we can assign values to Q.// 1-bit Register

```
module Reg1(clk, I enb, Q);
input clk;
input I, enb;
output Q;
reg Q;
endmodule
```

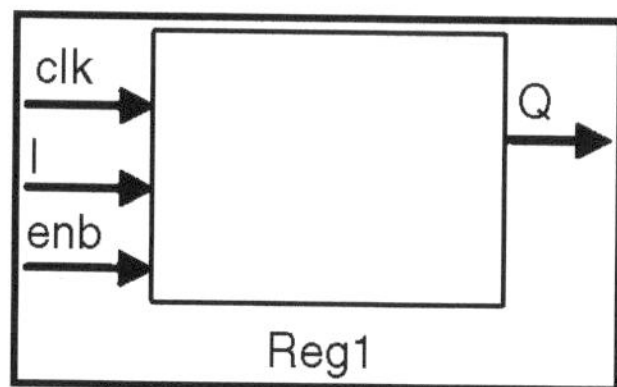

We then create a sensitivity list, which executes the code between the"begin" and"end" statements onv every rising edge of the clk.// 1-bit Register

```
module Reg1(clk, I enb, Q);
input clk;
input I, enb;
output Q;
```

```
reg Q;
always @ (posedge clk)
begin
end
endmodule
```

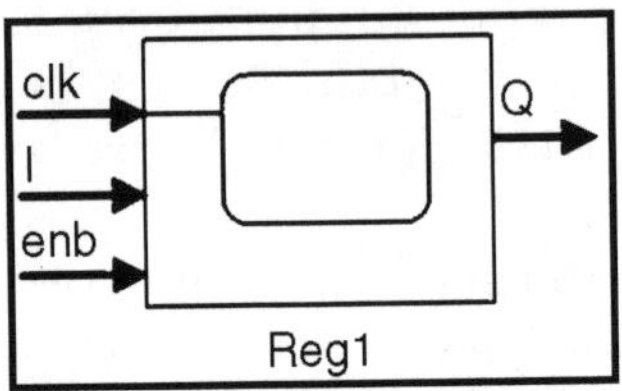

If enb=1 on the rising edge of the clk, then we store the value of input I.

```
// 1-bit Register
module Reg1(clk, I enb, Q);
input clk;
input I, enb;
output Q;
reg Q;
always @ (posedge clk)
begin
if(enb == 1)
Q ⇐ I;
end
endmodule
```

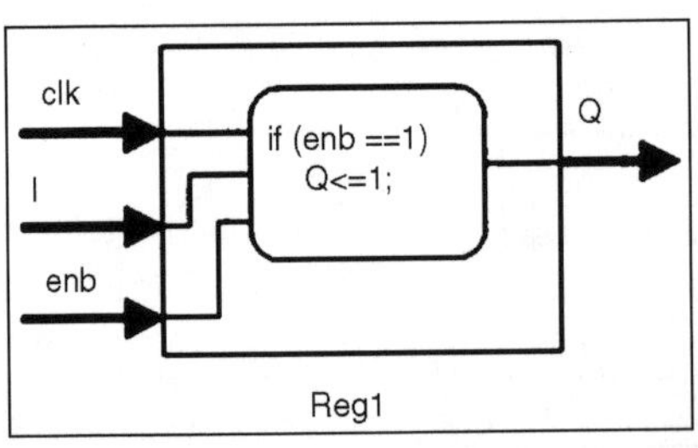

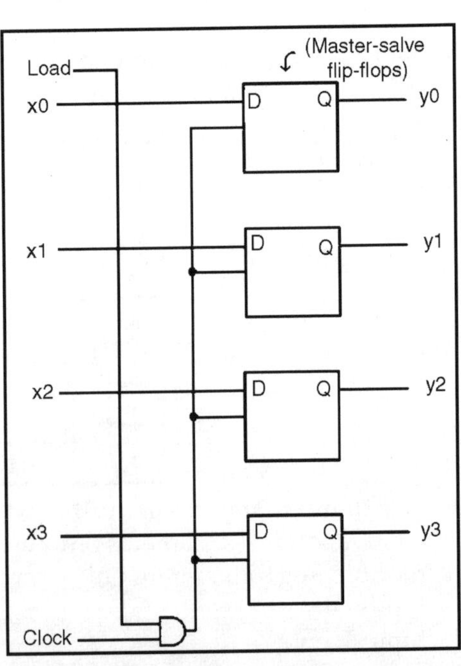

DESCRIPTION

A CPU contains very fast memory called registers. For example, a MIPS ISA stores 32 32-bit registers.

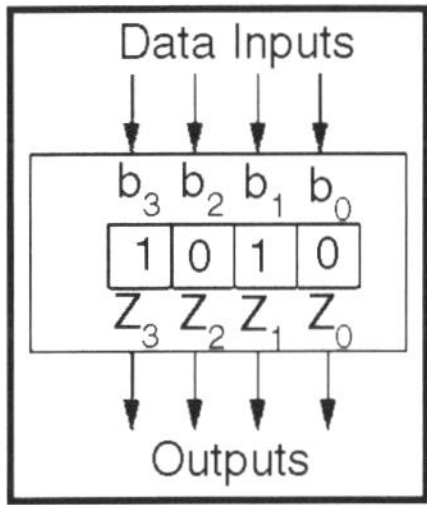

In some sense, this is not really a black box, because we have some idea of what's in it. However, let's not worry about that. The black box shown above has four inputs. The register can do one of two things: it can parallel load the inputs, that is, it can read in all four inputs and store the four bits into the array. Each element of the array can store a single bit. It can also choose to ignore the inputs, and the values of the arrays remain unchanged. In this case, we say the array is"holding" its values.

The array's values is continuously being sent as output. Thus, if we could measure the output of the register (say, with a voltmeter), the four bits being output would correspond to the four bits currently being stored in the array. This output is sent to the outside world where other devices can read the values, if they want to. A register can have more than 4 bits stored in it. In fact, modern CPUs either store 32 or 64 bits in registers. In reality, one bit of a register is really a flip flop, which is a device that can store one bit. We will discuss flip flops at some later point.

Parallel Load Register

This register has a name. It's called a parallel load register. It supports two operations. The operation is controlled by a single control bit, called c.

c	Operation
0	Hold
1	Parallel Load

Like combinational logic circuits, the c refers to a control bit. Recall that control bits are used to determine which operation or function a device performs. Unlike combinational logic circuits, registers use clocks. Here's an important feature of registers (and sequential logic devices in general).

A device that is edge-triggered (which this register is) can only change values when there is a clock edge. In particular, we assume a register is positive-edge triggered, so it can only change values whenever a positive edge occurs. Specifically, when the clock reaches a positive edge, the register detects this event. It then reads in the value of c, the control bit, and based on the value, either parallel loads (i.e., sets $z_3z_2z_1z_0 = x_3x_2x_1x_0$) or it holds. If a clock is not at a positive edge, then the value of the registers is held, and does not change.

Hold

You may wonder why we have a"hold" operation? After all, when there's no positive edge, the register seems to hold the value already. Here's an analogy. Many years ago, milk was delivered by a milkman, much in the same way newspapers are delivered today. (I guess in those days, they were always men, though if milkmen existed today they'd probably be called"milk attendants"). Imagine that the milkman comes by your house once a day during the weekdays. You don't always want to have milk dropped off each day. When you are done with a milk bottle (usually made of glass), you leave it outside your door. The milkman sees the bottle, picks it up, and replaces it with a new bottle. However, when the milkman does not see the bottle then no new bottle is dropped off. He assumes that you have not yet completed a bottle of milk. Similarly, a positive edge occurs on a regular basis (just like the milkman appears on a daily basis), and it's not always the case that you want to parallel load every time this happens. You want to control when a register is loaded. Sometimes you want it to parallel load, sometimes to hold its value. Thus, the need for a control signal.

Registers are Memory

Registers store information over time, which makes them devices with memory. There are other kinds of circuits (combinational logic circuits and wires) which merely process inputs and produce outputs, but do not store any values.

Clock

When we study combinational logic, we won't use a clock there. Combinational logic circuits are made from AND gates, OR gates, etc. These circuits can be modeled using a graph. Such graphs are acyclic, i.e., there are no cycles. However, in general, it's useful to design circuits with feedback (i.e., cycles). Circuits with feedback are easier to design if we can control how often the circuit is updated. Thus, the need for a clock. For now, just believe that it's important for a register to have a clock.

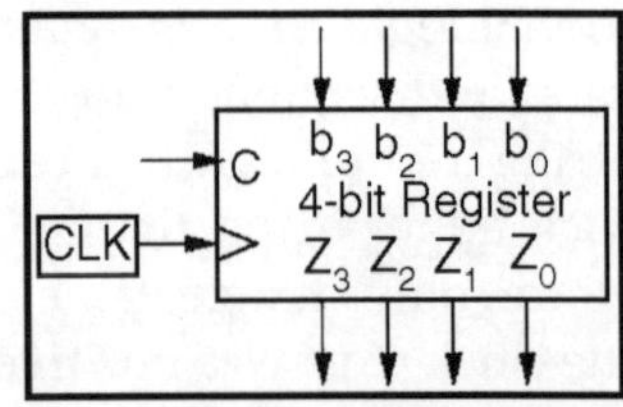

Diagram of a 4-bit Register

Let's look at a picture of a 4-bit register.

A 4-bit register has the following inputs and outputs:

- 4 bits of data input b3-0
- 1 control bit, c, which indicates whether to perform a parallel load or hold.
- A clock (which is an input), used for timing. The register will only perform the operation when there is a positive edge from the clock.

Example

Let's see an example of how a register behaves.

In particular, let's see what happens when a c = 0. I've added boxes inside the register diagram so you can see the"bits".

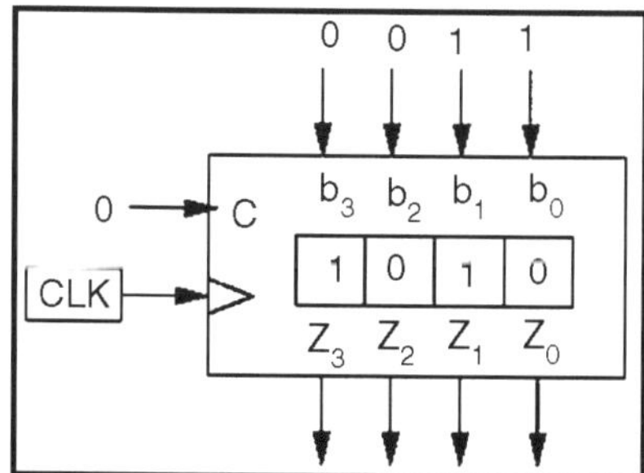

Suppose the register contains bits 1010. The inputs are 0011. When c == 0 and there is a positive clock edge, then the register maintains the value 1010. That is, itholds this value. The bits being fed into the input, i.e. 0011, are ignored. If the clock is not currently at a positive edge (i.e., it's at a lèvel 1, or a level 0, or a negative edge), then the register also holds the value. That is, it ignores the inputsb3-0, and maintains the value stored in the register, in this case, 1010. A register always outputs the bits it stores. Thus, z3-0 = 1010, since those are the bits stored in the register. This register has"memory". That is, it stores information. In this case, it stores 4 bits of information.

Parallel Load

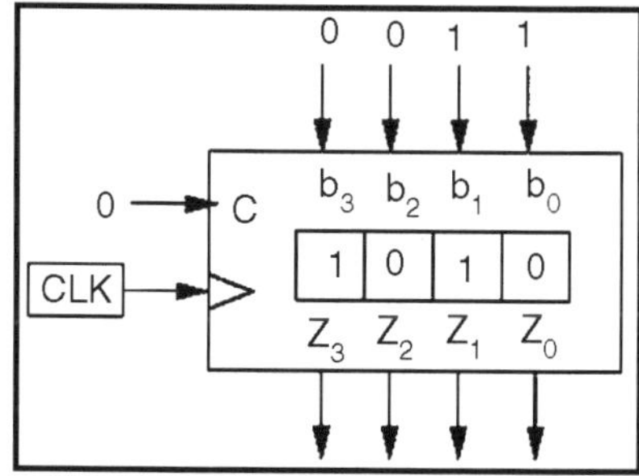

Now suppose the circuit is attempting a parallel load. That is, c == 1. Assume the positive edge has not yet arrived, and the register looks like:

After a positive edge occurs, the 4 bits are read in from the input, in parallel. This overwrites the previous bits, and replaces it with the new bits. In this case, those bits are 0011.

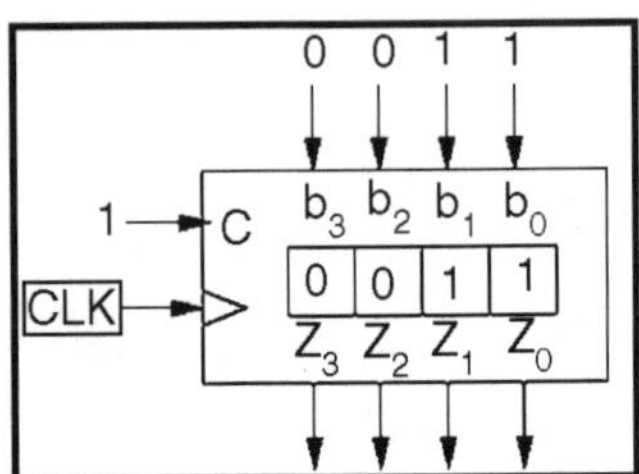

Once the bits are loaded, the outputs z3-0 is 0011.

Points to Remember

- A k-bit register contains k flip flops.
- Each flip flop can store a single bit.
- A positive edge-triggered flip flop (thus, a register) can only change the value of the bit on a positive edge.
- When a positive edge occurs, a register can either hold its value (if c == 0) or parallel load (if c == 1).
- A parallel load means the bits are read in from the input bits (b(k-1)-0) and stored in the k-bits within the register.
- A hold means the registers do not read in the bits, but maintains the current values of the bits.
- A register can only change its value at most once per positive edge.
- When the clock is not at a positive edge, the register maintains ("holds") its value.
- A k-bit register always outputs its values through z(k-1)-0.

CATEGORIES OF COUNTERS

Counters can be classified into two broad categories according to the way they are clocked:

- Asynchronous (Ripple) Counters - the first flip-flop is clocked by the external clock pulse, and then each successive flip-flop is clocked by the Q or Q' output of the previous flip-flop.
- Synchronous Counters - all memory elements are simultaneously triggered by the same clock.

ASYNCHRONOUS (RIPPLE) COUNTERS

A two-bit asynchronous counter is shown on the left. The external clock is connected to the clock input of the first flip-flop (FF0) only. So, FF0 changes state at the falling edge of each clock pulse, but FF1 changes only when triggered by the falling edge of the Q output of FF0. Because of the inherent propagation delay through a flip-flop, the transition of the input clock pulse and a transition of the Q output of FF0 can never occur at exactly the same time. Therefore, the flip-flops cannot be triggered simultaneously, producing an asynchronous operation.

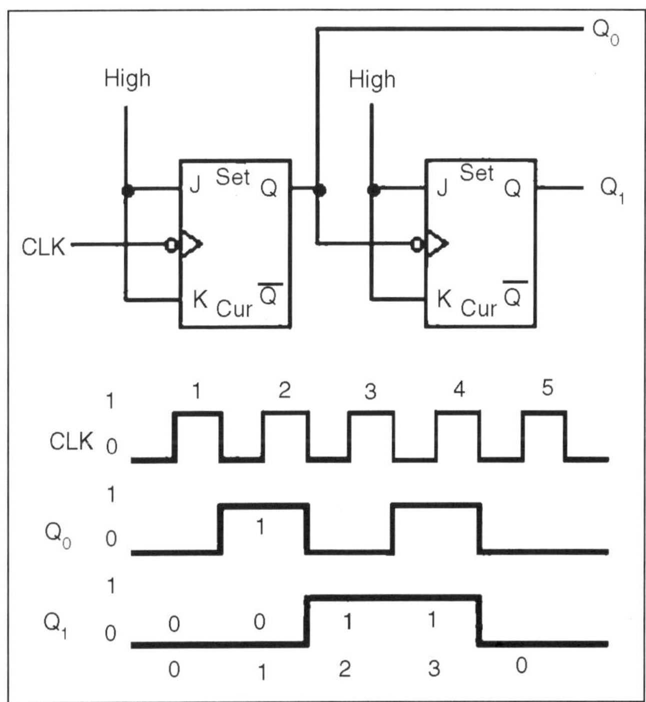

Note that for simplicity, the transitions of Q0, Q1 and CLK in the timing diagram above are shown as simultaneous even though this is an asynchronous counter. Actually, there is some small delay between the CLK, Q0 and Q1 transitions. 0 Usually, all the CLEAR inputs are connected together, so that a single pulse can clear all the flip-flops before counting starts. The clock pulse fed into FF0 is rippled through the other counters after propagation delays, like a ripple on water, hence the name Ripple Counter.

The 2-bit ripple counter circuit above has four different states, each one corresponding to a count value. Similarly, a counter with n flip-flops can have 2 to the power n states. The number of states in a counter is known as its mod (modulo) number. Thus a 2-bit counter is amod-4 counter. A mod-n counter may also described as a divide-by-n counter. This is because the most significant flip-flop (the furthest flip-flop from the original clock pulse) produces one pulse for every n pulses at the clock input of the least significant flip-flop (the one triggers by the clock pulse). Thus, the above counter is an example of a divide-by-4 counter.

The following is a three-bit asynchronous binary counter 0 and its timing diagram for one cycle. It works exactly the same way as a two-bit asynchronous binary counter mentioned above, except it has eight states due to the third flip-flop.

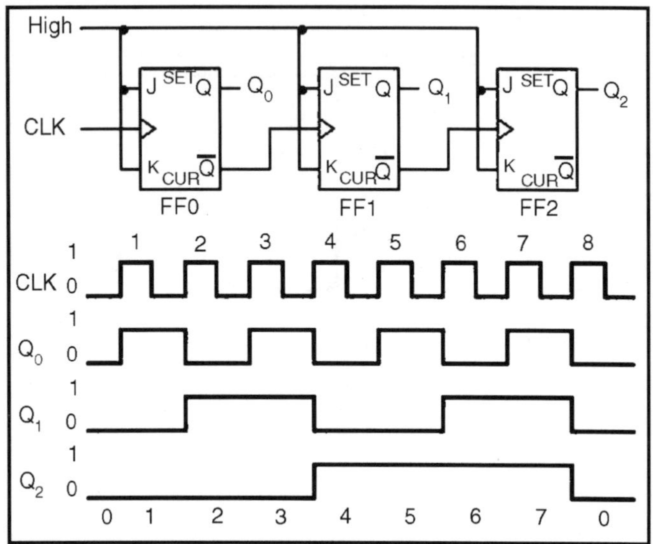

ASYNCHRONOUS DECADE COUNTERS

The binary counters previously introduced have two to the power n states. But counters with states less than this number are also possible. They are designed to have the number of states in their sequences, which are called truncated sequences. These sequences are achieved by forcing the counter to recycle before going through all of its normal states. A common modulus for counters with truncated sequences is ten.

A counter with ten states in its sequence is called a decade counter. The circuit below is an implementation of a decade counter.

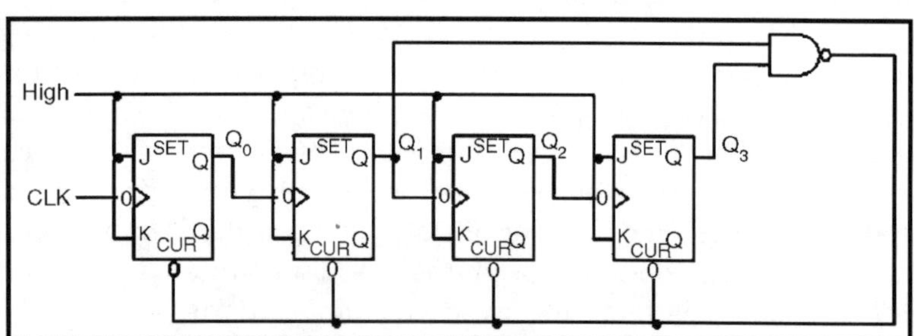

Once the counter counts to ten (1010), all the flip-flops are being cleared. Notice that only Q1 and Q3 are used to decode the count of ten.

This is called partial decoding, as none of the other states (zero to nine) have both Q1 and Q3 HIGH at the same time.

The sequence of the decade counter is shown in the table below:

Clock Pulse	Q3	Q2	Q1	Q0
0	0	0	0	0
1	0	0	0	1
2	0	0	1	0
3	0	0	1	1
4	0	1	0	0
5	0	1	0	1
6	0	1	1	0
7	0	1	1	1
8	1	0	0	0
9	1	0	0	1

Asynchronous Up-Down Counters

In certain applications a counter must be able to count both up and down. The circuit below is a 3-bit up-down counter. It counts up or down depending on the status of the control signals UP and DOWN.

When the UP input is at 1 and the DOWN input is at 0, the NAND network between FF0 and FF1 will gate the non-inverted output (Q) of FF0 into the clock input of FF1. Similarly, Q of FF1 will be gated through the other NAND network into the clock input of FF2. Thus the counter will count up.

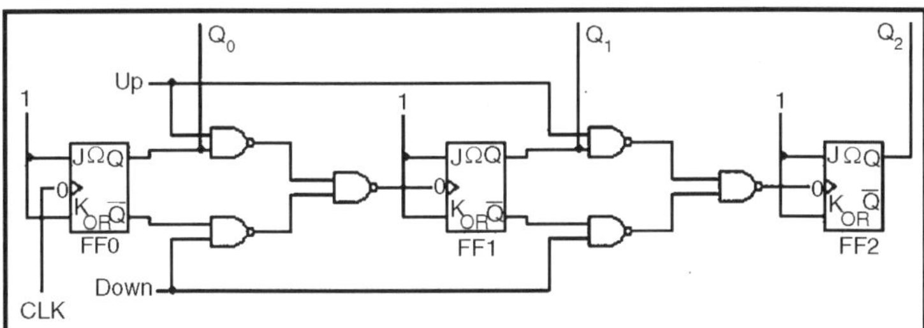

When the control input UP is at 0 and DOWN is at 1, the inverted outputs of FF0 and FF1 are gated into the clock inputs of FF1 and FF2 respectively. If the flip-flops are initially reset to 0's, then the counter will go through the following sequence as input pulses are applied.

FF2	FF1	FF0
0	0	0
1	1	1
1	1	0
1	0	1
1	0	0
0	1	1
0	1	0
0	0	1

Notice that an asynchronous up-down counter is slower than an up counter or a down counter because of the additional propagation delay introduced by the NAND networks.

SYNCHRONOUS COUNTERS

In synchronous counters, the clock inputs of all the flip-flops are connected together and are triggered by the input pulses. Thus, all the flip-flops change state simultaneously (in parallel). The circuit below is a 3-bit synchronous counter. The J and K inputs of FF0 are connected to HIGH. FF1 has its J and K inputs connected to the output of FF0, and the J and K inputs of FF2 are connected to the output of an AND gate that is fed by the outputs of FF0 and FF1.

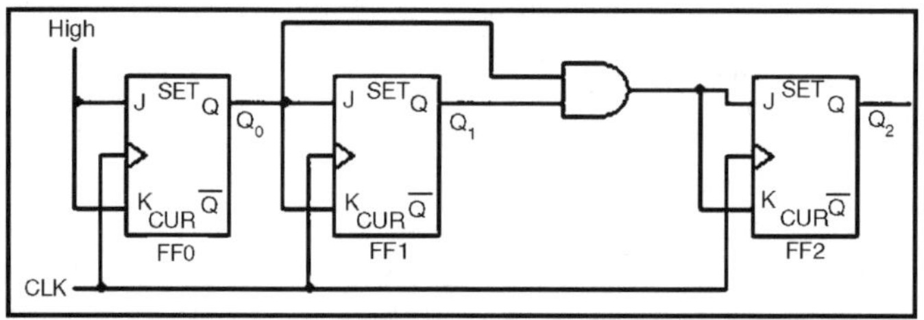

Pay attention to what happens after the 3rd clock pulse. Both outputs of FF0 and FF1 are HIGH. The positive edge of the 4th clock pulse will cause FF2 to change its state due to the AND gate.

SYNCHRONOUS DECADE COUNTERS

Similar to an asynchronous decade counter, a synchronous decade counter counts from 0 to 9 and then recycles to 0 again. This is done by forcing the 1010 state back to the 0000 state. This so called truncated sequence can be constructed by the following circuit.

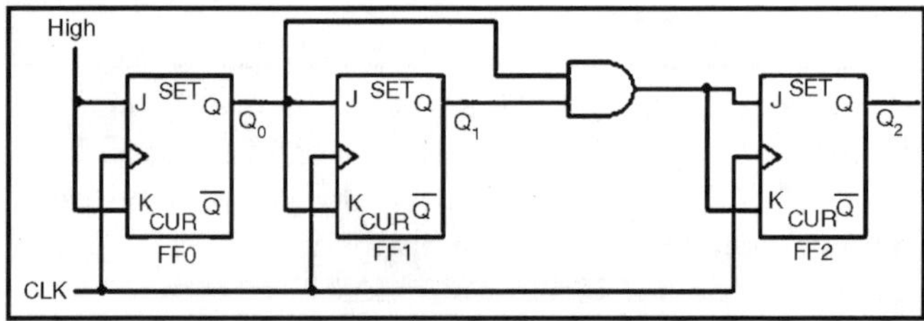

Clock Pulse	Q3	Q2	Q1	Q0
0	0	0	0	0
1	0	0	0	1
2	0	0	1	0
3	0	0	1	1
4	0	1	0	0
5	0	1	0	1
6	0	1	1	0
7	0	1	1	1
8	1	0	0	0
9	1	0	0	1

From the sequence on the left, we notice that:

- Q0 toggles on each clock pulse.
- Q1 changes on the next clock pulse each time Q0=1 and Q3=0.
- Q2 changes on the next clock pulse each time Q0=Q1=1.
- Q3 changes on the next clock pulse each time Q0=1, Q1=1 and Q2=1 (count 7), or when Q0=1 and Q3=1 (count 9).

Synchronous Up-Down Counters

A circuit of a 3-bit synchronous up-down counter and a table of its sequence are shown below.

Similar to an asynchronous up-down counter, a synchronous up-down counter also has an up-down control input. It is used to control the direction of the counter through a certain sequence.

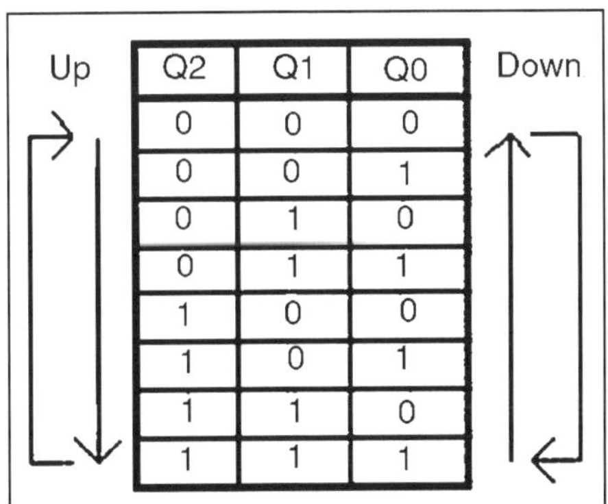

Q2	Q1	Q0
0	0	0
0	0	1
0	1	0
0	1	1
1	0	0
1	0	1
1	1	0
1	1	1

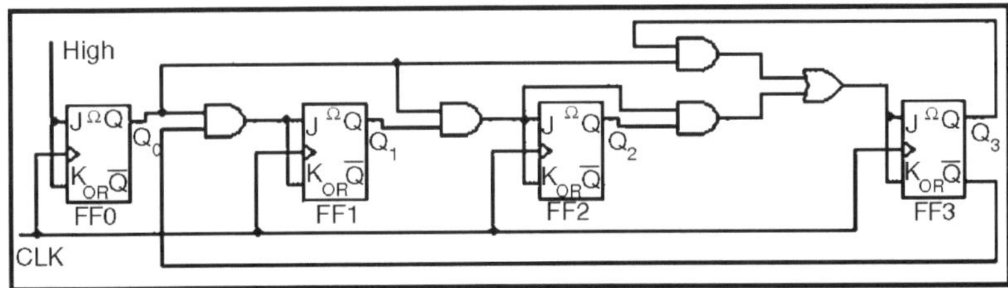

An examination of the sequence table shows:

- For both the UP and DOWN sequences, Q0 toggles on each clock pulse.
- For the UP sequence, Q1 changes state on the next clock pulse when Q0=1.
- For the DOWN sequence, Q1 changes state on the next clock pulse when Q0=0.
- For the UP sequence, Q2 changes state on the next clock pulse when Q0=Q1=1.
- For the DOWN sequence, Q2 changes state on the next clock pulse when Q0=Q1=0.

DEFINITION OF SHIFT REGISTERS

Shift registers are a type of sequential logic circuit, mainly for storage of digital data. They are a group of flip-flops connected in a chain so that the output from one flip-flop becomes the input of the next flip-flop. Most of the registers possess no characteristic internal sequence of states. All the flip-flops are driven by a common clock, and all are set or reset simultaneously.

In this chapter, the basic types of shift registers are studied, such as Serial In - Serial Out, Serial In - Parallel Out, Parallel In - Serial Out, Parallel In - Parallel Out, and bidirectional shift registers. A special form of counter - the shift register counter, is also introduced.

APPLICATIONS

Shift registers can be found in many applications. Here is a list of a few.

To produce time delay

The serial in -serial out shift register can be used as a time delay device. The amount of delay can be controlled by:

- The number of stages in the register
- The clock frequency

The ring counter technique can be effectively utilized to implement synchronous sequential circuits. A major problem in the realization of sequential circuits is the assignment of binary codes to the internal states of the circuit in order to reduce the complexity of circuits required. By assigning one flip-flop to one internal state, it is possible to simplify the combinational logic required to Realise the complete sequential circuit. When the circuit is in a particular state, the flip-flop corresponding to that state is set to HIGH and all other flip-flops remain LOW.

A computer or microprocessor-based system commonly requires incoming data to be in parallel format. But frequently, these systems must communicate with external devices that send or receive serial data. So, serial-to-parallel conversion is required. As shown in the previous sections, a serial in - parallel out register can achieve this.

PARALLEL IN - SERIAL OUT SHIFT REGISTERS

A four-bit parallel in - serial out shift register is shown below. The circuit uses D flip-flops and NAND gates for entering data (ie writing) to the register.

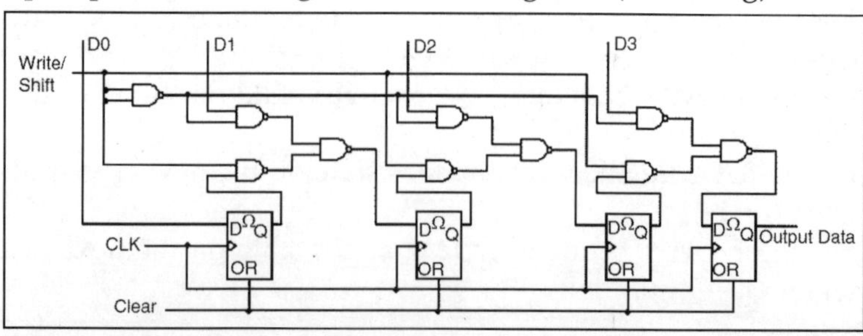

D0, D1, D2 and D3 are the parallel inputs, where D0 is the most significant bit and D3 is the least significant bit. To write data in, the mode control line is taken to LOW and the data is clocked in. The data can be shifted when the mode control line is HIGH as SHIFT is active high. The register performs right shift operation on the application of a clock pulse, as shown in the animation below.

	Q0	Q1	Q2	Q3
Clear	0	0	0	0

PARALLEL IN - PARALLEL OUT SHIFT REGISTERS

For parallel in - parallel out shift registers, all data bits appear on the parallel outputs immediately following the simultaneous entry of the data bits. The following circuit is a four-bit parallel in - parallel out shift register constructed by D flip-flops.

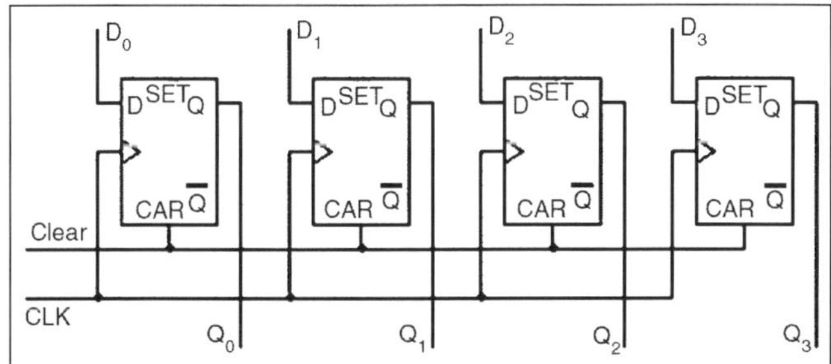

The D's are the parallel inputs and the Q's are the parallel outputs. Once the register is clocked, all the data at the D inputs appear at the corresponding Q outputs simultaneously.

BIDIRECTIONAL SHIFT REGISTERS

The registers discussed so far involved only right shift operations. Each right shift operation has the effect of successively dividing the binary number by two. If the operation is reversed (left shift), this has the effect of multiplying the number by two. With suitable gating arrangement a serial shift register can perform both operations. A bidirectional, or reversible, shift register is one in which the data can be shift either left or right. A four-bit bidirectional shift register using D flip-flops is shown below.

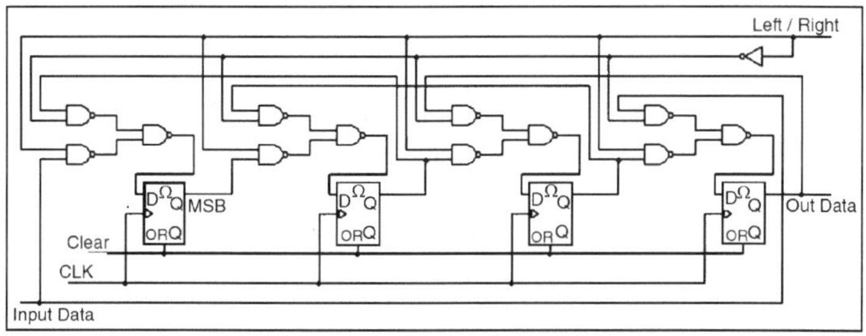

Here a set of NAND gates are configured as OR gates to select data inputs from the right or left adjacent bistables, as selected by the LEFT/RIGHT control line. The animation below performs right shift four times, then left shift four times. Notice the order of the four output bits are not the same as the order of the original four input bits. They are actually reversed!

Right	FF0	FF1	FF2	FF3
11111001	0	0	0	0

SERIAL IN - SERIAL OUT SHIFT REGISTERS

A basic four-bit shift register can be constructed using four D flip-flops, as shown below. The operation of the circuit is as follows. The register is first cleared, forcing all four outputs to zero. The input data is then applied sequentially to the D input of the first flip-flop on the left (FF0). During each clock pulse, one bit is transmitted from left to right. Assume a data word to be 1001. The least significant bit of the data has to be shifted through the register from FF0 to FF3.

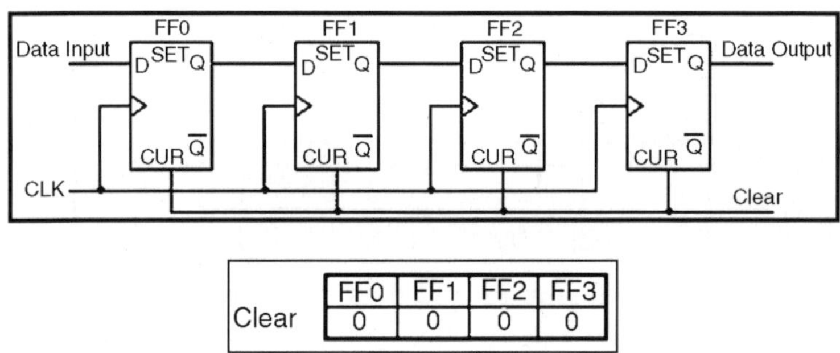

	FF0	FF1	FF2	FF3
Clear	0	0	0	0

In order to get the data out of the register, they must be shifted out serially. This can be done destructively or non-destructively. For destructive readout, the original data is lost and at the end of the read cycle, all flip-flops are reset to zero.

	FF0	FF1	FF2	FF3	
0000	1	0	0	1	0000

To avoid the loss of data, an arrangement for a non-destructive reading can be done by adding two AND gates, an OR gate and an inverter to the system. The construction of this circuit is shown below.

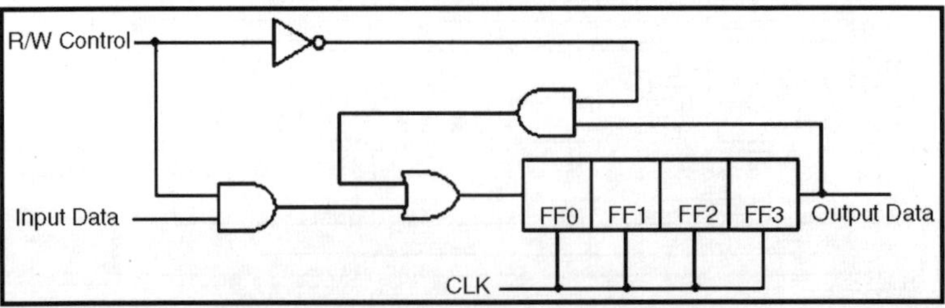

The data is loaded to the register when the control line is HIGH (ie WRITE). The data can be shifted out of the register when the control line is LOW (ie READ). This is shown in the animation below.

Write 1001	FF0	FF1	FF2	FF3
	0	0	0	0

SERIAL IN - PARALLEL OUT SHIFT REGISTERS

For this kind of register, data bits are entered serially in the same manner as discussed in the last section. The difference is the way in which the data bits are taken out of the register. Once the data are stored, each bit appears on its respective output line, and all bits are available simultaneously. A construction o f a four-bit serial in - parallel out register is shown below.

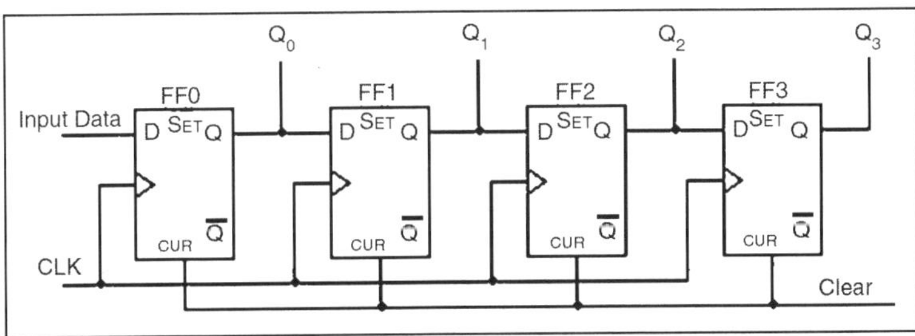

In the animation below, we can see how the four-bit binary number 1001 is shifted to the Q outputs of the register.

Clear 1001	Q0	Q1	Q2	Q3
	0	0	0	0

DEFINITION OF RIPPLE COUNTER

An n-stage counter that is formed from n cascaded flip-flops. The clock input to each of the individual flip-flops, with the exception of the first, is taken from the output of the preceding one. The count thus ripples along the counter's length due to the propagation delay associated with each stage of counting. A ripple counter contains a chain of flip-flops with the output of each one feeding the input of the next. A flip-flop output changes state every time the input changes from high to low (on the falling-edge). This simple arrangement works well, but there is a slight delay as the effect of the clock'ripples' through the chain of flip-flops.

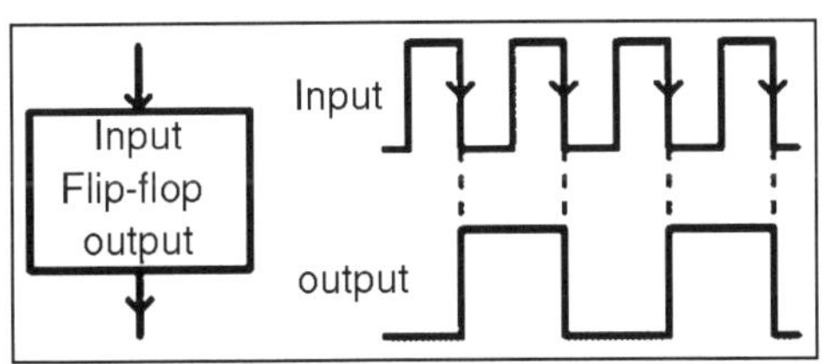

In most circuits the ripple delay is not a problem because it is far too short to be seen on a display. However, a logic system connected to ripple counter outputs will briefly see false counts which may produce'glitches' in the logic system and may disrupt its operation. For example a ripple counter changing from 0111 (7) to 1000 (8) will very briefly show 0110, 0100 and 0000 before 1000!

LINKING RIPPLE COUNTERS

The diagram below shows how to link standard ripple counters. Notice how the highest output QD of each counter drives the clock (CK) input of the next counter. This works because ripple counters have clock inputs that are'active-low' which means that the count advances as the clock input becomes low, on the falling-edge.

Remember that with all ripple counters there will be a slight delay before the later outputs respond to the clock signal, especially with a long counter chain. This is not a problem in simple circuits driving displays, but it may cause glitches in logic systems connected to the counter outputs.

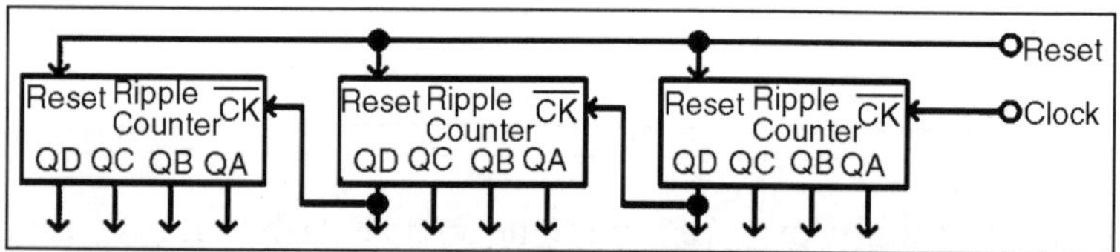

DESCRIPTION OF SYNCHRONOUS COUNTER

A synchronous counter, in contrast to an asynchronous counter, is one whose output bits change state simultaneously, with no ripple.

The only way we can build such a counter circuit from J-K flip-flops is to connect all the clock inputs together, so that each and every flip-flop receives the exact same clock pulse at the exact same time:

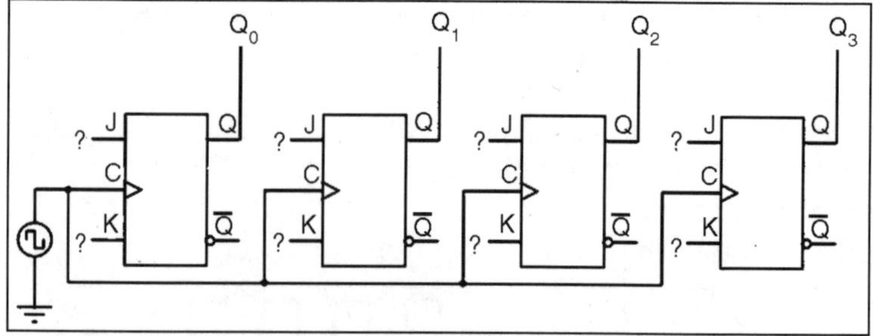

OPERATION

Now, the question is, what do we do with the J and K inputs? We know that we still have to maintain the same divide-by-two frequency pattern in order to count in a binary sequence, and that this pattern is best achieved utilizing the"toggle" mode of the flip-flop, so the fact that the J and K inputs must both be (at times)"high" is clear.

However, if we simply connect all the J and K inputs to the positive rail of the power supply as we did in the asynchronous circuit, this would clearly not work because all the flip-flops would toggle at the same time: with each and every clock pulse!

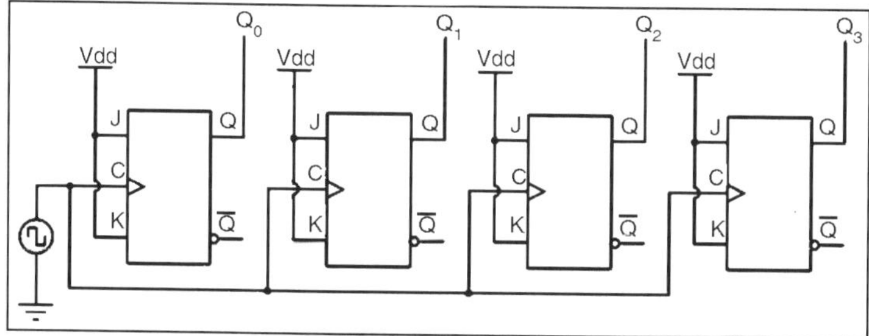

Fig. This Circuit will not Function as a counter!

EXAMPLE

Let's examine the four-bit binary counting sequence again, and see if there are any other patterns that predict the toggling of a bit.

0 0 0 0
0 0 0 1
0 0 1 0
0 0 1 1
0 1 0 0
0 1 0 1
0 1 1 0
0 1 1 1
1 0 0 0
1 0 0 1
1 0 1 0
1 0 1 1
1 1 0 0
1 1 0 1
1 1 1 0
1 1 1 1

Asynchronous counter circuit design is based on the fact that each bit toggle happens at the same time that the preceding bit toggles from a"high" to a"low" (from 1 to 0). Since we cannot clock the toggling of a bit based on the toggling of a previous bit in a synchronous counter circuit (to do so would create a ripple effect) we must find some other pattern in the counting sequence that can be used to trigger a bit toggle:

Examining the four-bit binary count sequence, another predictive pattern can be seen. Notice that just before a bit toggles, all preceding bits are"high:"

This pattern is also something we can exploit in designing a counter circuit. If we enable each J-K flip-flop to toggle based on whether or not all preceding flip-flop outputs (Q) are"high," we can obtain the same counting sequence as the asynchronous circuit without the ripple effect, since each flip-flop in this circuit will be clocked at exactly the same time:

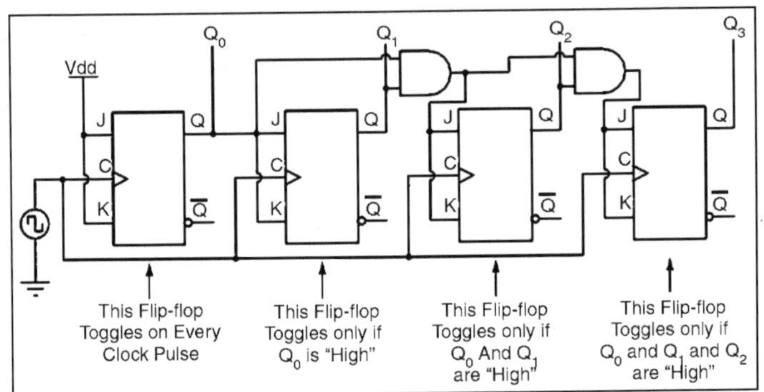

Fig. A Four-bit Synchronous "up" Counter

0 0 0 0
0 0 0 1
0 0 1 0
0 0 1 1
0 1 0 0
0 1 0 1
0 1 1 0
0 1 1 1
1 0 0 0
1 0 0 1
1 0 1 0
1 0 1 1
1 1 0 0
1 1 0 1
1 1 1 0
1 1 1 1

The result is a four-bit synchronous"up" counter. Each of the higher-order flip-flops are made ready to toggle (both J and K inputs"high") if the Q outputs of all previous flip-flops are"high." Otherwise, the J and K inputs for that flip-flop will both be"low," placing it into the"latch" mode where it will maintain its present output state at the next clock pulse. Since the first (LSB) flip-flop needs to toggle at every clock pulse, its J and K inputs are connected to Vcc or Vdd, where they will be"high" all the time. The next flip-flop need only"recognize" that the first flip-flop's Q output is high to be made ready to toggle, so no AND gate is needed. However, the remaining flip-flops should be made ready to toggle only when all lower-order output bits are"high," thus the need for AND gates. To make a synchronous"down" counter, we need to build the circuit to recognize the appropriate bit patterns predicting each toggle state while counting down. Not surprisingly, when we examine the four-bit binary count sequence, we see that all preceding bits are"low" prior to a toggle (following the sequence from bottom to top):

Since each J-K flip-flop comes equipped with a Q' output as well as a Q output, we can use the Q' outputs to enable the toggle mode on each succeeding flip-flop, being that each Q' will be"high" every time that the respective Q is"low:"

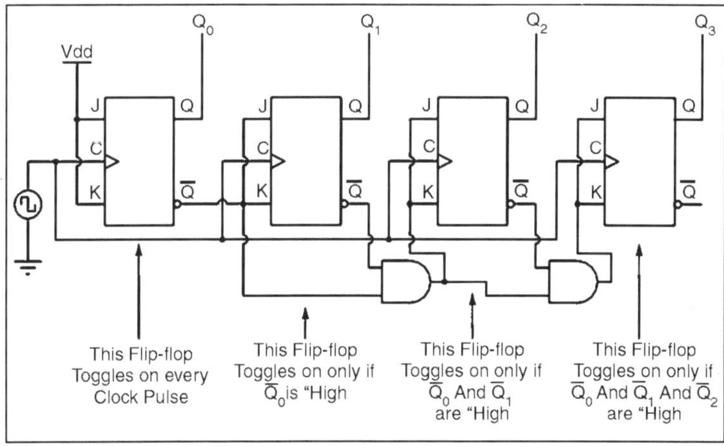

Fig. A Four-bit Synchronous "Down" Counter

Taking this idea one step further, we can build a counter circuit with selectable between"up" and"down" count modes by having dual lines of AND gates detecting the appropriate bit conditions for an"up" and a"down" counting sequence, respectively, then use OR gates to combine the AND gate outputs to the J and K inputs of each succeeding flip-flop:

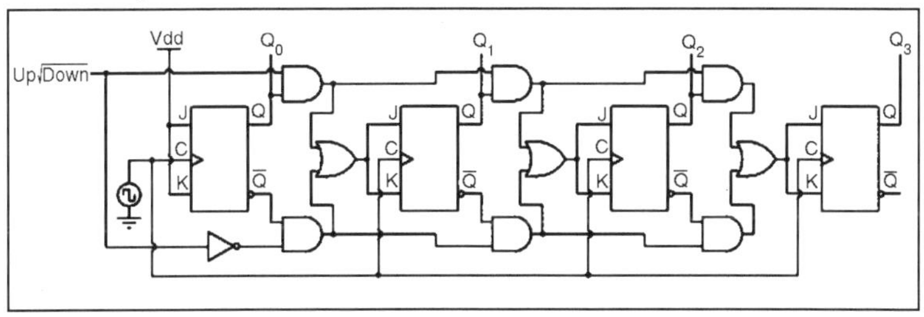

Fig. A Fout-bit Synchronous "up/down" Counter

This circuit isn't as complex as it might first appear. The Up/Down control input line simply enables either the upper string or lower string of AND gates to pass the Q/Q' outputs to the succeeding stages of flip-flops.

If the Up/Down control line is"high," the top AND gates become enabled, and the circuit functions exactly the same as the first ("up") synchronous counter circuit shown in this section.

If the Up/Down control line is made"low," the bottom AND gates become enabled, and the circuit functions identically to the second ("down" counter) circuit shown in this section.

To illustrate, here is a diagram showing the circuit in the"up" counting mode (all disabled circuitry shown in grey rather than black):

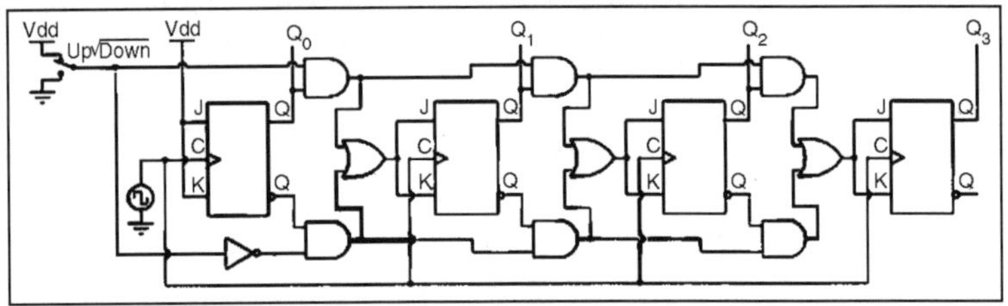

Fig. Counter in "up" Counting Mode

Here, shown in the"down" counting mode, with the same grey colouring representing disabled circuitry:

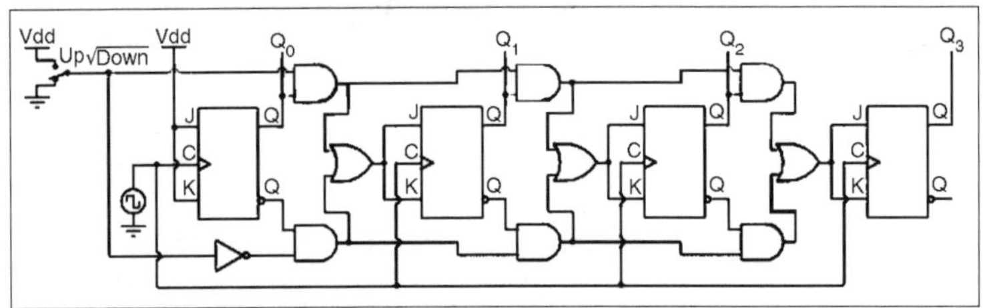

Fig.Counter in "Down" Counting Mode

Up/down counter circuits are very useful devices. A common application is in machine motion control, where devices called rotary shaft encoders convert mechanical rotation into a series of electrical pulses, these pulses"clocking" a counter circuit to track total motion:

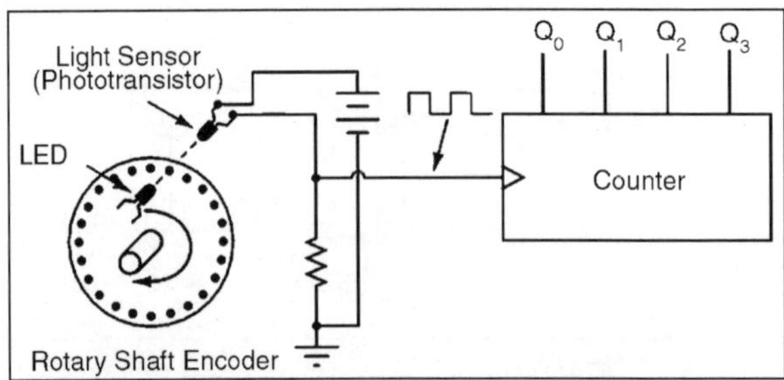

As the machine moves, it turns the encoder shaft, making and breaking the light beam between LED and phototransistor, thereby generating clock pulses to increment the counter circuit. Thus, the counter integrates, or accumulates, total motion of the shaft, serving as an electronic indication of how far the machine has moved. If all we care about is tracking total motion, and do not care to account for changes in the direction of motion, this arrangement will suffice. However, if we wish the counter to increment with one direction of motion and decrement with the reverse direction of motion, we must use an up/down counter, and an encoder/decoding circuit having the ability to discriminate between different directions. If we re-design the encoder to have two sets of LED/phototransistor pairs, those pairs aligned such that their square-wave output signals are 90o out of phase with each other, we have what is known as aquadrature output encoder (the word"quadrature" simply refers to a 90o angular separation). A phase detection circuit may be made from a D-type flip-flop, to distinguish a clockwise pulse sequence from a counter-clockwise pulse sequence:

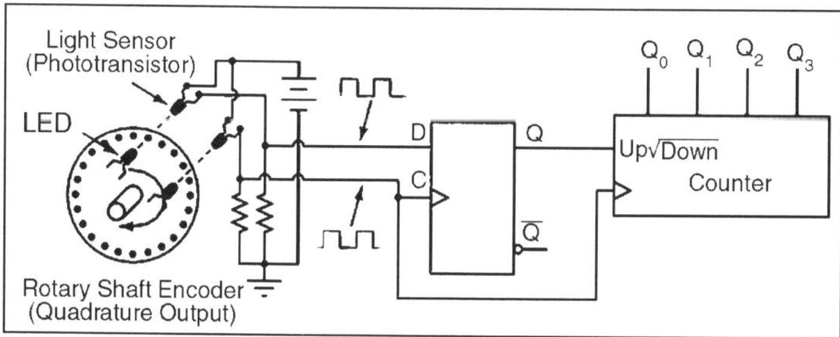

When the encoder rotates clockwise, the"D" input signal square-wave will lead the"C" input square-wave, meaning that the"D" input will already be"high" when the"C" transitions from"low" to"high," thus settingthe D-type flip-flop (making the Q output"high") with every clock pulse. A"high" Q output places the counter into the"Up" count mode, and any clock pulses received by the clock from the encoder (from either LED) will increment it.

Conversely, when the encoder reverses rotation, the"D" input will lag behind the"C" input waveform, meaning that it will be"low" when the"C" waveform transitions from"low" to"high," forcing the D-type flip-flop into the reset state (making the Q output"low") with every clock pulse. This"low" signal commands the counter circuit to decrement with every clock pulse from the encoder.

This circuit, or something very much like it, is at the heart of every position-measuring circuit based on a pulse encoder sensor. Such applications are very common in robotics, CNC machine tool control, and other applications involving the measurement of reversible, mechanical motion.

ANALYSIS OF CLOCKED SEQUENTIAL CIRCUITS AND DESIGN

Analysis consists of obtaining a state-table or a state-diagram from a given sequential circuit implementation. In other words analysis closes the loop by forming state-table from a given circuit-implementation. We will show the analysis procedure by deriving the state table of the example circuit we considered in synthesis. The circuit is shown in Figure.

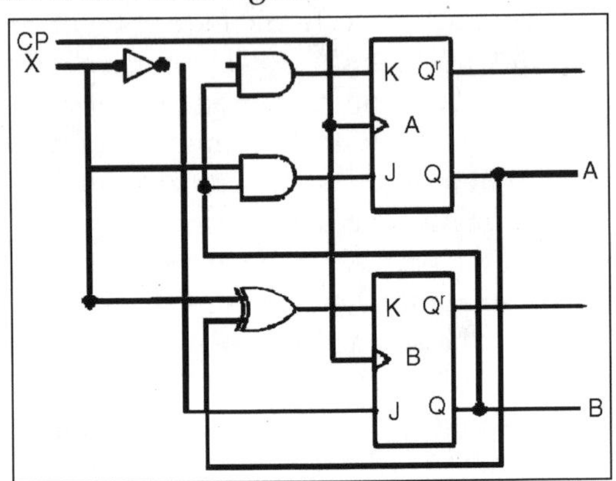

Fig.A Clocked Sequential Circuit

The circuit has

- Clock input, CP.
- One input x
- One output y
- One clocked JK flip-flop
- One clocked D flip-flop (the machine can be in maximum of 4 states)

A State table is representation of sequence of inputs, outputs, and flip-flop states in a tabular form. Two forms of state tables are shown.

Present State		Next State		Output	
	x = 0	x = 1	x = 0	x = 1	
A	B	A	B	y	y
0	0				
0	1				
1	0				
1	1				

Analysis is the generation of state table from the given sequential circuit.

The number of rows in the state table is equal to 2 (number of flip-flops+ number of inputs). For the circuit under consideration, number of rows = $2^{(2+1)}$ = $2^{(3)}$ = 8

Present State	Input	NextState	Output	
A(t)	B(t)	x	A(t+1)B(t+1)	y
0	0	0		
0	0	1		
0	1	0		
0	1	1		

1	0	0
1	0	1
1	1	0
1	1	1

In the present case there are two flip-flops and one input, thus a total of 8 rows as shown in the table.

Present State	Input	NextState		Output	
A(t)	B(t)	x		A(t+1)B(t+1)	y
0	0	0	0	0	0
0	0	1	0	1	0
0	1	0	1	0	0
0	1	1	0	1	0
1	0	0	0	0	0
1	0	1	0	0	0
1	1	1	1	1	0
1	1	1	0	1	1

The analysis can start from any arbitrary state. Let us start deriving the state table from the initial state 00.

As a first step, the input equations to the flip-flops and to the combinational circuit must be obtained from the given logic diagram. These equations are:

$J_A = BX'$

$K_A = BX + B'X'$

$D_B = X$

$y = ABX$

The first row of the state-table is obtained as follows:

When input X = 0; and present states A = 0 and B = 0 (as in the first row); then, using the above equations we get:

$$y = 0,\ J_A = 0,\ K_A = 1, \text{ and } D_B = 0.$$

The resulting state table is exactly same from which we started our design example. Thus analysis is opposite to design and combined they act as a closed loop.

WAYS OF REDUCING THE NUMBER OF STATES IN A SEQUENTIAL CIRCUIT

STATE REDUCTION

The problem of state reduction is to find ways of reducing the number of states in a sequential circuit without altering the input-output relationships. In other words, to reduce the number of states, redundant states should be eliminated. A redundant state Si is a state which is equivalent to another state Sj.

Two states are said to be equivalent if, for each member of the set of inputs, they give exactly the same output and send the circuit either to the same state or to an equivalent state.

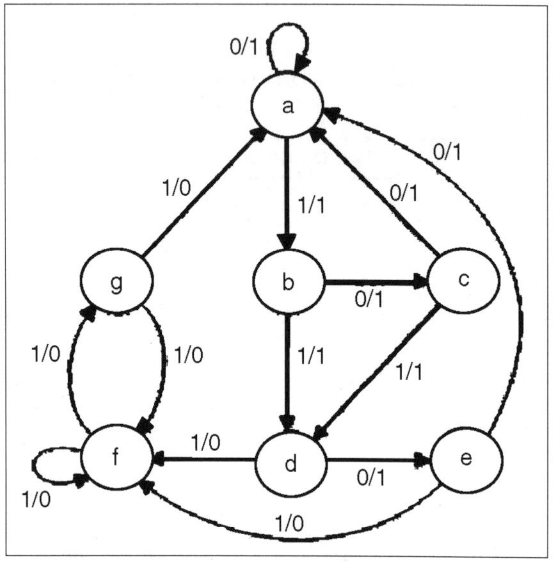

Fig. State Diagram

Since'm' flip-flops can describe a state machine of up to 2m states, reducing the number of states may (or may not) result in a reduction in the number of flip-flops. For example, if the number of states are reduced from 8 to 5, we still need 3 flip-flops. However, state reduction will result in more don't care states. The increased number of don't care states can help obtain a simplified circuit for the state machine.

The state reduction proceeds by first tabulating the information of the state diagram into its equivalent state-table form (as shown in the table) The problem of state reduction requires identifying equivalent states. Each N states is replaced by 1 state.

Consider the following state table.

Present State	Next State		Output	
	x = 0	x = 1	x = 0	x = 1
a	a	b	1	1
b	c	d	1	1
c	a	d	1	1
d	e	f d	1	0
e	a	f d	1	0
f	g e	f	1	0
g	a	f	1	0

States'g' and'e' produce the same outputs, i.e.'1' and'0', and take the state machine to same next-states,'a' and'f', on inputs'0' and'1' respectively. Thus, states'g' and'e' are equivalent states.

We can now remove state'g' and replace it with'e' as shown. We next note that the above change has caused the states'd' and'f' to be equivalent. Thus in the next step, we remove state'f' and replace it with'd'. There are no more equivalent states remaining. The reduced state table results in the following reduced state diagram.

STATE REDUCTION IN STATES ASSIGNMENT

When constructing a state diagram, variable names are used for states as the final number of states is not known a priori. Once the state diagram is constructed, prior to implementation (using gates and flip-flops), we need to perform the step of'state reduction'. The step that follows state reduction is state assignment. In state assignment, binary patterns are assigned to state variables.

Table. Possible State Assignments

State	Assignment 1	Assignment 2	Assignment 3
a	001	000	000
b	010	010	100
c	011	011	010
d	100	101	101
e	101	111	011

For a given machine, there are several state assignments possible. Different state assignments may result in different combinational circuits of varying complexities.

State assignment procedures try to assign binary values to states such that the cost (complexity) of the combinational circuit is reduced.

There are several heuristics that attempt to choose good state assignments (also known as state encoding) that try to reduce the required combinational logic complexity, and hence cost.

As mentioned earlier, for the reduced state machine obtained in the previous example, there can be a number of possible assignments. As an example, three different state assignments are shown in the table for the same machine.

SEQUENTIAL CIRCUIT IN DESIGN PROCEDURE

DESIGN WITH UNUSED STATES

There are occasions when a sequential circuit, implemented using m flip-flops, may not utilize all the possible 2m states

Table. Reduced Table with binary Assignment

Present State	Next State		Output	
	x = 0	x = 1	x = 0	x = 1
a	a	b	1	1
b	c	d	1	1
c	a	d	1	1
d	e	d	1	0
e	a	d	1	0

In the previous example of machine with 5 states, we need three flip-flops. Let us choose assignment 1, which is binary assignment for our sequential machine example (shown in the table).

The unspecified states can be used as don't-cares and will therefore help in simplifying the logic. The excitation table of previous example is shown. There are three states, 000, 110, and 111 that are not listed in the table under present state and input.

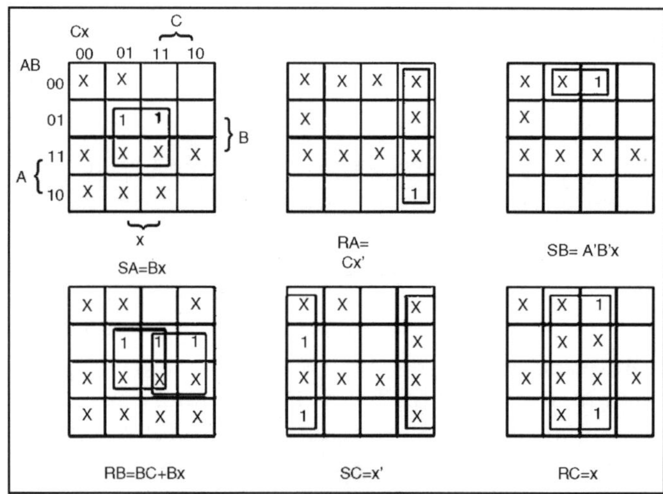

Fig. Exciation Table

The K-maps of SA and RA is shown in the figure. Other K-Maps can be obtained similarly and the equations derived are shown in the figure.

Present State			Input	Next state			Flip-flop inputs						Output
A	B	C	X	A	B	C	SA	RA	SB	RB	SC	RC	Y
0	0	1	0	0	0	1	0	X	0	X	X	0	1
0	0	1	1	0	0	1	0	X	1	0	0	1	1
0	1	0	0	0	1	0	0	X	X	0	1	0	1
0	1	0	1	0	1	0	1	0	0	1	0	X	1

0	1	1	0	0 1 1	0	X	0	1	X	1	1
0	1	1	1	0 1 1	1	0	0	1	0	0	1
1	0	0	0	0 0 0	1	0	0	1	0	1	1
1	0	0	0	1 0 0	x	0	0	X	1	0	1
1	0	1	0	1 0 1	0	1	0	X	X	0	1
1	0	1	1	1 0 1	X	0	0	X	0	1	0

With the inclusion of input 1 or 0, we obtain six don't-care minterms: 0, 1, 12, 13, 14, and 15.

The logic diagram thus obtained is shown in the figure.

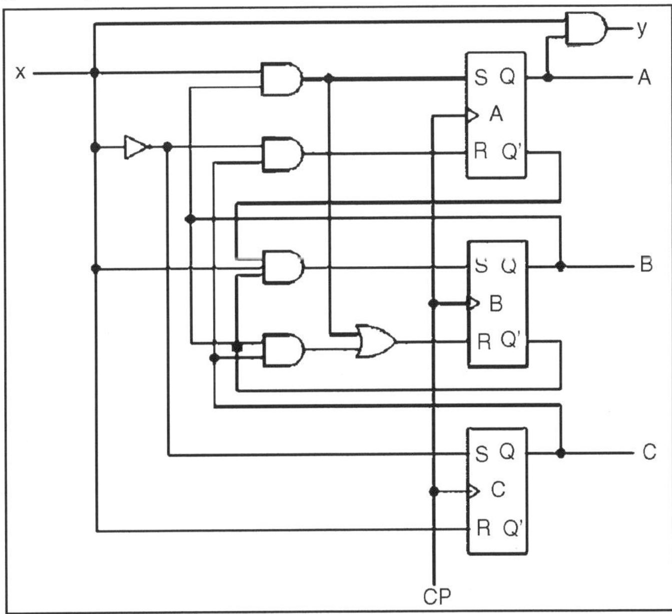

Fig. Logic Diagram

SA = BX

RA = Cx′

SB = A′B′X

RB = BC +Bx

SC = x′

RC = x

y = A′ + x′ = (Ax)′

Note that the design of the sequential circuit is dependent on binary codes for states. A different binary state codes set may have resulted in some different combinational circuit.

FINITE STATE MACHINES (FSM)

Strictly speaking a finite state machine (FSM) is a device that can exist in one of a finite number of states. Associated with an FSM is a memory that is used to store an identifier of the state, so that the machine may process its input (if any) and move to the next state. Due to this coincidence, finite state machines are often studied in conjunction with flip-flops.

We are all familiar with finite state machines, although we rarely think of them as such. Consider a washing machine. The states for this machine are: off, fill with water, wash, spin, and rinse. The control unit for the FSM is the knob on the washer that we turn to start it. A traffic light is also a finite state machine. We normally think of a traffic light as having only three states: Green, Yellow, and Red. The truth is a bit more complex, in that the physical unit must display at least two sets of lights, one for each intersecting street. Nevertheless, a standard traffic light can be modeled with no more than eight states, although the introduction of advanced green lights and turn signals complicates things a bit.

A standard digital clock that displays only hour, minute, and second, can be said to have 24 • 60 • 60 = 86,400 states - still a finite number. Normally the FSM construct is used to model systems with far fewer states; in our work we shall normally limit a FSM to either eight or sixteen states; that is N 2^3 or N 2^4. If a finite state machine does not have too many states, we may represent its operation by a state diagram. The following is a state diagram for a circuit called a sequence detector. For those who are interested, this is the state diagram for a 11011 sequence detector; it has five states because it is detecting a five-bit sequence.

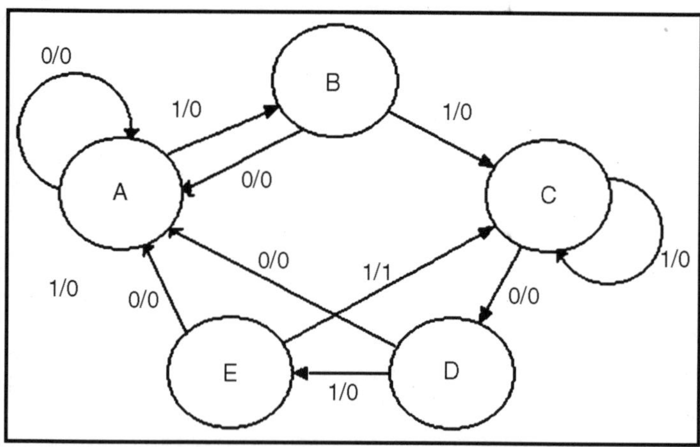

Fig. State Diagram for a 11011 Sequence Detector

At this stage of the presentation, we focus not on the details of generating the state diagram, but just use it as an example of a generic state diagram. What do we note about this one?

- In terms of discrete mathematics, it is a directed graph with loops. Thus, it is not a simple graph. In simple graphs, arcs do not connect any vertex with itself.

- The arcs each have direction and a label of the form X/Z. What we see here is the FSM reacting to input by moving between states and producing output. In the X/Z labeling scheme, X is the binary input (0 or 1) and Z is the binary output.
- There is output associated with the transitions. Not all FSM have output associated with the transitions between states. This one does.
- This and all typical FSM represents a synchronous machine; transitions between states and production of output (if any) takes place at a fixed phase of the clock, depending on the flip-flops used to implement the circuit.

Not all finite state machines have such complex state diagrams. The figure at left is the state diagram for a modulo-4 up-counter. It just counts 0, 1, 2, 3 and repeats, continually counting up (modulo 4). There is no input (other than the clock, which we almost never mention) and no output directly associated with the transitions. For this type of FSM, the output is associated with the states and not with the transitions.

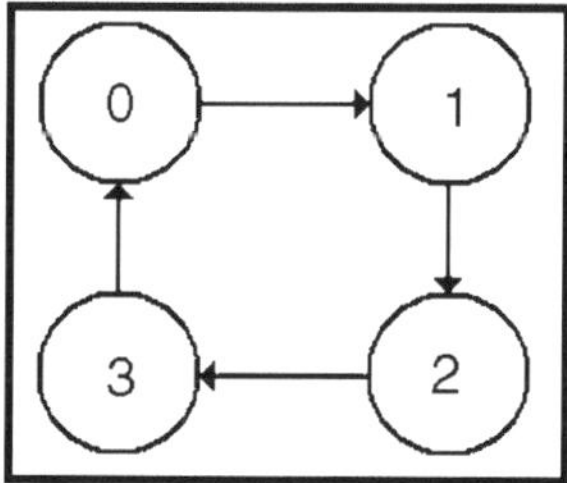

Fig. State Diagram for a Modulo-4 Up-Counter

Many mathematical models of FSM focus on the state diagram. For most of our work, it is more convenient to work with the state table of the FSM, a tabular representation of the state diagram. Translation between the state diagram and state table is automatic. The state table presents the data in terms of present state Q(t) and next state Q(t+1) using the labeling that most naturally fits the problem. Here are the state tables for the two FSM above. Note that eh state table contains exactly the same information as the state diagram.

Fig. State Table for 11011 Sequence Detector, Showing Output

Present State	Next State/ Output	
	X = 0	X = 1
A	A/ 0	B/ 0
B	A/ 0	C/ 0
C	D/ 0	C/ 0
D	A/ 0	E/ 0
E	A/ 0	C/ 1

Figure. State Table for a Modulo-Four Up-Counter

Present State	Next State
0	1
1	2
2	3
3	0

One notes immediately that the second state table is simpler than the first; this is expected it represents a simpler state diagram.

Specifically there is no input, so there is only one column for the next state. In general, for K inputs there are 2K next state columns in the table.

Another tool in the design and analysis of sequential circuits is the transition table. It contains the same information as the state table, except that all labels have been replaced by binary numbers.

There are many creative ways to assign binary numbers to state labels, here we just do the obvious. For the sequence detector, let A = 000, B = 001, C = 010, D = 011, and E = 100 (as there are five states). The following is the sequence detector transition table.

Table. Transition/Output Table for the 11011 Sequence Detector

Present State	Next State/ Output	
	X = 0	X = 1
A = 000	000/ 0	001/ 0
B = 001	000/ 0	010/ 0
C = 010	011/ 0	010/ 0
D = 011	000/ 0	100/ 0
E = 100	000/ 0	010/ 1

What we have in the above figure is a special type of truth table. We shall now investigate the table in a bit more detail. Note that the state of the machine is represented by a 3-bit binary number. We shall use the notation Y2Y1Y0 for that number. Given this notation, we write the table as shown below.

Table. The Transition/Output Table as a Modified Truth-Table

Present State			Next State/Output							
			X = 0				X = 1			
Y_2	Y_1	Y_0	Y_2	Y_1	Y_0	Z	Y_2	Y_1	Y_0	Z
0	0	0	0	0	0	0	0	0	1	0
0	0	1	0	0	0	0	0	1	0	0
0	1	0	0	1	1	0	0	1	0	0
0	1	1	0	0	0	0	1	0	0	0
1	0	0	0	0	0	0	0	1	0	1

The table above can be viewed as a truth table that has been"folded over". Another way to represent this table is as a standard truth table depending on Y2, Y1, Y0, and X.

Table. The Transition/Output Table as a Standard Truth Table

Y2	Y1	Y0	X	Y2	Y1	Y0	Z
0	0	0	0	0	0	0	0
0	0	0	1	0	0	1	0
0	0	1	0	0	0	0	0
0	0	1	1	0	1	0	0
0	1	0	0	0	1	1	0
0	1	0	1	0	1	0	0
0	1	1	0	0	0	0	0
0	1	1	1	1	0	0	0
1	0	0	0	0	0	0	0
1	0	0	1	0	1	0	1

Students are invited to use either form of the transition/output table that suit them. This instructor prefers to use the folded-over version, but that is not necessary.

We now have three equivalent representations of a FSM.

- The state diagram,
- The state table, and
- The transition/output table (probably not a standard name).

6

Digital Signals and Gates Design

INTRODUCTION

While the binary numeration system is an interesting mathematical abstraction, we haven't yet seen its practical application to electronics. This chapter is devoted to just that: practically applying the concept of binary bits to circuits. What makes binary numeration so important to the application of digital electronics is the ease in which bits may be represented in physical terms. Because a binary bit can only have one of two different values, either 0 or 1, any physical medium capable of switching between two saturated states may be used to represent a bit. Consequently, any physical system capable of representing binary bits is able to represent numerical quantities, and potentially has the ability to manipulate those numbers. This is the basic concept underlying digital computing.

ELECTRONIC CIRCUITS

Electronic circuits are physical systems that lend themselves well to the representation of binary numbers. Transistors, when operated at their bias limits, may be in one of two different states: either cutoff (no controlled current) or saturation (maximum controlled current).

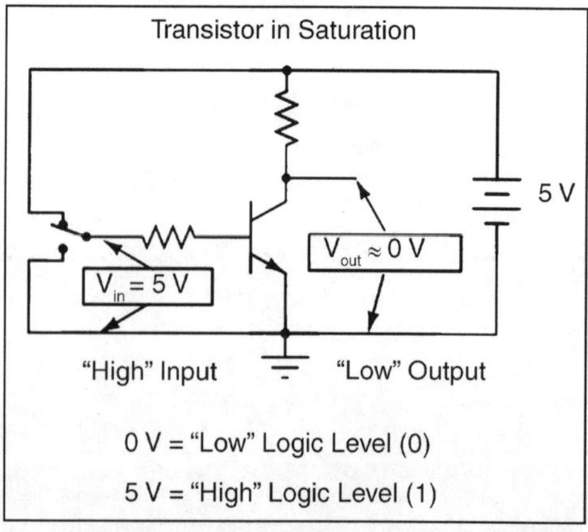

If a transistor circuit is designed to maximize the probability of falling into either one of these states (and not operating in the linear, or *active,* mode), it can serve as a physical representation of a binary bit. A voltage signal measured at the output of such a circuit may also serve as a representation of a single bit, a low voltage representing a binary "0" and a (relatively) high voltage representing a binary "1." Note the following transistor circuit:

In this circuit, the transistor is in a state of saturation by virtue of the applied input voltage (5 volts) through the two-position switch. Because its saturated, the transistor drops very little voltage between collector and emitter, resulting in an output voltage of (practically) 0 volts. If we were using this circuit to represent binary bits, we would say that the input signal is a binary "1" and that the output signal is a binary "0." Any voltage close to full supply voltage (measured in reference to ground, of course) is considered a "1" and a lack of voltage is considered a "0." Alternative terms for these voltage levels are *high* (same as a binary "1") and *low* (same as a binary "0"). A general term for the representation of a binary bit by a circuit voltage is *logic level.*

Moving the switch to the other position, we apply a binary "0" to the input and receive a binary "1" at the output:

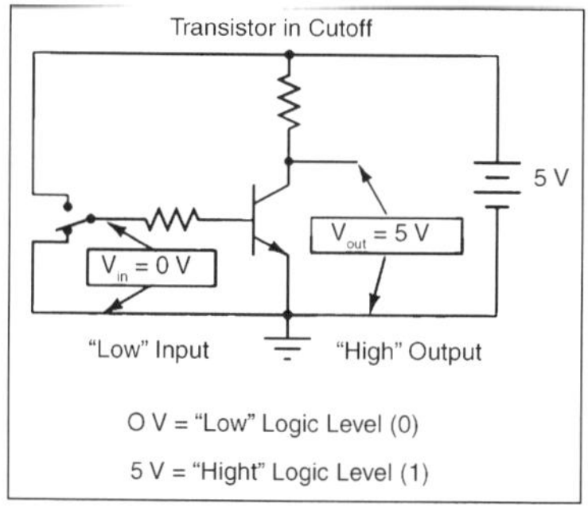

What we've created here with a single transistor is a circuit generally known as a *logic gate,* or simply *gate.* A gate is a special type of amplifier circuit designed to accept and generate voltage signals corresponding to binary 1's and 0's. As such, gates are not intended to be used for amplifying analog signals (voltage signals *between* 0 and full voltage). Used together, multiple gates may be applied to the task of binary number storage (memory circuits) or manipulation (computing circuits), each gate's output representing one bit of a multi-bit binary number. Right now it is important to focus on the operation of individual gates.

SINGLE TRANSISTOR

The gate shown here with the single transistor is known as an *inverter,* or NOT gate, because it outputs the exact opposite digital signal as what is input. For convenience, gate circuits are generally represented by their own symbols rather than by their constituent transistors and resistors. The following is the symbol for an inverter:

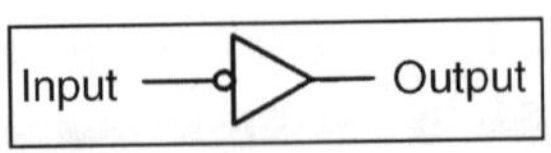

Fig. Invertor or NOT Gate

An alternative symbol for an inverter is shown here:

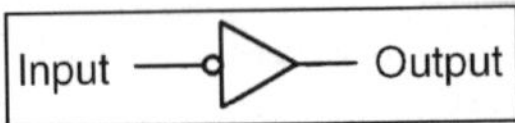

Notice the triangular shape of the gate symbol, much like that of an operational amplifier. As was stated before, gate circuits actually are amplifiers.

The small circle, or "bubble" shown on either the input or output terminal is standard for representing the inversion function. As you might suspect, if we were to remove the bubble from the gate symbol, leaving only a triangle, the resulting symbol would no longer indicate inversion, but merely direct amplification. Such a symbol and such a gate actually do exist, and it is called a *buffer*, the subject of the next section.

Like an operational amplifier symbol, input and output connections are shown as single wires, the implied reference point for each voltage signal being "ground." In digital gate circuits, ground is almost always the negative connection of a single voltage source (power supply). Dual, or "split," power supplies are seldom used in gate circuitry. Because gate circuits are amplifiers, they require a source of power to operate. Like operational amplifiers, the power supply connections for digital gates are often omitted from the symbol for simplicity's sake. If we were to show *all* the necessary connections needed for operating this gate, the schematic would look something like this:

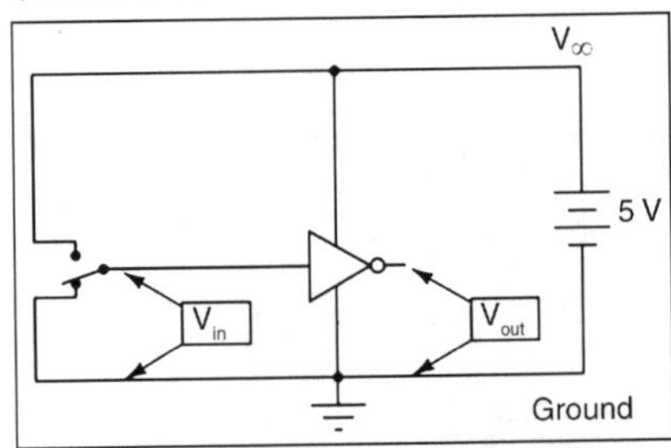

Power supply conductors are rarely shown in gate circuit schematics, even if the power supply connections at each gate are. Minimizing lines in our schematic, we get this:

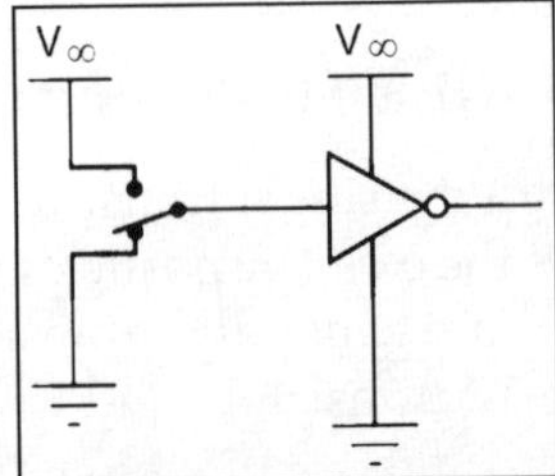

"V_{cc}" stands for the constant voltage supplied to the collector of a bipolar junction transistor circuit, in reference to ground. Those points in a gate circuit marked by the label "V_{cc}" are all connected to the same point, and that point is the positive terminal of a *DC* voltage source, usually 5 volts.

As we will see in other sections of this chapter, there are quite a few different types of logic gates, most of which have multiple input terminals for accepting more than one signal. The output of any gate is dependent on the state of its input(s) and its logical function.

One common way to express the particular function of a gate circuit is called a *truth table*. Truth tables show all combinations of input conditions in terms of logic level states (either "high" or "low," "1" or "0," for each input terminal of the gate), along with the corresponding output logic level, either "high" or "low." For the inverter, or NOT, circuit just illustrated, the truth table is very simple indeed:

Input —▷o— Output

Input	Output
0	1
1	0

Fig. Not Gate Truth Table

Truth tables for more complex gates are, of course, larger than the one shown for the NOT gate. A gate's truth table must have as many rows as there are possibilities for unique input combinations. For a single-input gate like the NOT gate, there are only two possibilities, 0 and 1. For a two input gate, there are*four* possibilities (00, 01, 10, and 11), and thus four rows to the corresponding truth table. For a three-input gate, there are *eight* possibilities (000, 001, 010, 011, 100, 101, 110, and 111), and thus a truth table with eight rows are needed. The mathematically inclined will realise that the number of truth table rows needed for a gate is equal to 2 raised to the power of the number of input terminals.

THE NOT GATE

The single-transistor inverter circuit illustrated earlier is actually too crude to be of practical use as a gate. Real inverter circuits contain more than one transistor to maximize voltage gain (so as to ensure that the final output transistor is either in full cutoff or full saturation), and other components designed to reduce the chance of accidental damage.

Shown here is a schematic diagram for a real inverter circuit, complete with all necessary components for efficient and reliable operation:

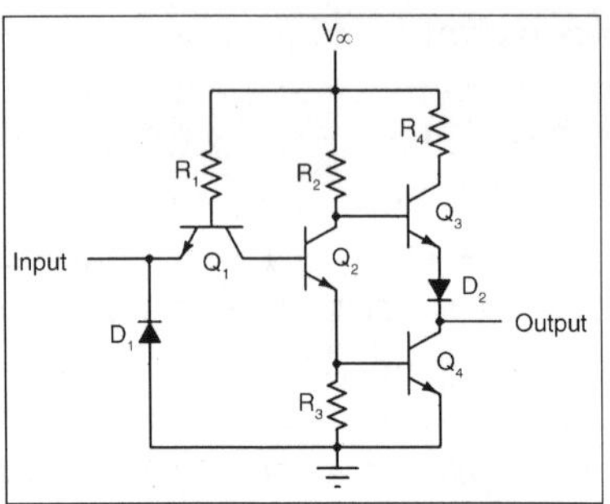

Fig. Practical Inverter (NOT) Circuit

This circuit is composed exclusively of resistors and bipolar transistors. Bear in mind that other circuit designs are capable of performing the NOT gate function, including designs substituting field-effect transistors for bipolar.

Let's analyse this circuit for the condition where the input is "high," or in a binary "1" state. We can simulate this by showing the input terminal connected to V_{cc} through a switch:

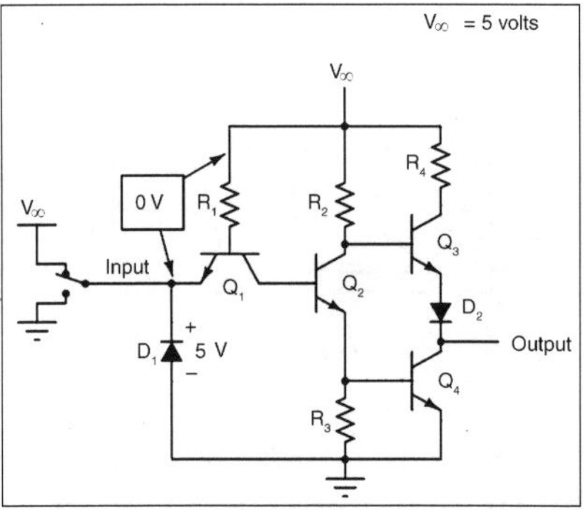

In this case, diode D_1 will be reverse-biased, and therefore not conduct any current. In fact, the only purpose for having D_1 in the circuit is to prevent transistor damage in the case of a *negative* voltage being impressed on the input (a voltage that is negative, rather than positive, with respect to ground). With no voltage between the base and emitter of transistor Q_1, we would expect no current through it, either.

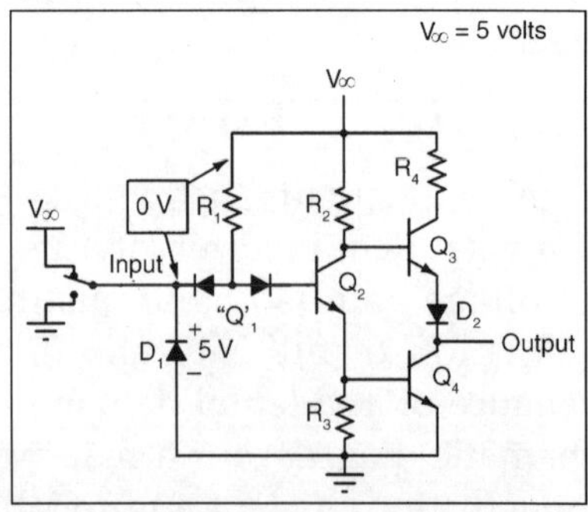

However, as strange as it may seem, transistor Q_1 is not being used as is customary for a transistor. In reality, Q_1 is being used in this circuit as nothing more than a back-to-back pair of diodes. The following schematic shows the real function of Q_1:

The purpose of these diodes is to "steer" current to or away from the base of transistor Q_2, depending on the logic level of the input. Exactly how these two diodes are able to "steer" current isn't exactly obvious at first inspection, so a short example may be necessary for understanding. Suppose we had the following diode/resistor circuit, representing the base-emitter junctions of transistors Q_2 and Q_4 as single diodes, stripping away all other portions of the circuit so that we can concentrate on the current "steered" through the two back-to-back diodes:

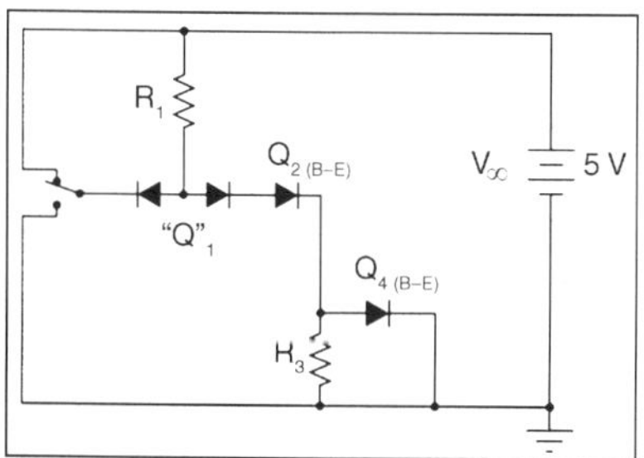

With the input switch in the *"up"* position (connected to V_{cc}), it should be obvious that there will be no current through the left steering diode of Q_1, because there isn't any voltage in the switch-diode-R_1-switch loop to motivate electrons to flow. However, there *will* be current through the right steering diode of Q_1, as well as through Q_2's base-emitter diode junction and Q_4's base-emitter diode junction:

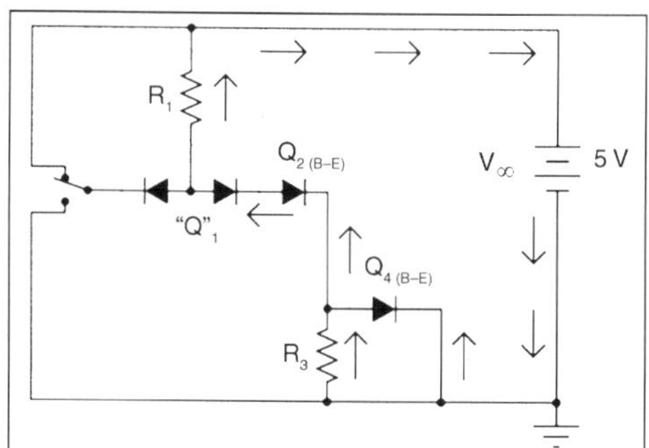

This tells us that in the real gate circuit, transistors Q_2 and Q_4 will have base current, which will turn them on to conduct collector current. The total voltage dropped between the base of Q_1 (the node joining the two back-to-back steering diodes) and ground will be about 2.1 volts, equal to the combined voltage drops of three *PN* junctions: the right steering diode, Q_2's base-emitter diode, and Q_4's base-emitter diode. Now, let's move the input switch to the "down" position and see what happens:

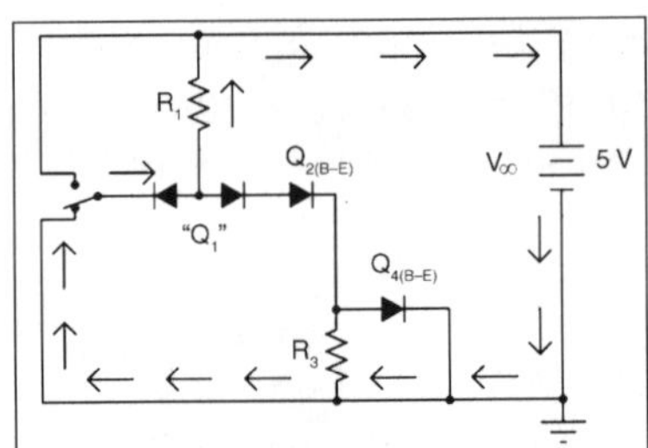

If we were to measure current in this circuit, we would find that *all* of the current goes through the left steering diode of Q_1 and *none* of it through the right diode. Why is this? It still appears as though there is a complete path for current through Q_4's diode, Q_2's diode, the right diode of the pair, and R_1, so why will there be no current through that path?

NON-LINEAR DEVICES

Remember that PN junction diodes are very non-linear devices: they do not even begin to conduct current until the forward voltage applied across them reaches a certain minimum quantity, approximately 0.7 volts for silicon and 0.3 volts for germanium. And then when they begin to conduct current, they will not drop substantially more than 0.7 volts. When the switch in this circuit is in the "down" position, the left diode of the steering diode pair is fully conducting, and so it drops about 0.7 volts across it and no more.

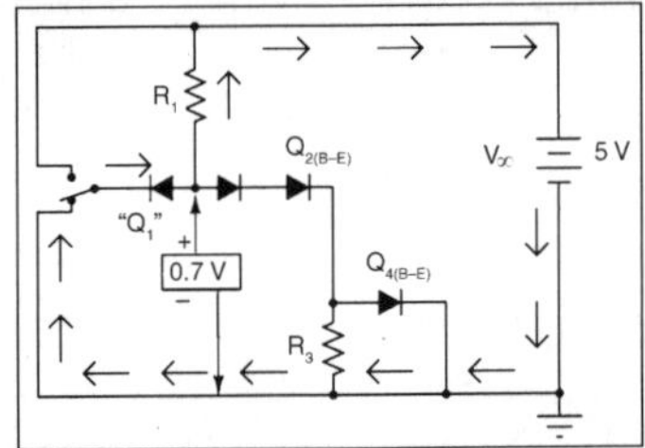

Recall that with the switch in the "*up*" position (transistors Q_2 and Q_4 conducting), there was about 2.1 volts dropped between those same two points (Q_1's base and ground), which also happens to be the *minimum* voltage necessary to forward-bias three series-connected silicon PN junctions into a state of conduction. The 0.7 volts provided by the left diode's forward voltage drop is simply insufficient to allow any electron flow through the series string of the right diode, Q_2's diode, and the $R_3//Q_4$ diode parallel subcircuit, and so no electrons flow through that path. With no current through the bases of either transistor Q_2 or Q_4, neither one will be able to conduct collector current: transistors Q_2 and Q_4 will both be in a state of cutoff.

CIRCUIT CONFIGURATION

Consequently, this circuit configuration allows 100 percent switching of Q_2 base current (and therefore control over the rest of the gate circuit, including voltage at the output) by diversion of current through the left steering diode. In the case of our example gate circuit, the input is held "high" by the switch (connected to V_{cc}), making the left steering diode (zero voltage dropped across it). However, the right steering diode is conducting current through the base of Q_2, through resistor R_1:

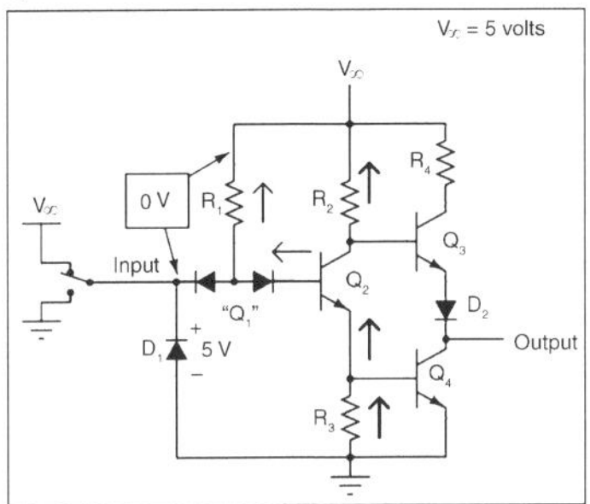

With base current provided, transistor Q_2 will be turned "on." More specifically, it will be *saturated* by virtue of the more-than-adequate current allowed by R_1 through the base. With Q_2 saturated, resistor R_3 will be dropping enough voltage to forward-bias the base-emitter junction of transistor Q_4, thus saturating it as well:

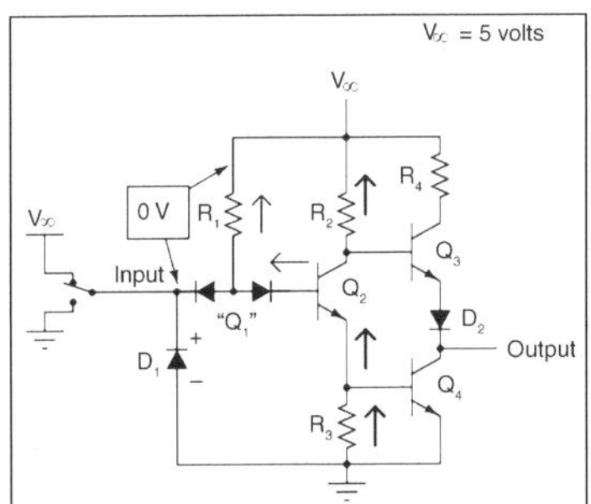

With Q_4 saturated, the output terminal will be almost directly shorted to ground, leaving the output terminal at a voltage (in reference to ground) of almost 0 volts, or a binary "0" ("low") logic level. Due to the presence of diode D_2, there will not be enough voltage between the base of Q_3 and its emitter to turn it on, so it remains in cutoff.

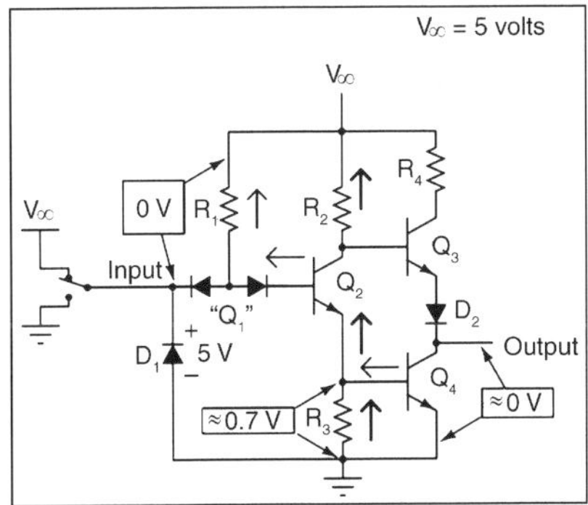

Let's see now what happens if we reverse the input's logic level to a binary "0" by actuating the input switch:

Now there will be current through the left steering diode of Q_1 and no current through the right steering diode. This eliminates current through the base of Q_2, thus turning it off. With Q_2 off, there is no longer a path for Q_4 base current, so Q_4 goes into cutoff as well. Q_3, on the other hand, now has sufficient voltage dropped between its base and ground to forward-bias its base-emitter junction and saturate it, thus raising the output terminal voltage to a "high" state. In actuality, the output voltage will be somewhere around 4 volts depending on the degree of saturation and any load current, but still high enough to be considered a "high" (1) logic level.

With this, our simulation of the inverter circuit is complete: a "1" in gives a "0" out, and vice versa.

The astute observer will note that this inverter circuit's input will assume a "high" state of left floating (not connected to either V_{cc} or ground). With the input terminal left unconnected, there will be no current through the left steering diode of Q_1, leaving all of R_1's current to go through Q_2's base, thus saturating Q_2 and driving the circuit output to a "low" state:

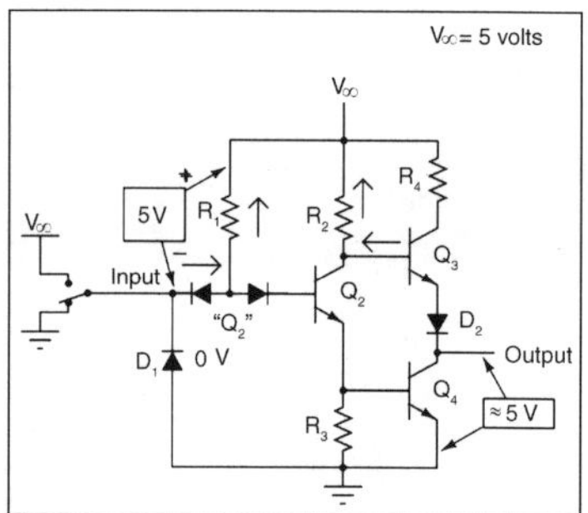

The tendency for such a circuit to assume a high input state if left floating is one shared by all gate circuits based on this type of design, known asTransistor-to-Transistor Logic, or *TTL*. This characteristic may be taken advantage of in simplifying the design of a gate's *output* circuitry, knowing that the outputs of gates typically drive the inputs of other gates. If the input of a *TTL* gate circuit assumes a high state when floating, then the output of any gate driving a *TTL* input need only provide a path to ground for a low state and be floating for a high state. This concept may require further elaboration for full understanding. A gate circuit as we have just analysed has the ability to handle output current in two directions: in and out. Technically, this is known as *sourcing* and*sinking* current, respectively. When the gate output is high, there is continuity from the output terminal to V_{cc} through the top output transistor (Q_3), allowing electrons to flow from ground, through a load, into the gate's output terminal, through the emitter of Q_3, and eventually up to the V_{cc} power terminal (positive side of the DC power supply):

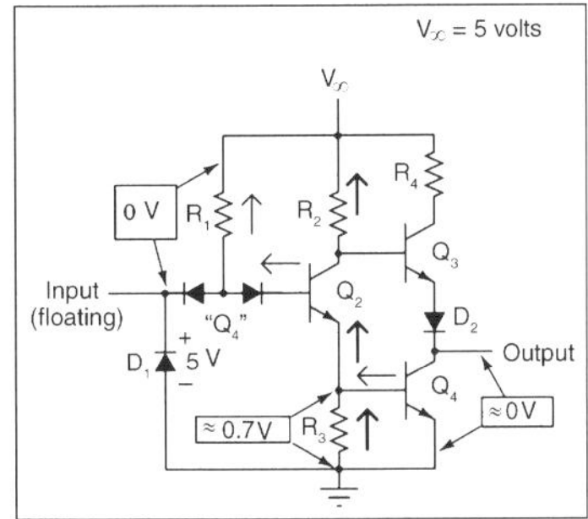

THE OUTPUT OF A GATE CIRCUIT

To simplify this concept, we may show the output of a gate circuit as being a double-throw switch, capable of connecting the output terminal either to V_{cc} or ground, depending on its state. For a gate outputting a "high" logic level, the combination of Q_3 saturated and Q_4 cutoff is analogous to a double-throw switch in the "V_{cc}" position, providing a path for current through a grounded load:

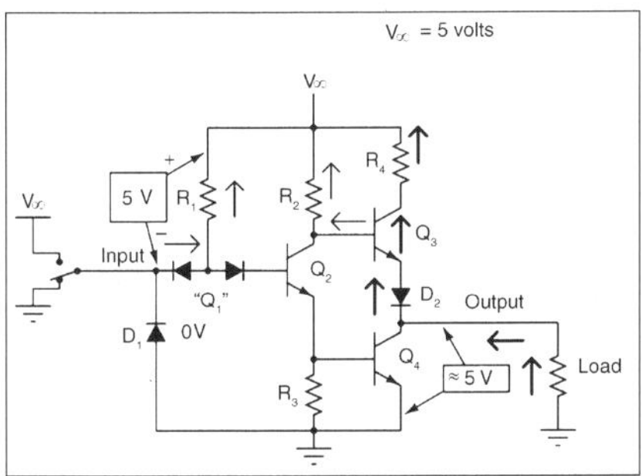

Please note that this two-position switch shown inside the gate symbol is representative of transistors Q_3 and Q_4 alternately connecting the output terminal to V_{cc} or ground, *not* of the switch previously shown sending an input signal to the gate! Conversely, when a gate circuit is outputting a "low" logic level to a load, it is analogous to the double-throw switch being set in the "ground" position.

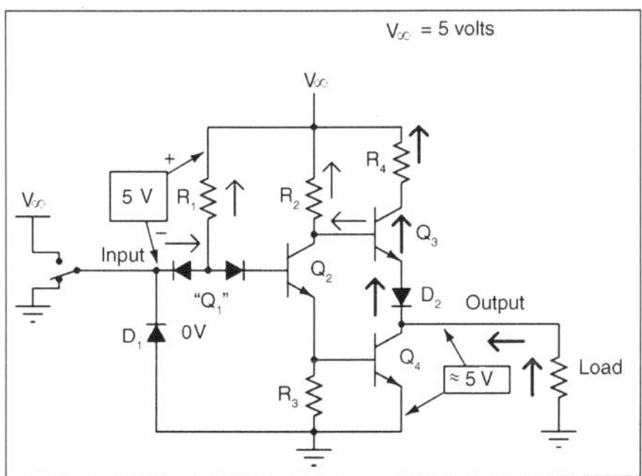

Fig. Inverter Gate Sinking Current

Current will then be going the other way if the load resistance connects to V_{cc}: from ground, through the emitter of Q_4, out the output terminal, through the load resistance, and back to V_{cc}. In this condition, the gate is said to be *sinking* current:

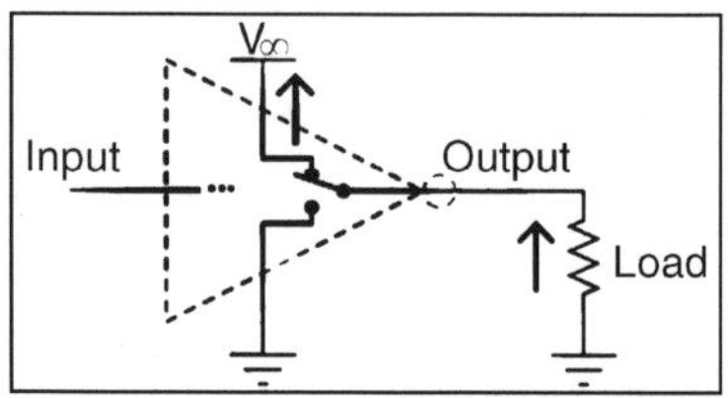

Fig. Simplified Gate Cuircuit Sinking Current

The combination of Q_3 and Q_4 working as a "push-pull" transistor pair (otherwise known as a *totem pole output*) has the ability to either source current (draw in current to V_{cc}) or sink current (output current from ground) to a load. However, a standard *TTL* gate *input* never needs current to be sourced, only sunk. That is, since a *TTL* gate input naturally assumes a high state if left floating, any gate output driving a *TTL* input need only sink current to provide a "0" or "low" input, and need not source current to provide a "1" or a "high" logic level at the input of the receiving gate:

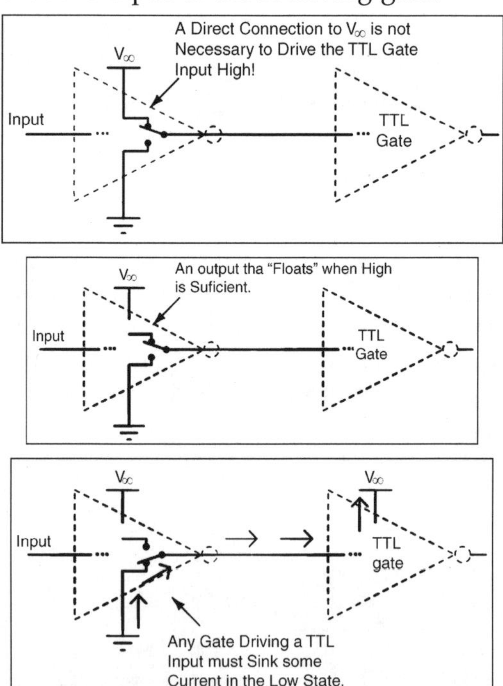

This means we have the option of simplifying the output stage of a gate circuit so as to eliminate Q_3 altogether. The result is known as an *open-collector output*:

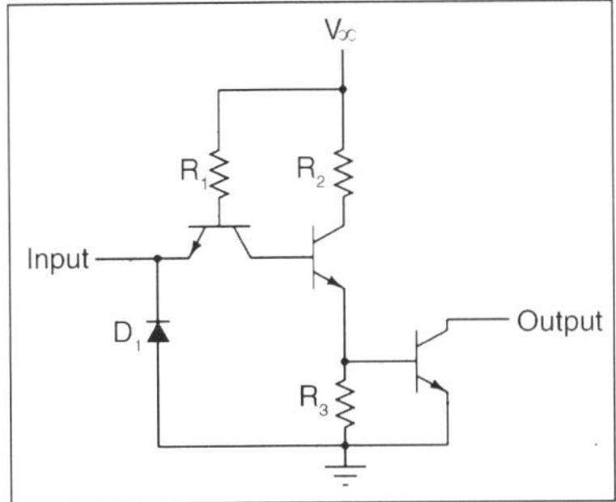

Fig. Invertor With Open Collector Output

To designate open-collector output circuitry within a standard gate symbol, a special marker is used. Shown here is the symbol for an inverter gate with open-collector output:

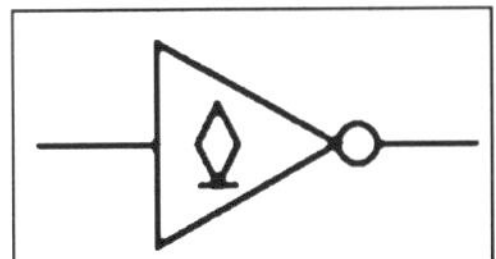

Fig. Invertor With Open Collector Output

Please keep in mind that the "high" default condition of a floating gate input is only true for TTL circuitry, and not necessarily for other types, especially for logic gates constructed of field-effect transistors.

THE "BUFFER" GATE

If we were to connect two inverter gates together so that the output of one fed into the input of another, the two inversion functions would "cancel" each other out so that there would be no inversion from input to final output:

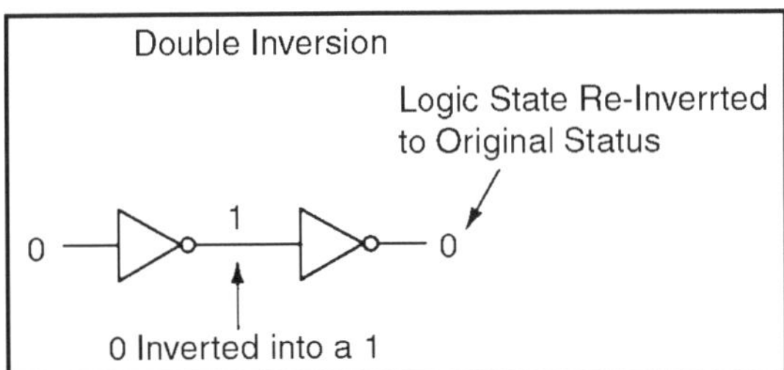

While this may seem like a pointless thing to do, it does have practical application. Remember that gate circuits are signal*amplifiers*, regardless of what logic function they may perform. A weak signal source (one that is not capable of sourcing or sinking very much current to a load) may be boosted by means of two inverters like the pair shown in the previous illustration. The logic level is unchanged, but the full current-sourcing or -sinking capabilities of the final inverter are available to drive a load resistance if needed.

For this purpose, a special logic gate called a *buffer* is manufactured to perform the same function as two inverters. *Its symbol is simply a triangle, with no inverting "bubble" on the output terminal*:

"Buffer" Gate

Input —▷— Output

Input	Output
0	0
1	1

The internal schematic diagram for a typical open-collector buffer is not much different from that of a simple inverter: only one more common-emitter transistor stage is added to re-invert the output signal.

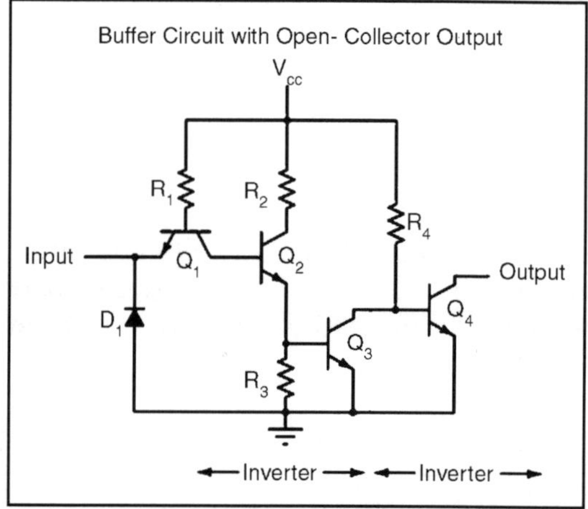

Let's analyse this circuit for two conditions: an input logic level of "1" and an input logic level of "0." First, a "high" (1) input:

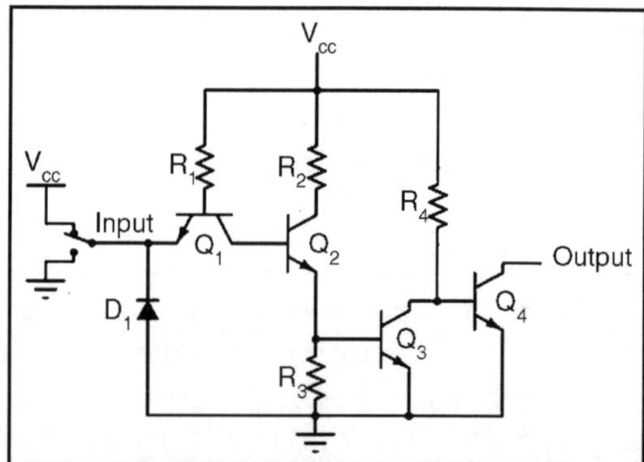

As before with the inverter circuit, the "high" input causes no conduction through the left steering diode of Q_1 (emitter-to-base PN junction). All of R_1's current goes through the base of transistor Q_2, saturating it:

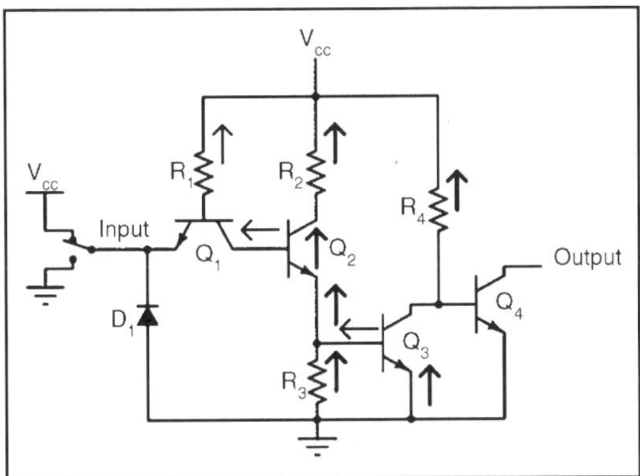

Having Q_2 saturated causes Q_3 to be saturated as well, resulting in very little voltage dropped between the base and emitter of the final output transistor Q_4. Thus, Q_4 will be in cutoff mode, conducting no current. The output terminal will be floating (neither connected to ground nor V_{cc}), and this will be equivalent to a "high" state on the input of the next TTL gate that this one feeds in to. Thus, a "high" input gives a "high" output.

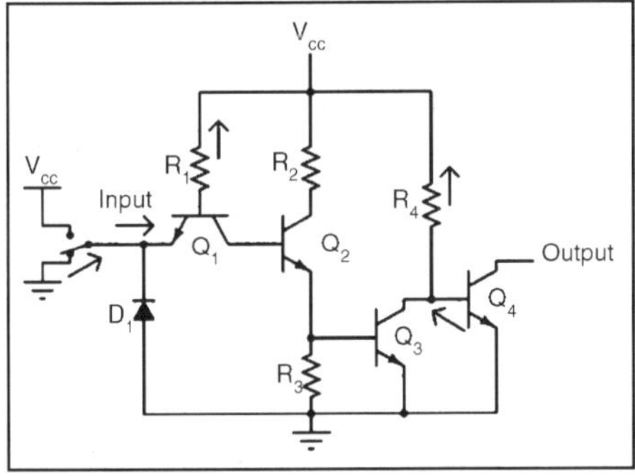

With a "low" input signal (input terminal grounded), the analysis looks something like this:

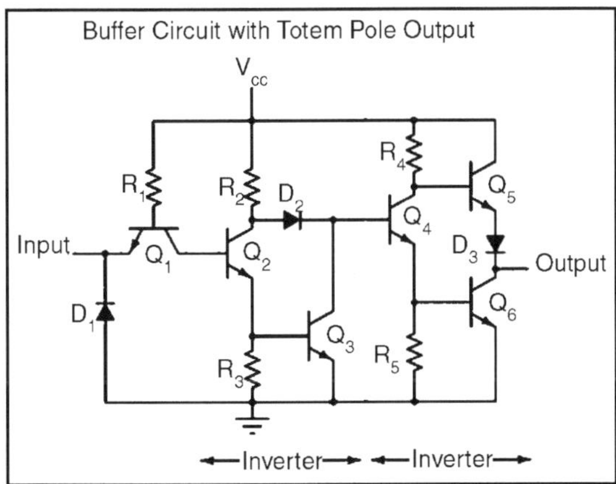

All of R_1's current is now diverted through the input switch, thus eliminating base current through Q_2. This forces transistor Q_2 into cutoff so that no base current goes through Q_3 either. With Q_3 cutoff as well, Q_4 is will be saturated by the current through resistor R_4, thus connecting the output terminal to ground, making it a "low" logic level. Thus, a "low" input gives a "low" output.

The schematic diagram for a buffer circuit with totem pole output transistors is a bit more complex, but the basic principles, and certainly the truth table, are the same as for the open-collector circuit:

LOGIC DIAGRAM DESIGN USING GATE EQUIVALENCIES

Keeping in mind the preceding introduction to the concept of gate equivalencies, we now turn to the design of a logic diagram using this concept.

Logic Diagram Design Using NAND Gates

This diagram is easily converted to use just NAND gates by first replacing all OR gates with the OR form of the NAND gate and replacing all AND gates with the AND form of the NAND gate.

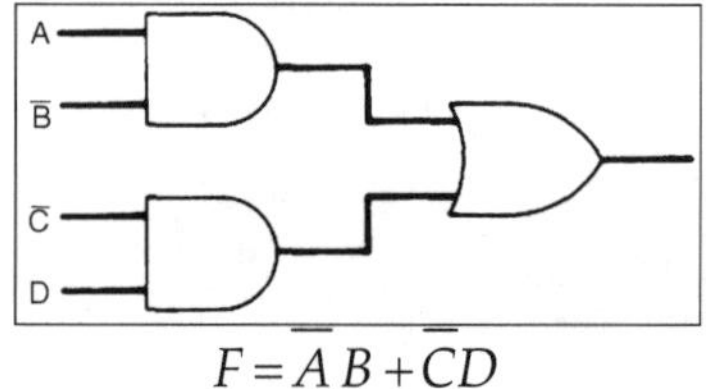

$$F = \overline{A}B + \overline{C}D$$

Fig. AND-OR Implementation of Equation

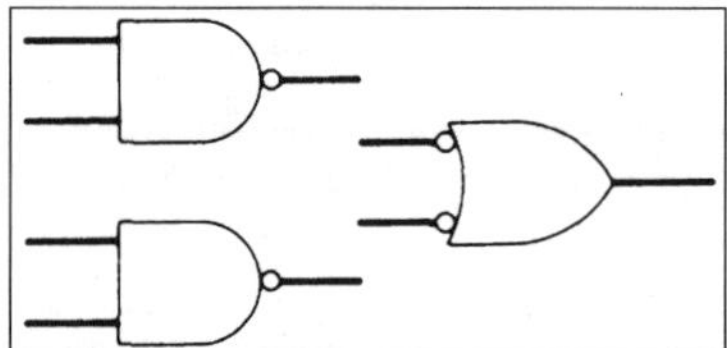

Fig. Replacing with AND and OR Form of NAND Gate

Since we deal with active high inputs and outputs in the "real" world, the initial inputs and final outputs of a circuit should generally be active high. The next step is connec-tion of the AND forms to the OR form. Since the OR form of the NAND gate requires active low inputs and the output of the AND form is already active low, no level conversion is needed, and a direct connection can be made between the outputs of the AND forms and the inputs of the OR form.

In general, a direct connection can be made between an active low output and input or between an active high output and input. If one is active high and the other is active low, then an inverter must be inserted to convert from one level to the other.

Making connections as just described, the final logic diagram using only NAND gates is illustrated in Figure below. When the AND and OR forms of the NAND gates on the logic diagram are shown, the in-tended function of the circuit is more obvious. Looking at Figure below, we can readily see it is a SOP implementation—a series of ANDs feeding an OR—while looking at the equivalent circuit in Figure above, we find that the actual function of the circuit is not clear.

A second example will illustrate a condition where level conversion is necessary. Consider implementing a three-input OR gate using only two-input NAND gates. If

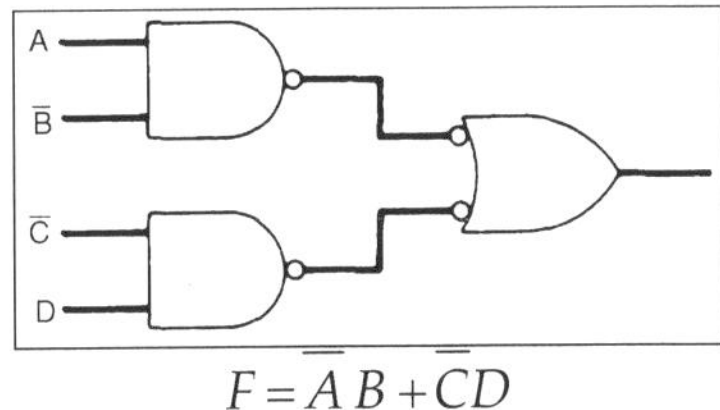

$$F = \overline{A}B + \overline{C}D$$

Fig. AND Implementation of Equation

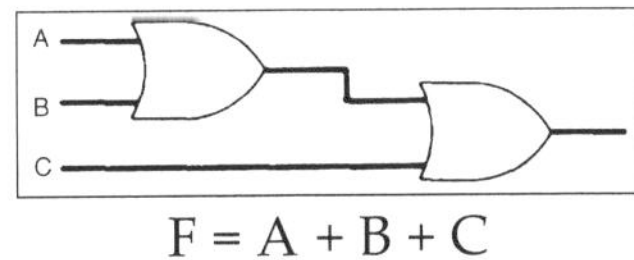

$$F = A + B + C$$

Fig. Three-Input OR with Two-Input Gates two-input OR gates were available, this function could be implemented as shown in Figure above

To implement using only two-input NAND gates we proceed as before replacing all OR gates with the OR form of the NAND gate. This yields Figure below. To complete the diagram we must first insure active high inputs and outputs to the real world. Since the inputs of the OR forms are active low, we must invert the input variables. Since we are assuming double-rail inputs, input inverters are not shown in Figure. The final output of the circuit is already active high; so no conversion is needed there.

The final connection is that between the first and second OR form. Since the output of the first OR form is active high and the input of the second OR form is active low, a direct connection cannot be made. To match the levels, a NAND gate used as an inverter is inserted. Since the inverter is converting from active high to active low, we show the AND form of the NAND gate as an inverter since it has active high inputs and an active low input.

Logic Diagram Design Using NOR Gates

As with the NAND gates, the first step in implementing a circuit using only NOR gates is replacing all OR gates with the OR form of the NOR gate and all AND gates with the AND form of the NOR gate. First we must provide active high inputs and outputs to the real world. The AND form of the NOR gate has active low inputs; so the input variables must be comple-mented. Since we are assuming double-rail inputs, the input. The final output of the circuit is also active low; so an inverter must be added to convert this output to active high.

A NOR gate is used as an inverter, and to show the active low input and active high output required of the inverter, the AND form of the NOR gate is shown. Finally, connections can be made directly between the outputs of the AND forms and the inputs of the OR form since both are active high. The final logic diagram using only NOR gates is shown in Figure.

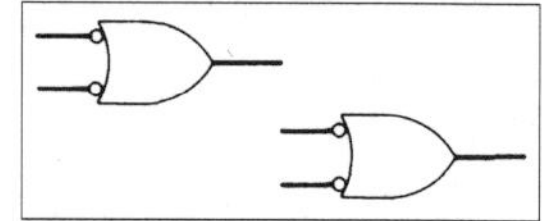

Fig. Replacing with OR Form of NAND Gate

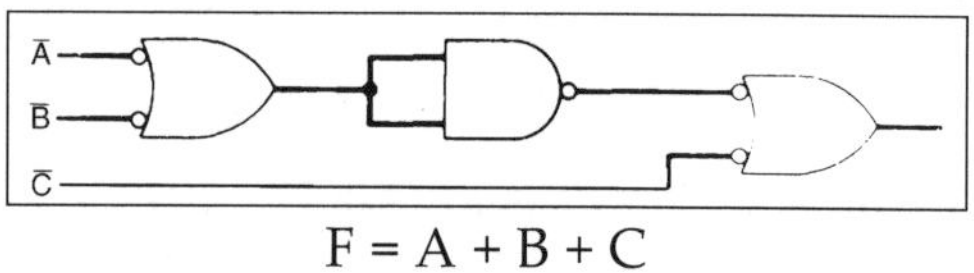

F = A + B + C

Fig. Two-Input NAND Implementation of Three-Input OR

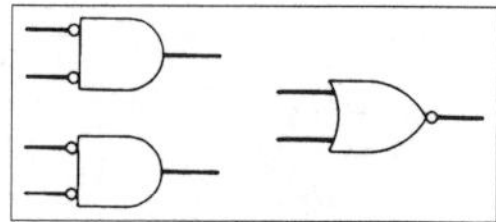

Fig. Replacing with AND and OR Form of NOR Gate

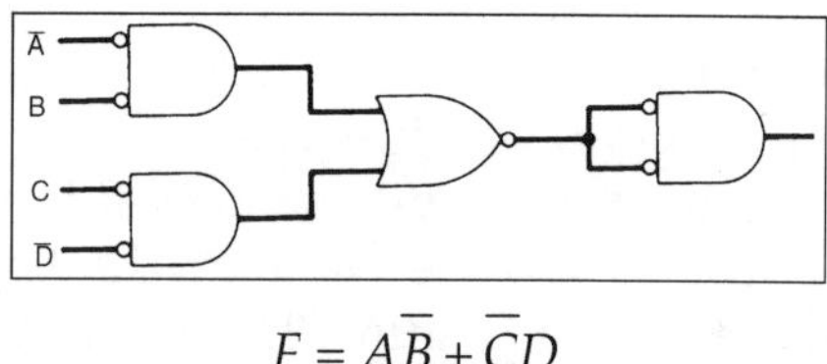

$$F = A\overline{B} + \overline{C}D$$

Fig. NOR Implementation of Equation

GATE EQUIVALENCIES

Boolean algebra is composed of just three operators: AND, OR, and NOT. Any expression can be implemented using combinations of these three functions. To imple-ment a circuit with just one gate type, all three of these operations must somehow be performed.

Performing the NOT Function with NAND and NOR

TThe problem of performing the NOT function with NAND and NOR gates as an exercise. NAND and NOR can function as inverters by tying the inputs together.

Performing the AND Function with NAND and NOR

Through simple boolean transformations we can determine how to perform the AND function with a NAND or NOR gate.

The AND function on two inputs is illustrated in boolean algebra as:

$$F = AB$$

Fig. 4.21 Invert Function with NAND and NOR

The NAND function is expressed as:

$$F = \overline{AB}$$

This equation can be reduced to the AND function by complementing the entire equation (*i.e.*, the output) to remove the NOT bar. Thus a NAND gate will perform the AND function if the output is inverted. The NOR function on two inputs is illustrated as:

$$F = \overline{A + B}$$

By applying De Morgan's theorem, we obtain an equivalent expression:

$$F = \overline{A}\,\overline{B}$$

This can be reduced to the AND function by complementing the input variables. Thus a NOR gate will perform the AND function if the inputs are inverted.

In summary, a NAND gate will perform the AND function if the output is inverted. A NOR gate will perform the AND function if the inputs are inverted.

Performing the OR Function with NAND and NOR

The OR function on two inputs is illustrated in boolean algebra as:

$$F = A + B$$

The NAND function is expressed as:

$$F = \overline{AB}$$

By applying De Morgan's theorem we obtain an equivalent expression:

$$F = \overline{A} + \overline{B}$$

This can be reduced to the OR function by complementing the input variables. Thus a NAND gate performs the OR function if the input variables are complemented. The NOR function on two inputs is illustrated as:

$$F = \overline{A + B}$$

This can be reduced to the OR function by complementing the entire equation (*i.e.*, the output) to remove the NOT bar. Thus a NOR gate performs the OR function if the output is complemented.

In summary, a NAND gate will perform the OR function if the inputs are inverted. A NOR gate will perform the OR function if the output is inverted.

Symbolic Definitions

To clarify this dual nature of gates, each can be represented symbolically in two forms. The representations of a NAND gate and two representations of a NOR gate respectively. The intended function (AND or OR) is represented by the basic shape of the symbol. Input or output requirements are dictated by the presence of circles at the inputs or outputs. We note that the presence of a circle at an input indicates that the input must be inverted to properly perform the function dictated by the symbol shape, that is, the AND or OR function. Also, the presence of a circle at an output indicates that the output must be inverted to properly perform the function dictated by the symbol shape, that is, the AND or OR function.

It is important to note that these symbols do not represent new gates, just two ways of viewing the NAND and NOR gates. Thus we can speak of the *AND form* of a NAND gate and the *OR form* of a NAND gate. Likewise we can speak of the *AND form* of a NOR gate and the *OR form* of a NOR gate.

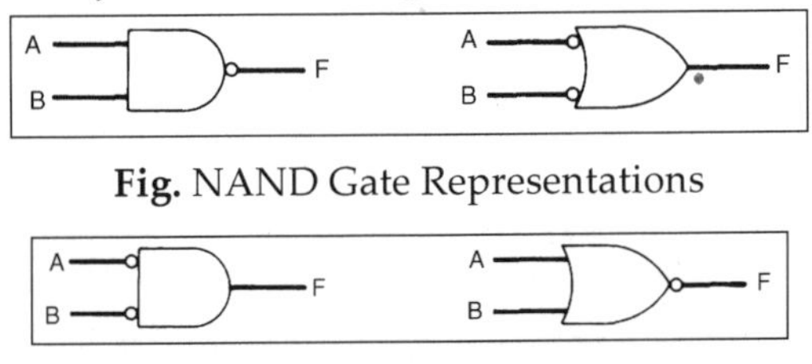

Fig. NAND Gate Representations

Fig. NOR Gate Representations

Active High and Active Low Inputs and Outputs

To further define the meaning of the circles on the gate symbols, we say a gate form with a circle at the output has an *active low output.* A gate form with circles at the inputs is said to have *active low inputs.* Conversely, a gate form without a circle at the output is said to have an *active high output,* and a gate form without circles at the inputs is said to have *active high inputs.* For example, the AND form of the NAND gate can be thought of as an AND gate with active high inputs and an active low output. The OR form of the NAND gate can be thought of as an OR gate with active low inputs and an active high output.

The term *active high* refers to logic levels as we have dealt with them so far: a true is represented by a high logic level. *Active low* indicates that a true is represented by a low logic level rather than a high logic level. To illustrate, consider performing the AND function with a NAND gate. The AND function provides a true output only if all inputs are true. Looking at the AND form of the NAND gate, we see it too performs the AND function, but since it has an active low output, a true output is indicated by a low logic level rather than a high logic level. Thus to see a *normal* active high output, we must invert the output of the NAND gate just.

An OR gate produces a true output if any of its inputs are true. Looking at the OR form of the NAND gate, we see it also performs the OR function, but since it has active low inputs, a true input is indicated by a low logic level rather than a high logic level. To provide these active low inputs from *normal* active high inputs, we must invert each of the inputs.

MULTIPLE-INPUT GATES

Inverters and buffers exhaust the possibilities for single-input gate circuits. What more can be done with a single logic signal but to buffer it or invert it? To explore more logic gate possibilities, we must add more input terminals to the circuit(s).

Adding more input terminals to a logic gate increases the number of input state possibilities. With a single-input gate such as the inverter or buffer, there can only be two possible input states: either the input is "high" (1) or it is "low" (0). As was mentioned previously in this chapter, a two input gate has *four* possibilities (00, 01, 10, and 11). A three-input gate has*eight* possibilities (000, 001, 010, 011, 100, 101, 110, and 111) for input states. The number of possible input states is equal to two to the power of the number of inputs:

$$\text{Number of Possible Input States} = 2^n$$

Where:

n=Number of Inputs

This increase in the number of possible input states obviously allows for more complex gate behaviour. Now, instead of merely inverting or amplifying (buffering) a single "high" or "low" logic level, the output of the gate will be determined by whatever *combination* of 1's and 0's is present at the input terminals. Since so many combinations are possible with just a few input terminals, there are many different types of multiple-input gates, unlike single-input gates which can only be inverters or buffers. Each basic gate type will be presented in this section, showing its standard symbol, truth table, and practical operation. The actual TTL circuitry of these different gates will be explored in subsequent sections.

THE AND GATE

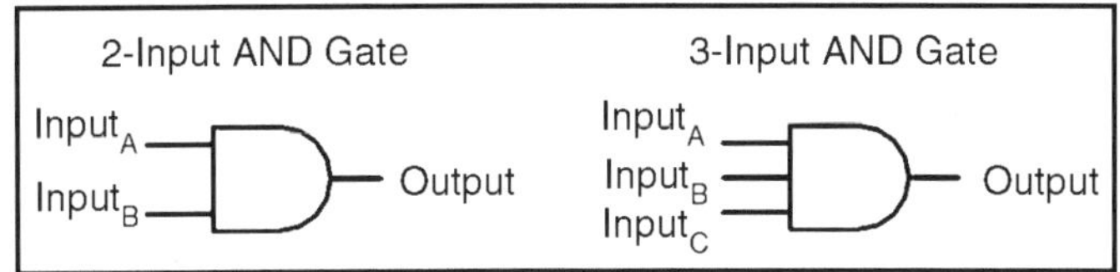

One of the easiest multiple-input gates to understand is the AND gate, so-called because the output of this gate will be "high" (1) if and only if *all* inputs (first input *and* the second input *and* ...) are "high" (1). If any input(s) are "low" (0), the output is guaranteed to be in a "low" state as well.

In case you might have been wondering, AND gates are made with more than three inputs, but this is less common than the simple two-input variety.

***A two-input and Gate's Truth Table Looks Like this*:**

2-Input AND Gate

$Input_A$, $Input_B$ — Output

A	B	Output
0	0	0
0	1	0
1	0	0
1	1	1

What this truth table means in practical terms is shown in the following sequence of illustrations, with the 2-input AND gate subjected to all possibilities of input logic levels. An LED (Light-Emitting Diode) provides visual indication of the output logic level:

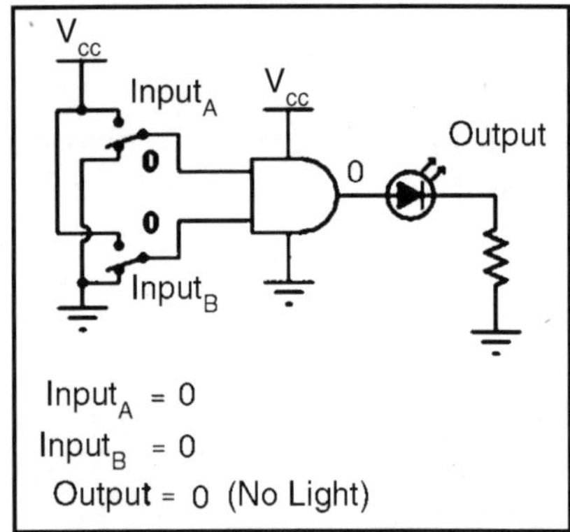

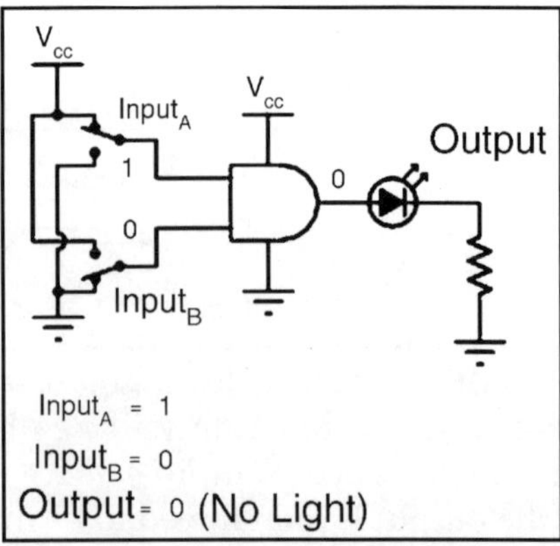

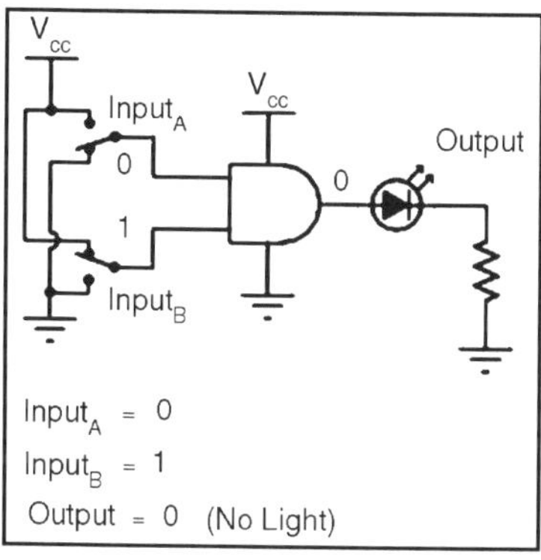

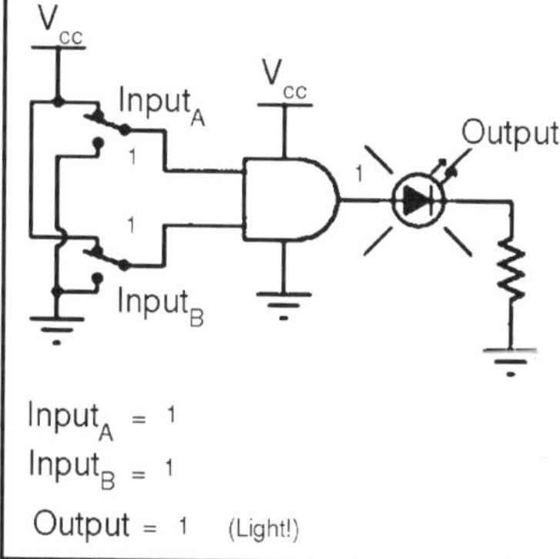

It is only with all inputs raised to "high" logic levels that the AND gate's output goes "high," thus energizing the LED for only one out of the four input combination states.

THE NAND GATE

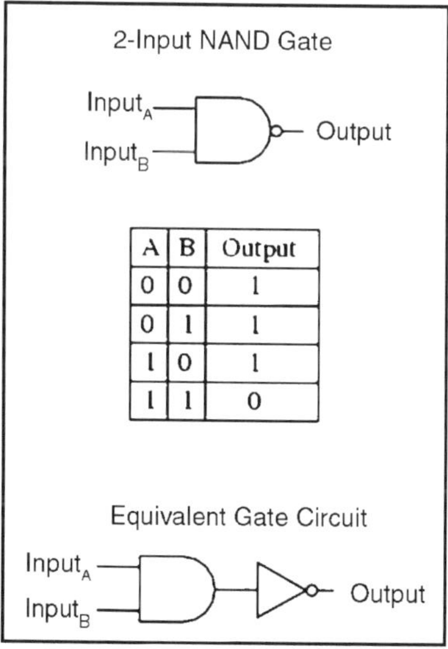

A	B	Output
0	0	1
0	1	1
1	0	1
1	1	0

A variation on the idea of the AND gate is called the NAND gate. The word "NAND" is a verbal contraction of the words NOT and AND. Essentially, a NAND gate behaves the same as an AND gate with a NOT (inverter) gate connected to the output terminal. To symbolize this output signal inversion, the NAND gate symbol has a bubble on the output line. The truth table for a NAND gate is as one might expect, exactly opposite as that of an AND gate: As with AND gates, NAND gates are made with more than two inputs. In such cases, the same general principle applies: the output will be "low" (0) if and only if all inputs are "high" (1). If any input is "low" (0), the output will go "high" (1).

THE OR GATE

Our next gate to investigate is the OR gate, so-called because the output of this gate will be "high" (1) if *any* of the inputs (first input *or* the second input *or* ...) are "high" (1). The output of an OR gate goes "low" (0) if and only if all inputs are "low" (0).

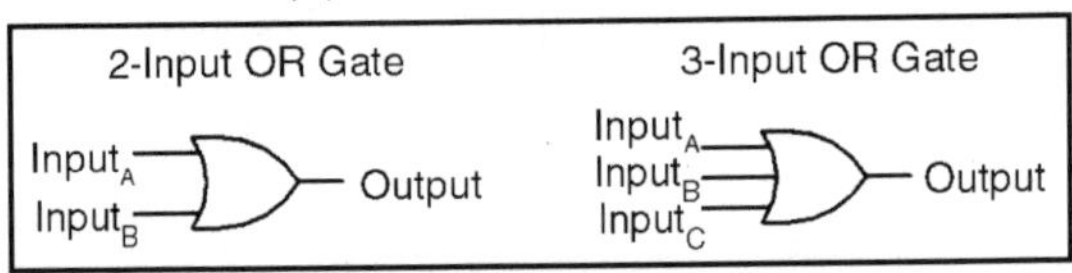

A two-input OR gate's truth table looks like this:

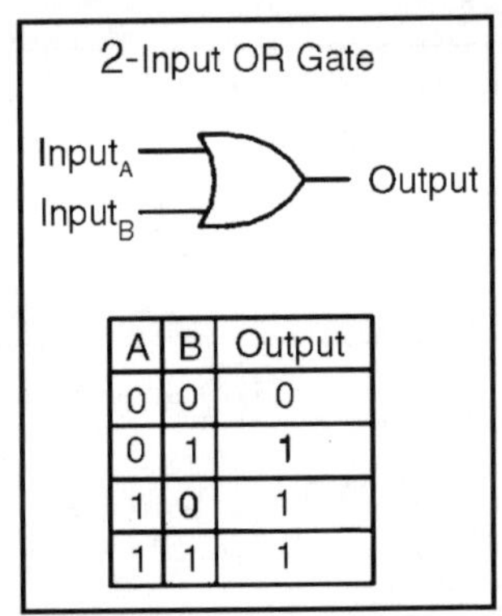

A	B	Output
0	0	0
0	1	1
1	0	1
1	1	1

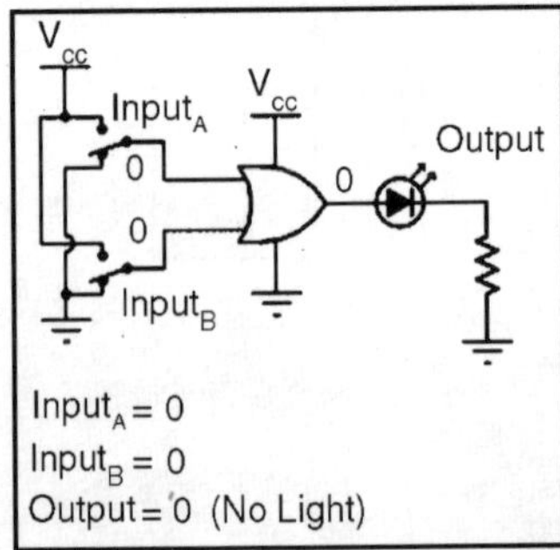

The following sequence of illustrations demonstrates the OR gate's function, with the 2-inputs experiencing all possible logic levels. An LED (Light-Emitting Diode) provides visual indication of the gate's output logic level:

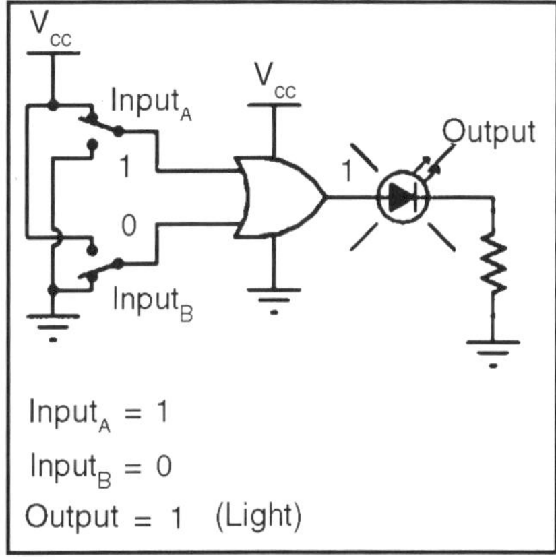

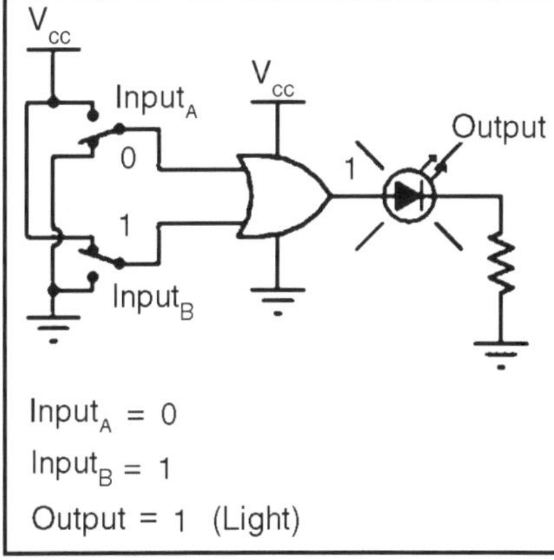

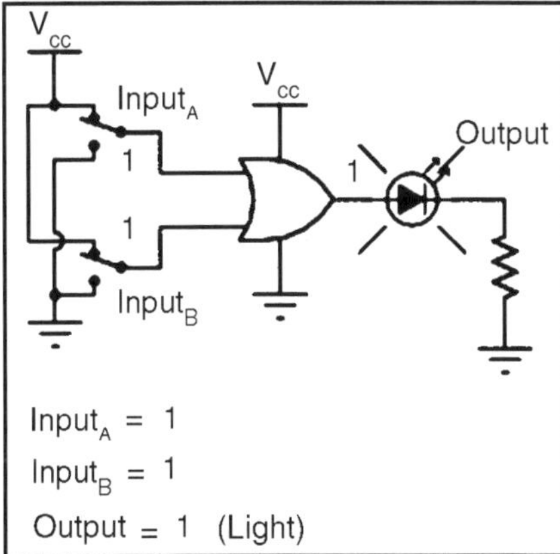

A condition of any input being raised to a "high" logic level makes the OR gate's output go "high," thus energizing the LED for three out of the four input combination states.

THE NOR GATE

As you might have suspected, the NOR gate is an OR gate with its output inverted, just like a NAND gate is an AND gate with an inverted output.

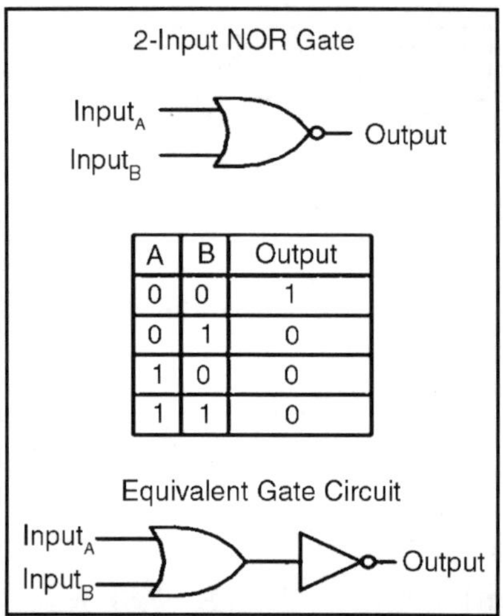

A	B	Output
0	0	1
0	1	0
1	0	0
1	1	0

NOR gates, like all the other multiple-input gates seen thus far, can be manufactured with more than two inputs. Still, the same logical principle applies: the output goes "low" (0) if any of the inputs are made "high" (1). The output is "high" (1) only when all inputs are "low" (0).

THE NEGATIVE-AND GATE

Its truth table, actually, is identical to a NOR gate:

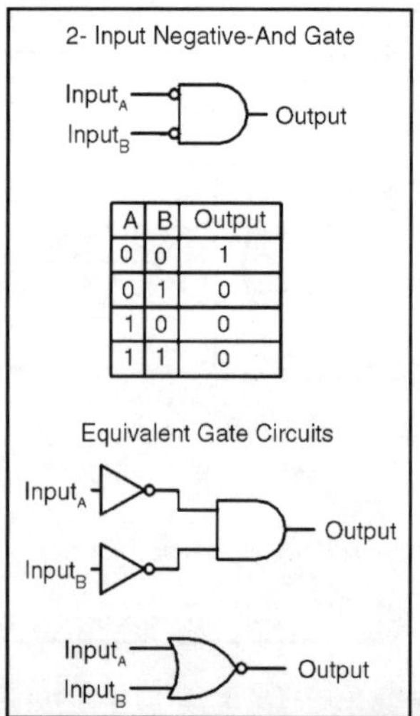

A	B	Output
0	0	1
0	1	0
1	0	0
1	1	0

A Negative-AND gate functions the same as an AND gate with all its inputs inverted (connected through NOT gates). In keeping with standard gate symbol convention, these inverted inputs are signified by bubbles. Contrary to most peoples' first instinct, the logical behaviour of a Negative-AND gate is *not* the same as a NAND gate.

THE NEGATIVE-OR GATE

Following the same pattern, a Negative-OR gate functions the same as an OR gate with all its inputs inverted. In keeping with standard gate symbol convention, these inverted inputs are signified by bubbles. The behaviour and truth table of a Negative-OR gate is the same as for a NAND gate:

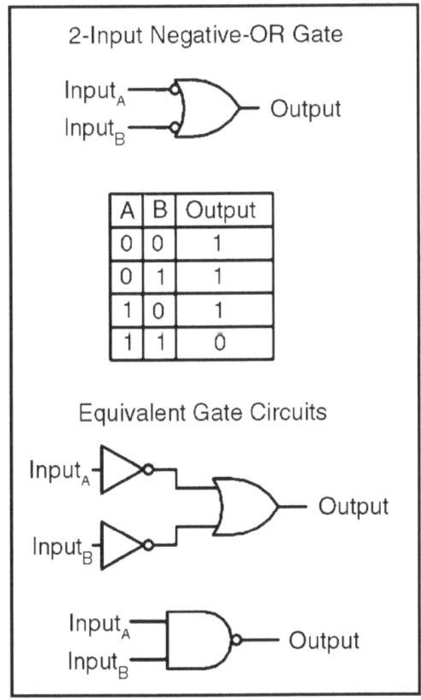

7

Boolean Algebra

INTRODUCTION

Mathematical rules are based on the defining limits we place on the particular numerical quantities dealt with. When we say that 1 + 1 = 2 or 3 + 4 = 7, we are implying the use of integer quantities: the same types of numbers we all learned to count in elementary education. What most people assume to be self-evident rules of arithmetic valid at all times and for all purposes actually depend on what we define a number to be.

For instance, when calculating quantities in AC circuits, we find that the "real" number quantities which served us so well in DC circuit analysis are inadequate for the task of representing AC quantities. We know that voltages add when connected in series, but we also know that it is possible to connect a 3-volt AC source in series with a 4-volt AC source and end up with 5 volts total voltage (3 + 4 = 5)! Does this mean the inviolable and self-evident rules of arithmetic have been violated? No, it just means that the rules of "real" numbers do not apply to the kinds of quantities encountered in AC circuits, where every variable has both a magnitude and a phase.

Consequently, we must use a different kind of numerical quantity, or object, for AC circuits (*complex* numbers, rather than *real* numbers), and along with this different system of numbers comes a different set of rules telling us how they relate to one another. An expression such as "3 + 4 = 5" is nonsense within the scope and definition of real numbers, but it fits nicely within the scope and definition of complex numbers (think of a right triangle with opposite and adjacent sides of 3 and 4, with a hypotenuse of 5).

Because complex numbers are two-dimensional, they are able to "add" with one another trigonometrically as single-dimension "real" numbers cannot. Logic is much like mathematics in this respect: the so-called "Laws" of logic depend on how we define what a proposition is.

The Greek philosopher Aristotle founded a system of logic based on only two types of propositions: true and false. His bivalent (two-mode) definition of truth led to the four foundational laws of logic: the Law of Identity (A is A); the Law of Non-contradiction (A is not non-A); the Law of the Excluded

Middle (either A or non-A); and the Law of Rational Inference. These so-called Laws function within the scope of logic where a proposition is limited to one of two possible values, but may not apply in cases where propositions can hold values other than "true" or "false." In fact, much work has been done and continues to be done on "multivalued," or *fuzzy* logic, where propositions may be true or false *to a limited degree.*

In such a system of logic, "Laws" such as the Law of the Excluded Middle simply do not apply, because they are founded on the assumption of bivalence. Likewise, many premises which would violate the Law of Non-contradiction in Aristotelian logic have validity in "fuzzy" logic. Again, the defining limits of propositional values determine the Laws describing their functions and relations. The English mathematician George Boole sought to give symbolic form to Aristotle's system of logic. Boole wrote a treatise on the subject in 1854, titled *An Investigation of the Laws of Thought, on Which Are Founded the Mathematical Theories of Logic and Probabilities,* which codified several rules of relationship between mathematical quantities limited to one of two possible values: true or false, 1 or 0. His mathematical system became known as Boolean algebra.

All arithmetic operations performed with Boolean quantities have but one of two possible outcomes: either 1 or 0. There is no such thing as "2" or "–1" or "1/2" in the Boolean world. It is a world in which all other possibilities are invalid by fiat. As one might guess, this is not the kind of math you want to use when balancing a checkbook or calculating current through a resistor. However, Claude Shannon of MIT fame recognized how Boolean algebra could be applied to on-and-off circuits, where all signals are characterized as either "high" (1) or "low" (0). The thesis, titled *A Symbolic Analysis of Relay and Switching Circuits,* put Boole's theoretical work to use in a way Boole never could have imagined, giving us a powerful mathematical tool for designing and analyzing digital circuits.

In this chapter, you will find a lot of similarities between Boolean algebra and "normal" algebra, the kind of algebra involving so-called real numbers. Just bear in mind that the system of numbers defining Boolean algebra is severely limited in terms of scope, and that there can only be one of two possible values for any Boolean variable: 1 or 0. Consequently, the "Laws" of Boolean algebra often differ from the "Laws" of real-number algebra, making possible such statements as 1 + 1 = 1, which would normally be considered absurd. Once you comprehend the premise of all quantities in Boolean algebra being limited to the two possibilities of 1 and 0, and the general philosophical principle of Laws depending on quantitative definitions, the "nonsense" of Boolean algebra disappears.

It should be clearly understood that Boolean numbers are not the same as *binary* numbers. Whereas Boolean numbers represent an entirely different system of mathematics from real numbers, binary is nothing more than an alternative *notation*for real numbers. The two are often confused because both Boolean math and binary notation use the same two ciphers: 1 and 0. The difference is that Boolean quantities are restricted to a single bit (either 1 or 0), whereas binary numbers may be composed of many bits adding up in place-weighted form to a value of any finite size. The binary number 10011_2("nineteen") has no more place in the Boolean world than the decimal number 2_{10} ("two") or the octal number 32_8 ("twenty-six").

BASIC OPERATIONS OF BOOLEAN ALGEBRA

Boolean Algebra was introduced by George Boole in the year 1854. Like other algebras, it uses variables (called statements) and operations (called relations). Variables in Boolean algebra is called logic variable that only has 2 possible values, either true (1) or false (0) whereas its operation is called logical operations.

There are 3 basic logical operations, *i.e.* AND (•), OR (+) and NOT (–). These operations are used to combine operands (logical constants and variables) to form logical expressions. Following are a few logical expressions with *X*, *Y* and *Z* as the logical variables that can only assume the value of FALSE (or 0) or TRUE (or 1).

X or NOT (*X*)

$\overline{X.Y} + Z$ or NOT(*X* AND *Y*) OR Z

$\left(X.\overline{Y}\right)+\left(Y.Z\right)$ or (*X* AND NOT(*Y*)) OR (*Y* AND *Z*)

Other than the basic logical operations of AND, OR and NOT, there are multiple of other logical operations, among them are NAND, NOR and XOR that are widely used in building logical circuits in computers. These logical operations are combination of a few of the basic logical operations, *e.g.*:

NAND is a combination of AND followed by NOT

NOR is a combination of OR followed by NOT

The following is a truth table for the logical operations of AND, OR, NOT, NAND, NOR and XOR.

P	Q	NOT P	P AND Q	P OR Q	P XOR Q	P NAND Q	P NORQ
0	0	1	0		0	0	1 1
0	1	1	0	1	1	1	0
1	0	0	0	1	1	1	0
1	1	0	1	1	0	0	0

BOOLEAN OPERATORS

In this section we will look at four different Boolean operators. You know how to handle arithmetic operators (+ – * / %). Addition, subtraction, multiplication, division and remainder operations are performed according to the rules for each operator. There are also a set of Boolean operators with their own set of rules. The rules of Boolean operators can be conveniently displayed in a truth table. This is a table, which shows the possible combinations of a Boolean statement and indicates the value (true or false) of each statement.

In the truth tables that follow, a single letter indicates a single, simple Boolean condition. Such a condition is either true or false. Boolean statement *A* is true or false. Likewise Boolean statement *B* is true or false. The truth tables will show the results of Boolean statements that use both *A* and *B* with a variety of Boolean operators.Employment requirements will be used to explain the logic of each truth table.In each case imagine that an accountant needs to be hired. Condition A determines if the applicant has a Degree and condition B determines if the applicant has at least five years experience.

Boolean Or

The or Operator

A
B
A or *B*
T
T
T
T
F
T
F
T
T
F
F
F

Notice that two conditions have four possible combinations. It is important that you know the results for each type of combination.

In this case the employment analogy requires a Degree OR Experience.This requirement is quite relaxed.You have a Degree, fine.You have Experience, that's also fine.You have both, definitely fine.You have neither, that's a problem.

Boolean And

The and Operator

A
B
A and *B*
T
T

T
T
F
F
F
T
F
F
F
F

Now employment requires a Degree AND Experience. This requirement is much more demanding than the or operator. You have a Degree, fine, provided you also have Experience. If you have only one qualification, or the other qualification, that is not good enough. If you have neither qualification, forget showing up.

Boolean Xor

The xor Operator

A
B
A xor *B*
T
T
F
T
F
T
F
T
T
F
F
F

Everyday human language, like English, uses both the or operator, as well as the and operator. How about the "exclusive or" *x*or? A peek at the truth table shows something pretty weird.If conditions *A* and *B* are both true, the compound result is false. We will try to explain this by using the "cheap boss" analogy. A manager wants to hire somebody and not pay much money.The advertisement states that a degree or experience is required.

Candidate *X* walks in and the boss says: "I'll hire you, but your pay will be low. You see, you have a degree but you have no experience at all."

Candidate *Y* walks in and the boss says: "I'll hire you, but your pay will be low. You see, you have experience but you have no degree." Candidate *Z* walks in and the boss says: "I'm sorry I cannot hire you.You are over qualified since you have a degree and also experience." Let us try another example with a "Student Council" election analogy. The student council sponsor requires that a candidate shows school spirit by belonging to a musical organisation (band, orchestra, choir) or an athletic team. You can only run for student council if you belong to one or the other group.However, if you are active in both a musical organisation and an athletic team, the sponsor does not want you in Student Council.You are too active and won't have the time to do a proper job.

Boolean Not

The not Operator

A

not *A*

T

F

F

T

This section will finish with the simplest Boolean operator, not.This operator takes the condition that follows and changes true to false or false to true.There are special rules that need to be followed when a complex Boolean statement is used with not. Right now we want to understand the simple truth table shown earlier. In English we need to use "double negative" sentences to create an appropriate analogy. I can say "It is not true that Tom Smith is valedictorian." This statement results in Tom not being Valedictorian. On the other hand, if I say "It is not true that Tom Smith is not the Valedictorian." Now Tom is the Valedictorian.

EXPLORATION OF BOOLEAN ARITHMETIC

Let us begin our exploration of Boolean algebra by adding numbers together:

$$0+0=0$$

$$0+1=1$$

$$1+0=1$$

$$1+1=1$$

The first three sums make perfect sense to anyone familiar with elementary addition. The last sum, though, is quite possibly responsible for more confusion than any other single statement in digital electronics, because it seems to run contrary to the basic principles of mathematics. Well, it *does* contradict principles of addition for real numbers, but not for Boolean numbers.

Remember that in the world of Boolean algebra, there are only two possible values for any quantity and for any arithmetic operation: 1 or 0. There is no such thing as "2" within the scope of Boolean values. Since the sum "1 + 1" certainly isn't 0, it must be 1 by process of elimination.

It does not matter how many or few terms we add together, either. Consider the following sums:

$$0+1+1=1$$
$$1+1+1=1$$
$$0+1+1+1=1$$
$$1+0+1+1+1=1$$

Take a close look at the two-term sums in the first set of equations. Does that pattern look familiar to you? It should! It is the same pattern of 1's and 0's as seen in the truth table for an OR gate. In other words, Boolean addition corresponds to the logical function of an "OR" gate, as well as to parallel switch contacts:

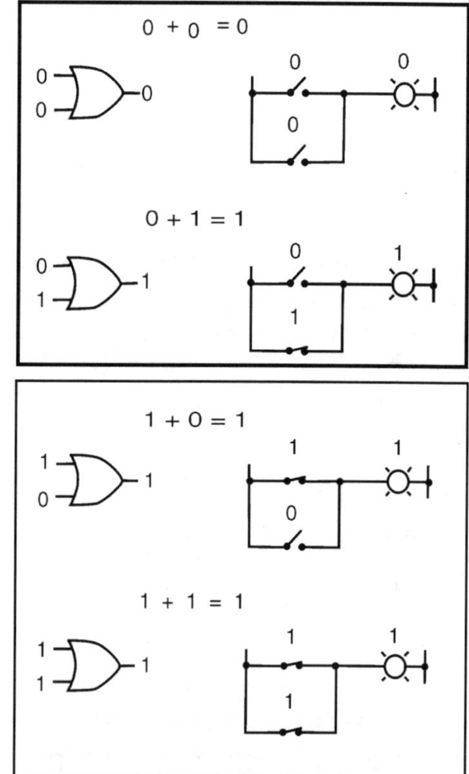

There is no such thing as subtraction in the realm of Boolean mathematics. Subtraction implies the existence of negative numbers: 5 – 3 is the same thing as 5 + (-3), and in Boolean algebra negative quantities are forbidden. There is no such thing as division in Boolean mathematics, either, since division is really nothing more than compounded subtraction, in the same way that multiplication is compounded addition. Multiplication is valid in Boolean algebra, and thankfully it is the same as in real-number algebra: anything multiplied by 0 is 0, and anything multiplied by 1 remains unchanged:

$$0 \times 0 = 0$$
$$0 \times 1 = 0$$
$$1 \times 0 = 0$$
$$1 \times 1 = 1$$

This set of equations should also look familiar to you: it is the same pattern found in the truth table for an AND gate. In other words, Boolean multiplication corresponds to the logical function of an "AND" gate, as well as to series switch contacts:

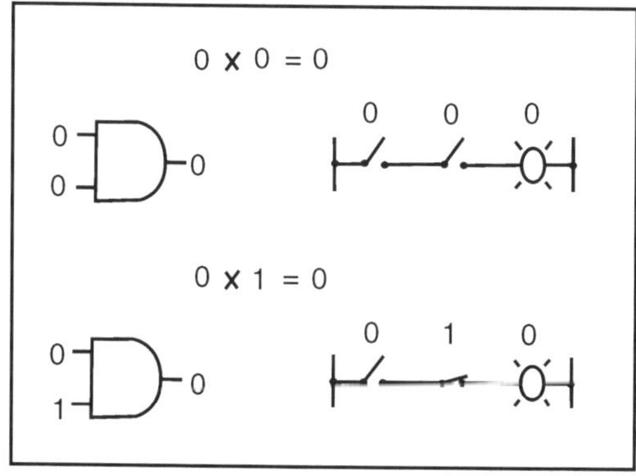

Like "normal" algebra, Boolean algebra uses alphabetical letters to denote variables. Unlike "normal" algebra, though, Boolean variables are always CAPITAL letters, never lower-case. Because they are allowed to possess only one of two possible values, either 1 or 0, each and every variable has a *complement*: the opposite of its value. For example, if variable "A" has a value of 0, then the complement of A has a value of 1. Boolean notation uses a bar above the variable character to denote complementation, like this:

$\text{If}: A = 0$

$\text{Then}: \bar{A} = 1$

$\text{If}: A = 0$

$\text{Then}: \bar{A} = 0$

In written form, the complement of "A" denoted as "A-not" or "A-bar". Sometimes a "prime" symbol is used to represent complementation. For example, A′ would be the complement of A, much the same as using a prime symbol to denote differentiation in calculus rather than the fractional notation d/dt. Usually, though, the "bar" symbol finds more widespread use than the "prime" symbol, for reasons that will become more apparent later in this chapter.

Boolean complementation finds equivalency in the form of the NOT gate, or a normally-closed switch or relay contact:

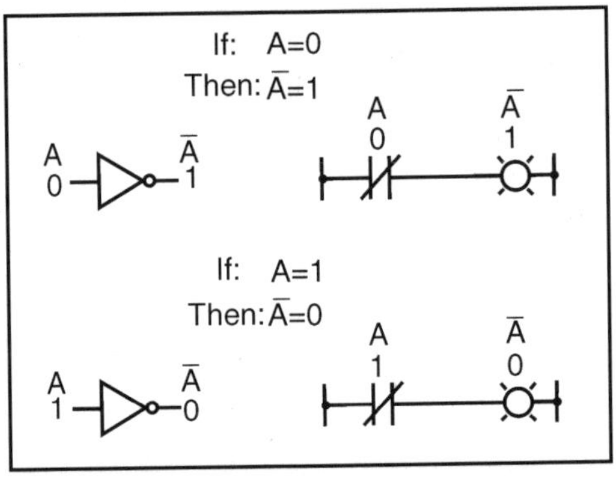

The basic definition of Boolean quantities has led to the simple rules of addition and multiplication, and has excluded both subtraction and division as valid arithmetic operations. We have a symbology for denoting Boolean variables, and their complements. In the next section we will proceed to develop Boolean identities.

IDENTITY OF BOOLEAN ALGEBRAIC

In mathematics, an *identity* is a statement true for all possible values of its variable or variables. The algebraic identity of $x + 0 = x$ tells us that anything (x) added to zero equals the original "anything," no matter what value that "anything" (x) may be. Like ordinary algebra, Boolean algebra has its own unique identities based on the bivalent states of Boolean variables. The first Boolean identity is that the sum of anything and zero is the same as the original "anything." This identity is no different from its real-number algebraic equivalent:

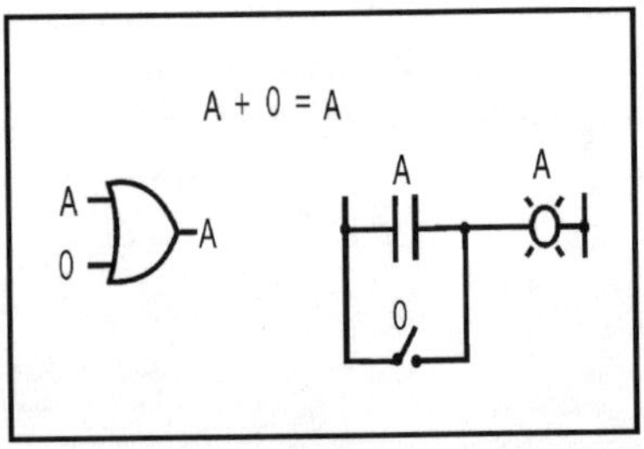

No matter what the value of A, the output will always be the same: when A=1, the output will also be 1; when A=0, the output will also be 0. The next identity is most definitely *different* from any seen in normal algebra.

Here we discover that the sum of anything and one is one:

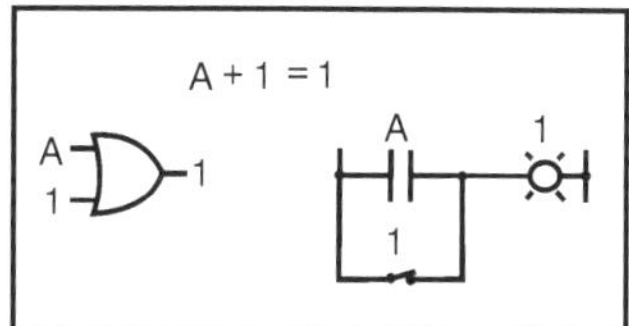

No matter what the value of A, the sum of A and 1 will always be 1. In a sense, the "1" signal *overrides* the effect of A on the logic circuit, leaving the output fixed at a logic level of 1. Next, we examine the effect of adding A and A together, which is the same as connecting both inputs of an OR gate to each other and activating them with the same signal:

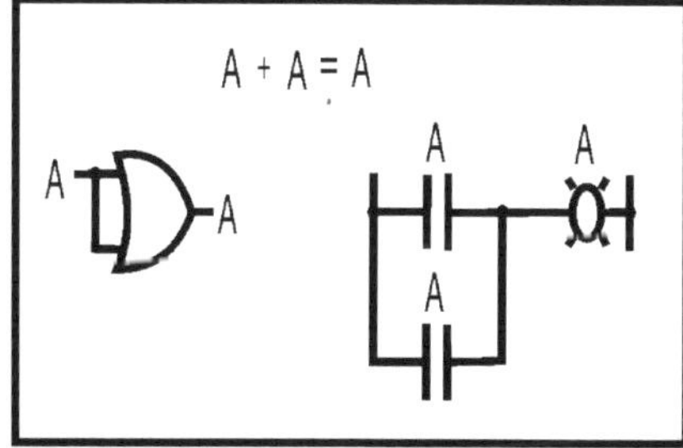

In real-number algebra, the sum of two identical variables is twice the original variable's value (x + x = 2x), but remember that there is no concept of "2" in the world of Boolean math, only 1 and 0, so we cannot say that A + A = 2A. Thus, when we add a Boolean quantity to itself, the sum is equal to the original quantity: 0 + 0 = 0, and 1 + 1 = 1.

Introducing the uniquely Boolean concept of complementation into an additive identity, we find an interesting effect. Since there must be one "1" value between any variable and its complement, and since the sum of any Boolean quantity and 1 is 1, the sum of a variable and its complement must be 1:

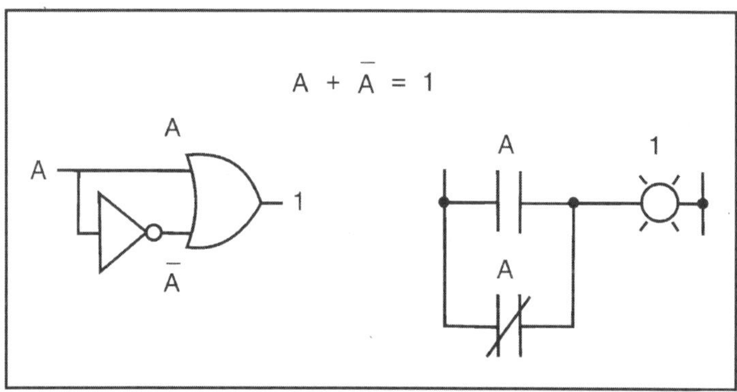

Just as there are four Boolean additive identities (A+0, A+1, A+A, and A+A'), so there are also four multiplicative identities: A × 0, A × 1, A × A, and A × A'. Of these, the first two are no different from their equivalent expressions in regular algebra:

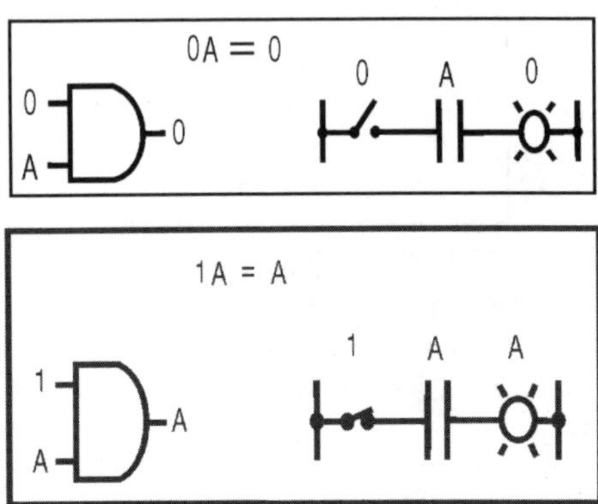

The third multiplicative identity expresses the result of a Boolean quantity multiplied by itself. In normal algebra, the product of a variable and itself is the *square* of that variable ($3 \times 3 = 3^2 = 9$). However, the concept of "square" implies a quantity of 2, which has no meaning in Boolean algebra, so we cannot say that,

$$A \times A = A^2.$$

Instead, we find that the product of a Boolean quantity and itself is the original quantity, since,

$$0 \times 0 = 0 \text{ and } 1 \times 1 = 1:$$

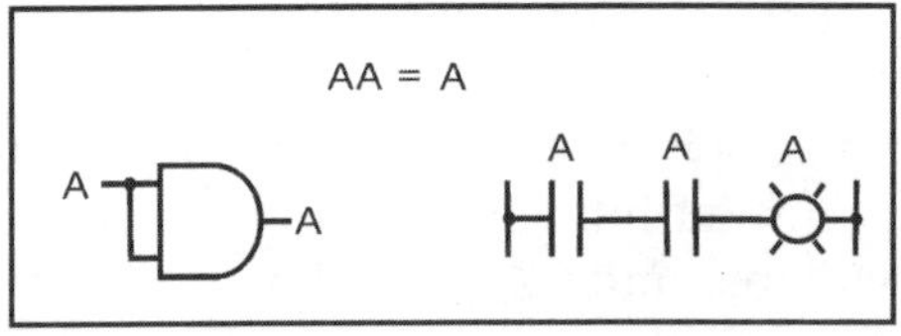

The fourth multiplicative identity has no equivalent in regular algebra because it uses the complement of a variable, a concept unique to Boolean mathematics. Since there must be one "0" value between any variable and its complement, and since the product of any Boolean quantity and 0 is 0, the product of a variable and its complement must be 0:

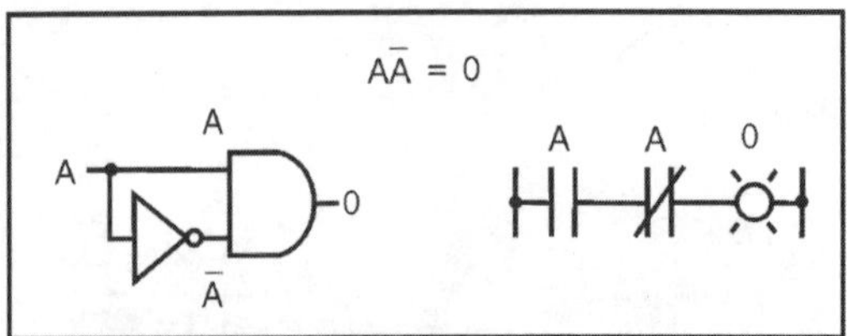

To summarize, then, we have four basic Boolean identities for addition and four for muSltiplication:

Table. Basic Boolean Algebraic Identities

Additive	Multiplicative
$A + 0 = A$	$0A = 0$
$A + 1 = 1$	$1A = A$
$A + A = A$	$AA = A$
$A + \overline{A} = 1$	$A\overline{A} = 0$

Another identity having to do with complementation is that of the *double complement*: a variable inverted twice. Complementing a variable twice (or any even number of times) results in the original Boolean value.

This is analogous to negating (multiplying by -1) in real-number algebra: an even number of negations cancel to leave the original value:

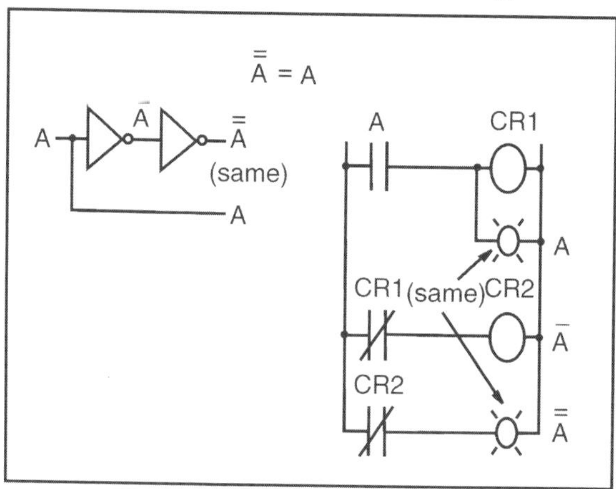

BOOLEAN ALGEBRA AND SOME COMBINATIONAL CIRCUITS

Boolean algebra is the algebra of variables that can assume two values: True and False. Conventionally we associate these as follows: True = 1 and False = 0. This association will become important when we consider the use of Boolean components to synthesise arithmetic circuits, such as a binary adder.

Formally, Boolean algebra is defined over a set of elements {0, 1}, two binary operators

{AND, OR}, and a single unary operator NOT. These operators are conventionally represented as follows:

• for AND

+ for OR

′ for NOT, thus X′ is Not(X).

The Boolean operators are completely defined by Truth Tables.

AND 0•0 = 0 OR 0+0 = 0 NOT′=1
0•1 = 0 0+1 = 1 1′ = 0
1•0 = 0 1+0 = 1
1•1 = 1 1+1 = 1

Note that the use of "+" for the OR operation is restricted to those cases in which addition is not being discussed. When addition is also important, we use different symbols for the binary Boolean operators, the most common being $\wedge$ for AND, and $\vee$ for OR.

There is another notation for the complement (NOT) function that is preferable. If X is a Boolean variable, then $\overline{X}$ is its complement, so that $\overline{X}$ = 1 and $\overline{X}$ = 0. The only reason that this author uses X′ to denote $\overline{X}$ is that the former notation is easier to create in MS-Word. There is another very handy function, called the XOR (Exclusive OR) function. Although it is not basic to Boolean algebra, it comes in quite handy in circuit design. The symbol for the Exclusive OR function is ⊕. Here is its complete definition using a truth table.

$0 \oplus 0 = 0$ $0 \oplus 1 = 1$
$1 \oplus 0 = 1$ $1 \oplus 1 = 0$

TRUTH TABLES

A truth table for a function of N Boolean variables depends on the fact that there are only 2^N different combinations of the values of these *N* Boolean variables. For small values of *N*, this allows one to list every possible value of the function. Consider a Boolean function of two Boolean variables X and Y. The only possibilities for the values of the variables are:

X = 0 and Y = 0
X = 0 and Y = 1
X = 1 and Y = 0
X = 1 and Y = 1

Similarly, there are eight possible combinations of the three variables *X*, *Y*, and *Z*, beginning with *X* = 0, *Y* = 0, *Z* = 0 and going through *X* = 1, *Y* = 1, *Z* = 1. Here they are.

X = 0, *Y* = 0, *Z* = 0 *X* = 0, *Y* = 0, *Z* = 1 *X* = 0, *Y* = 1, *Z* = 0 *X* = 0, *Y* = 1, *Z* = 1
X = 1, *Y* = 0, *Z* = 0 *X* = 1, *Y* = 0, *Z* = 1 *X* = 1, *Y* = 1, *Z* = 0 *X* = 1, *Y* = 1, *Z* = 1

As we shall see, we prefer truth tables for functions of not too many variables.

X	Y	F(X, Y)
0	0	1
0	1	0
1	0	0
1	1	1

The figure at left is a truth table for a two-variable function. Note that we have four rows in the truth table, corresponding to the four possible combinations of values for *X* and *Y*. Note also the standard order in which the values are written: 00, 01, 10, and 11. Other orders can be used when needed (it is done below), but one must list all combinations.

X	Y	Z	F1	F2
0	0	0	0	0
0	0	1	1	0
0	1	0	1	0
0	1	1	0	1
1	0	0	1	0
1	0	1	0	1
1	1	0	0	1
1	1	1	1	1

For another example of truth tables, we consider the figure at the right, which shows two Boolean functions of three Boolean variables. Truth tables can be used to define more than one function at a time, although they become hard to read if either the number of variables or the number of functions is too large. Here we use the standard shorthand of *F*1 for *F*1(*X*, *Y*, *Z*) and *F*2 for F2(*X*, *Y*, *Z*). Also note the standard ordering of the rows, beginning with 0 0 0 and ending with 1 1 1. This causes less confusion than other ordering schemes, which may be used when there is a good reason for them.

As an example of a truth table in which non-standard ordering might be useful, consider the following table for two variables. As expected, it has four rows.

X	Y	X • Y	X + Y
0	0	0	0
1	0	0	1
0	1	0	1
1	1	1	1

A truth table in this non-standard ordering would be used to prove the standard Boolean axioms:

X • 0 = 0 for all X　　　X + 0 = X for all X

X • 1 = X for all X　　　X + 1 = 1 for all X

In future lectures we shall use truth tables to specify functions without needing to consider their algebraic representations. Because 2^N is such a fast growing function, we give truth tables for functions of 2, 3, and 4 variables only, with 4, 8, and 16 rows, respectively.Truth tables for 4 variables, having 16 rows, are almost too big. For five or more variables, truth tables become unusable, having 32 or more rows.

Labeling Rows in Truth Tables

We now discuss a notation that is commonly used to identify rows in truth tables. The exact identity of the rows is given by the values for each of the variables, but we find it convenient to label the rows with the integer equivalent of the binary values.We noted above that for N variables, the truth table has 2^N rows.These are conventionally numbered from 0 through $2^N - 1$ inclusive to give us a handy way to reference the rows. Thus a two variable truth table would have four rows numbered 0, 1, 2, and 3. Here is a truth-table with labeled rows.

Row	A	B	G(A,B)
0	0	0	0
1	0	1	1
2	1	0	1
3	1	1	0

We can see that G(A, B) = A ⊕ B, but,

$$0 = 0 \bullet 2 + 0 \bullet 1$$
$$1 = 0 \bullet 2 + 1 \bullet 1$$
$$2 = 1 \bullet 2 + 0 \bullet 1$$
$$3 = 1 \bullet 2 + 1 \bullet 1$$

this value has nothing to do with the row numberings, which are just the decimal equivalents of the values in the A & B columns as binary. A three variable truth table would have eight rows, numbered 0, 1, 2, 3, 4, 5, 6, and 7. Here is a three variable truth table for a function F(X, Y, Z) with the rows numbered.

Row Number	X	Y	Z	F(X, Y, Z)
0	0	0	0	1
1	0	0	1	1
2	0	1	0	0
3	0	1	1	1
4	1	0	0	1
5	1	0	1	0
6	1	1	0	1
7	1	1	1	1

Note that the row numbers correspond to the decimal value of the three bit binary, thus

$$0 = 0 \bullet 4 + 0 \bullet 2 + 0 \bullet 1$$
$$1 = 0 \bullet 4 + 0 \bullet 2 + 1 \bullet 1$$
$$2 = 0 \bullet 4 + 1 \bullet 2 + 0 \bullet 1$$
$$3 = 0 \bullet 4 + 1 \bullet 2 + 1 \bullet 1$$
$$4 = 1 \bullet 4 + 0 \bullet 2 + 0 \bullet 1$$
$$5 = 1 \bullet 4 + 0 \bullet 2 + 1 \bullet 1$$
$$6 = 1 \bullet 4 + 1 \bullet 2 + 0 \bullet 1$$
$$7 = 1 \bullet 4 + 1 \bullet 2 + 1 \bullet 1$$

Truth tables are purely Boolean tables in which decimal numbers, such as the row numbers above do not really play a part. However, we find that the ability to label a row with a decimal number to be very convenient and so we use this. The row numberings can be quite important for the standard algebraic forms used in representing Boolean functions.

Where to Put the Ones and Zeroes

Every truth table corresponds to a Boolean expression. For some truth tables, we begin with a Boolean expression and evaluate that expression in order to find where to place the 0's and 1's. For other tables, we just place a bunch of 0's and 1's and then ask what Boolean expression we have created. The truth table just above was devised by selecting an interesting pattern of 0's and 1's. The author of these notes had no particular pattern in mind when creating it. Other truth tables are more deliberately generated.

Let's consider the construction of a truth table for the Boolean expression.

$$F(X, Y, Z) = X \bullet Y + Y \bullet Z + X \bullet \overline{Y} \bullet Z$$

Let's evaluate this function for all eight possible values of X, Y, Z.

$X = 0 \quad Y = 0 \quad Z = 0 \quad F(X, Y, Z) = 0 \bullet 0 + 0 \bullet 0 + 0 \bullet 1 \bullet 0 = 0 + 0 + 0 = 0$
$X = 0 \quad Y = 0 \quad Z = 1 \quad F(X, Y, Z) = 0 \bullet 0 + 0 \bullet 1 + 0 \bullet 1 \bullet 1 = 0 + 0 + 0 = 0$
$X = 0 \quad Y = 1 \quad Z = 0 \quad F(X, Y, Z) = 0 \bullet 1 + 1 \bullet 0 + 0 \bullet 0 \bullet 0 = 0 + 0 + 0 = 0$
$X = 0 \quad Y = 1 \quad Z = 1 \quad F(X, Y, Z) = 0 \bullet 1 + 1 \bullet 1 + 0 \bullet 0 \bullet 1 = 0 + 1 + 0 = 1$
$X = 1 \quad Y = 0 \quad Z = 0 \quad F(X, Y, Z) = 1 \bullet 0 + 0 \bullet 0 + 1 \bullet 1 \bullet 0 = 0 + 0 + 0 = 0$
$X = 1 \quad Y = 0 \quad Z = 1 \quad F(X, Y, Z) = 1 \bullet 0 + 0 \bullet 1 + 1 \bullet 1 \bullet 1 = 0 + 0 + 1 = 1$
$X = 1 \quad Y = 1 \quad Z = 0 \quad F(X, Y, Z) = 1 \bullet 1 + 1 \bullet 0 + 1 \bullet 0 \bullet 0 = 1 + 0 + 0 = 1$
$X = 1 \quad Y = 1 \quad Z = 1 \quad F(X, Y, Z) = 1 \bullet 1 + 1 \bullet 1 + 1 \bullet 0 \bullet 1 = 1 + 1 + 0 = 1$

From the above, we create the truth table for the function. Here it is.

X	Y	Z	F(X, Y, Z)
0	0	0	0
0	0	1	0
0	1	0	0
0	1	1	1
1	0	0	0
1	0	1	1
1	1	0	1
1	1	1	1

A bit later we shall study how to derive Boolean expressions from a truth table. Truth tables used as examples for this part of the course do not appear to be associated with a specific Boolean function. Often the truth tables are associated with well-known functions, but the point is to derive that function starting only with 0's and 1's.

Consider the truth table given below, with no explanation of the method used to generate the values of F1 and F2 for each row.

Table. Our Sample Functions F1 and F2

Row	X	Y	Z	F1	F2
0	0	0	0	0	0
1	0	0	1	1	0
2	0	1	0	1	0
3	0	1	1	0	1
4	1	0	0	1	0
5	1	0	1	0	1
6	1	1	0	0	1
7	1	1	1	1	1

Students occasionally ask how the author knew where to place the 0's and 1's in the above table. There are two answers to this, both equally valid. We reiterate the statement that a Boolean function is completely specified by its truth table. Thus, one can just make an arbitrary list of 2^N 0's and 1's and then decide what function of N Boolean variables has been represented. In that view, the function F2 is that function specified by the sequence

(0, 0, 0, 1, 0, 1, 1, 1) and nothing more. We can use methods described below to assign it a functional representation. Note that F2 is 1 if and only if two of X, Y, and Z are 1. Given this, we can give a functional description of the function as F2 = $X{\bullet}Y + X{\bullet}Z + Y{\bullet}Z$.

As the student might suspect, neither the pattern of 0's and 1's for F1 nor that for F2 were arbitrarily selected. The real answer is that the instructor derived the truth table from a set of known Boolean expressions, one for F1 and one for F2. The student is invited to compute the value of F2 = X•Y + X•Z + Y•Z for all possible values of X, Y, and Z; this will verify the numbers as shown in the truth table.

We have noted that a truth table of two variables has four rows (numbered 0, 1, 2, and 3) and that a truth table of three variables has eight rows (numbered 0 through 7). We now prove that a truth table of N variables has 2^N rows, numbered 0 through $2^N - 1$.

Here is an inductive proof, beginning with the case of one variable.

- Base case: a function of one variable X requires 2 rows,one row for $X = 0$ and one row for $X = 1$.
- If a function of N Boolean variables $X_1, X_2, \ldots, X_N$ requires 2^N rows, then the function of (N + 1) variables $X_1, X_2, \ldots, X_N, X_{N+1}$ would require
  2^N rows for $X_1, X_2, \ldots, X_N$ when $X_{N+1} = 0$
  2^N rows for $X_1, X_2, \ldots, X_N$ when $X_{N+1} = 1$
- $2^N + 2^N = 2^{N+1}$, so the function of $(N + 1)$ variables required 2^{N+1} rows.

While we are at it, we show that the number of Boolean functions of N Boolean variables is 2^R where $R = 2^N$, thus the number is 2^{2^N}. The argument is quite simple. We have shown that the number of rows in a truth table is given by $R = 2^N$. The value in the first row could be a 0 or 1; thus two choices. Each of the $R = 2^N$ rows could have two choices, thus the total number of functions is 2^R where $R = 2^N$.

For N = 1, R = 2, and $2^2 = 4$. A truth table for the function F(X) would have two rows, one for $X = 0$ and one for $X = 1$. There are four functions of a single Boolean variable.

$$F_1(X) = 0,\ F_2(X) = 1,\ F_3(X) = X, \text{ and } F_4(X) = \overline{X}.$$

It might be interesting to give a table of the number of rows in a truth table and number of possible Boolean functions for N variables. The number of rows grows quickly, but the number of functions grows at an astonishing rate.

N	$R = 2^N$	2^R
1	2	4
2	4	16
3	8	256
4	16	65 536
5	32	4 294 967 296

Note on computation: $\log 2 = 0.30103$, so $2^{64} = (10^{0.30103})^{64} = 10^{19.266}$
$\log 1.845 = 0.266$, so $10^{0.266} \approx 1.845$ and $10^{19.266} \approx 1.845 \bullet 10^{19}$

The number of Boolean functions of N Boolean variables is somewhat of interest. More to interest in this course is the number of rows in any possible truth-table representation of a function of N Boolean variables. For $N = 2$, 3, and 4, we have $2^N = 4$, 8, and 16 respectively, so that truth tables for 2, 3, and 4 variables are manageable. Truth tables for five variables are a bit unwieldy and truth tables for more than five variables are almost useless.

Truth Tables and Associated Tables with Don't Care Conditions

At this point, we mention a convention normally used for writing large truth tables and associated tables in which there is significant structure. This is called the "don't care" condition, denoted by a "d" in the table. When that notation appears, it indicates that the value of the Boolean variable for that slot can be either 0 or 1, but give the same effect. Let's look at two tables, each of which to be seen and discussed later in this textbook. We begin with a table that is used to describe control of memory; it has descriptive text.

Select	$R / \overline{W}$	Action
0	0	Memory not active
0	1	Memory not active
1	0	CPU writes to memory
1	1	CPU reads from memory

The two control variables are Select and. $R/\overline{W}$ But note that when Select = 0, the action of the memory is totally independent of the value of $R/\overline{W}$ .For this reason, we may write:

Select	$R/\bar{W}$	Action
0	d	Memory not active
1	0	CPU writes to memory
1	1	CPU reads from memory

Similarly, consider the truth table for a two–to–four decoder with Enable. The complete version is shown first; it has eight rows and describes four outputs: Y_0, Y_1, Y_2, and Y_3.

Enable	X_1	X_0	Y_0	Y_1	Y_2	Y_3
0	0	0	0	0	0	0
0	0	1	0	0	0	0
0	1	0	0	0	0	0
0	1	1	0	0	0	0
1	0	0	1	0	0	0
1	0	1	0	1	0	0
1	1	0	0	0	1	0
1	1	1	0	0	0	1

The more common description uses the "don't care" notation.

Enable	X_1	X_0	Y_0	Y_1	Y_2	Y_3
0	d	d	0	0	0	0
1	0	0	1	0	0	0
1	0	1	0	1	0	0
1	1	0	0	0	1	0
1	1	1	0	0	0	1

This latter form is simpler to read. The student should not make the mistake of considering the "*d*" as an algebraic value. What the first row says is that if Enable = 0, then I don't care what X_1 and X_0 are; even if they have different values, all outputs are 0.

The next section will discuss conversion of a truth table into a Boolean expression. The safest way to do this is to convert a table with "don't cares" back to the full representation.

PROPERTY OF BOOLEAN ALGEBRAIC

Another type of mathematical identity, called a "property" or a "law," describes how differing variables relate to each other in a system of numbers. One of these properties is known as the *commutative property*, and it applies equally to addition and multiplication. In essence, the commutative property tells us we can reverse the order of variables that are either added together or multiplied together without changing the truth of the expression:

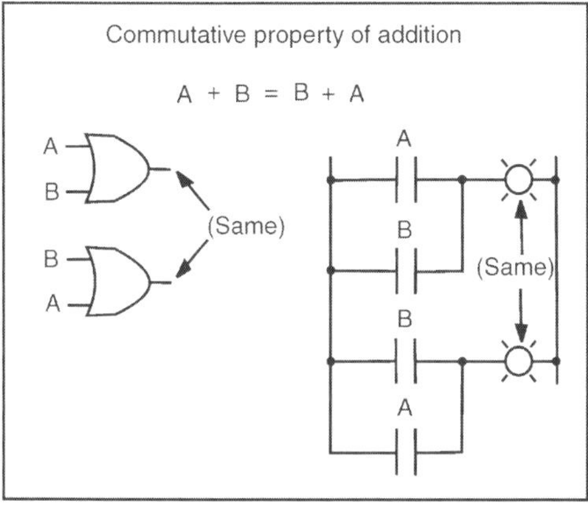

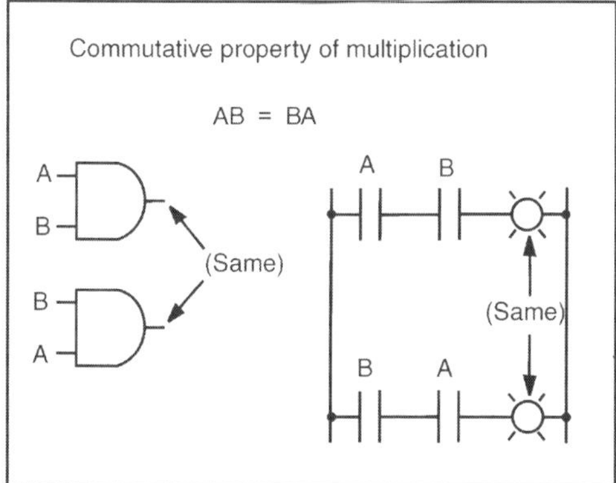

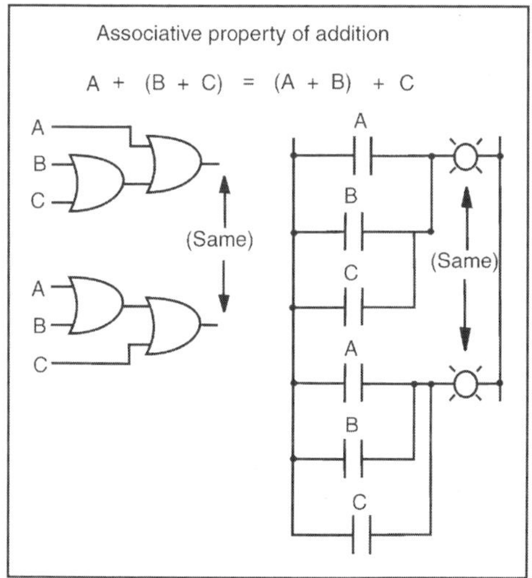

Along with the commutative properties of addition and multiplication, we have the *associative property*, again applying equally well to addition and multiplication. This property tells us we can associate groups of added or multiplied variables together with parentheses without altering the truth of the equations.

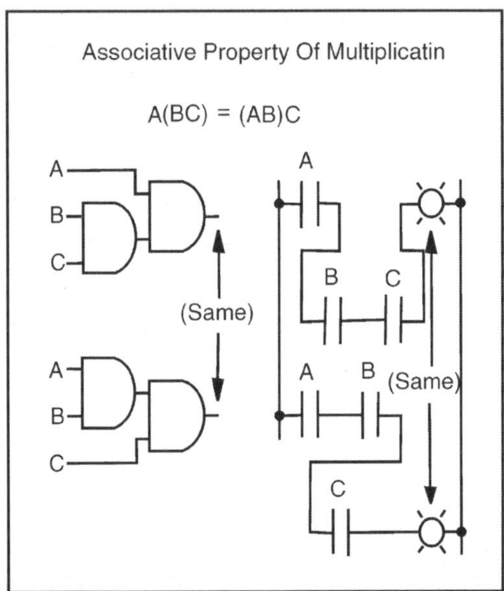

Lastly, we have the *distributive property*, illustrating how to expand a Boolean expression formed by the product of a sum, and in reverse shows us how terms may be factored out of Boolean sums-of-products:

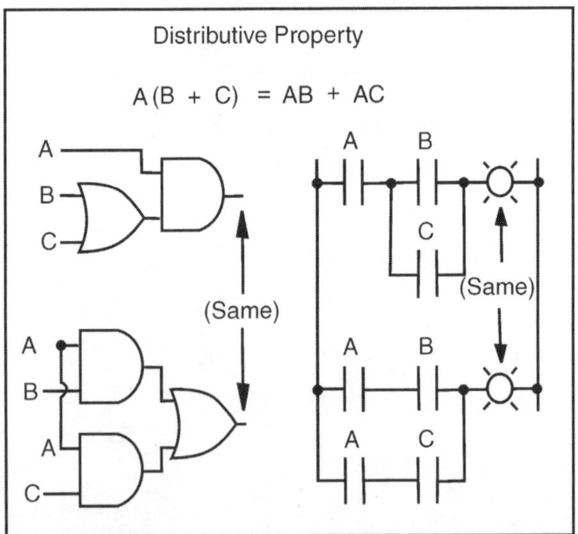

To summarize, here are the three basic properties: commutative, associative, and distributive.

Table. Basic Boolean Algebraic Properties

Additive	Multiplicative
A + B = B + A	AB = BA
A + (B + C) = (A + B) + C	A(BC) = (AB) C
A (B + C) = AB + AC	

SIMPLIFICATION OF LOGIC CIRCUITS IN BOOLEAN RULES

Boolean algebra finds its most practical use in the simplification of logic circuits. If we translate a logic circuit's function into symbolic (Boolean) form, and apply certain algebraic rules to the resulting equation to reduce the number of terms and/or arithmetic operations, the simplified equation may be translated back into circuit form for a logic circuit performing the same function with fewer components. If equivalent function may be achieved with fewer components, the result will be increased reliability and decreased cost of manufacture. To this end, there are several rules of Boolean algebra presented in this section for use in reducing expressions to their simplest forms. The identities and properties already reviewed in this chapter are very useful in Boolean simplification, and for the most part bear similarity to many identities and properties of "normal" algebra. However, the rules shown in this section are all unique to Boolean mathematics.

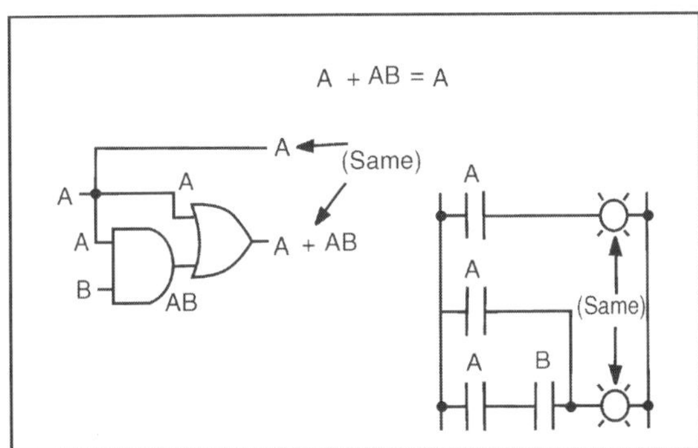

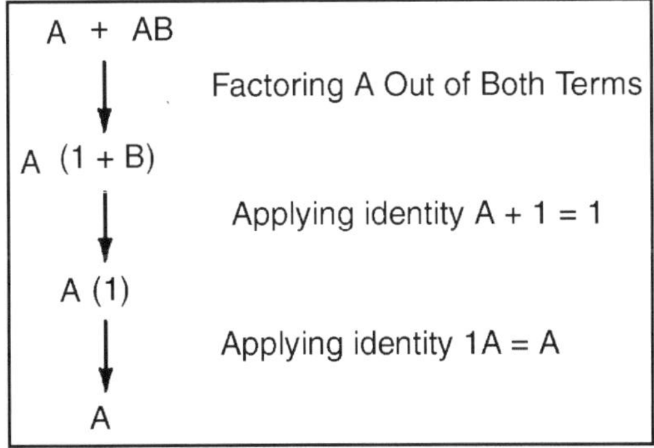

This rule may be proven symbolically by factoring an "A" out of the two terms, then applying the rules of A + 1 = 1 and 1A = A to achieve the final result:

Please note how the rule A + 1 = 1 was used to reduce the (B + 1) term to 1. When a rule like "A + 1 = 1" is expressed using the letter "A", it doesn't mean it only applies to expressions containing "A". What the "A" stands for in a rule like A + 1 = 1 is *any* Boolean variable or collection of variables. This is perhaps the most difficult concept for new students to master in Boolean simplification: applying standardized identities, properties, and rules to expressions not in standard form. For instance, the Boolean expression ABC + 1 also reduces to 1 by means of the "A + 1 = 1" identity. In this case, we recognize that the "A" term in the identity's standard form can represent the entire "ABC" term in the original expression. The next rule looks similar to the first on shown in this section, but is actually quite different and requires a more clever proof:

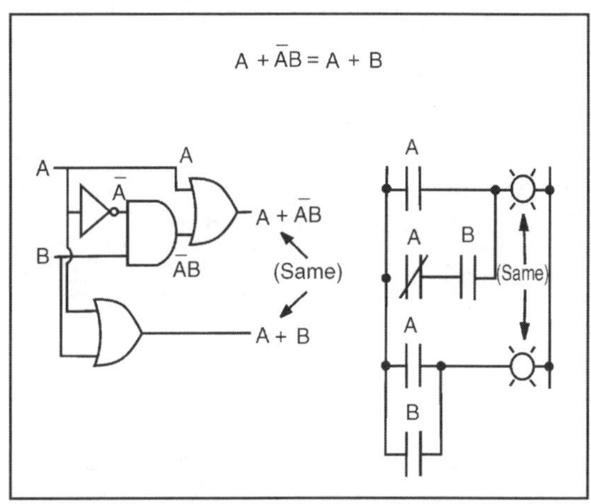

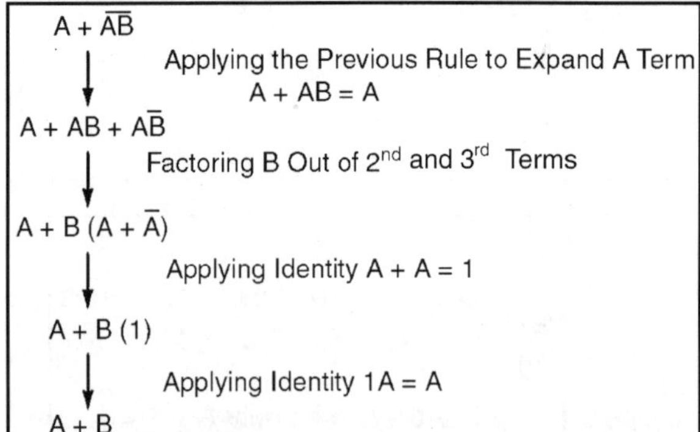

Note how the last rule (A + AB = A) is used to "un-simplify" the first "A" term in the expression, changing the "A" into an "A + AB".

While this may seem like a backward step, it certainly helped to reduce the expression to something simpler! Sometimes in mathematics we must take "backward" steps to achieve the most elegant solution. Knowing when to take such a step and when not to is part of the art-form of algebra, just as a victory in a game of chess almost always requires calculated sacrifices.

Another rule involves the simplification of a product-of-sums expression:

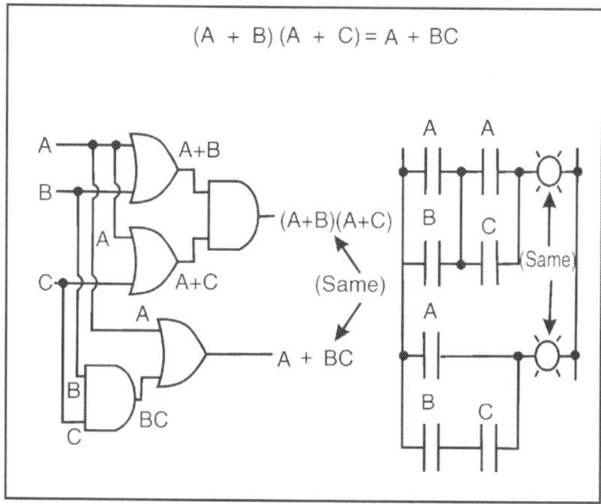

And, the corresponding proof:

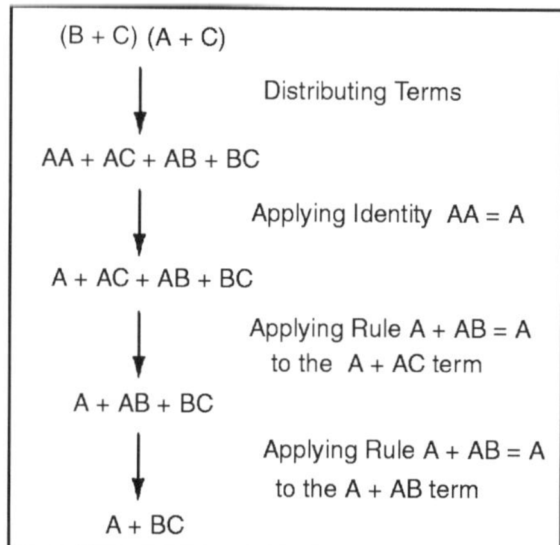

To summarize, here are the three new rules of Boolean simplification expounded in this section:

$$A + AB = A$$

$$A + \bar{A}B = A + B$$

$$(A + B)(A + C) = A + BC$$

CIRCUIT SIMPLIFICATION EXAMPLES

Let's begin with a semiconductor gate circuit in need of simplification. The "A," "B," and "C" input signals are assumed to be provided from switches, sensors, or perhaps other gate circuits.

Where these signals originate is of no concern in the task of gate reduction.

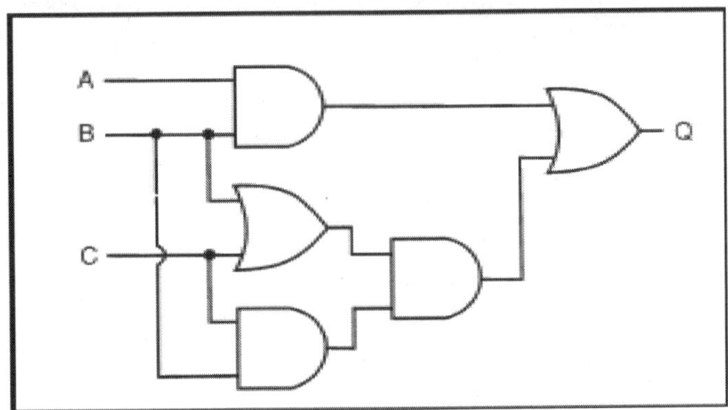

Our first step in simplification must be to write a Boolean expression for this circuit. This task is easily performed step by step if we start by writing sub-expressions at the output of each gate, corresponding to the respective input signals for each gate. Remember that OR gates are equivalent to Boolean addition, while AND gates are equivalent to Boolean multiplication.

For example, I'll write sub-expressions at the outputs of the first three gates:

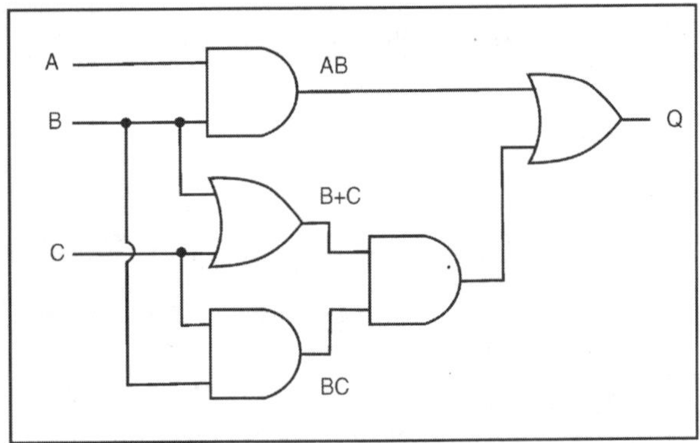

... then another sub-expression for the next gate:

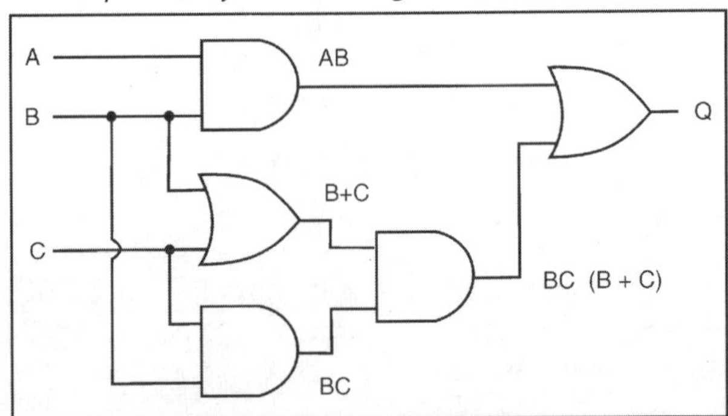

Finally, the output ("Q") is seen to be equal to the expression AB + BC(B + C):

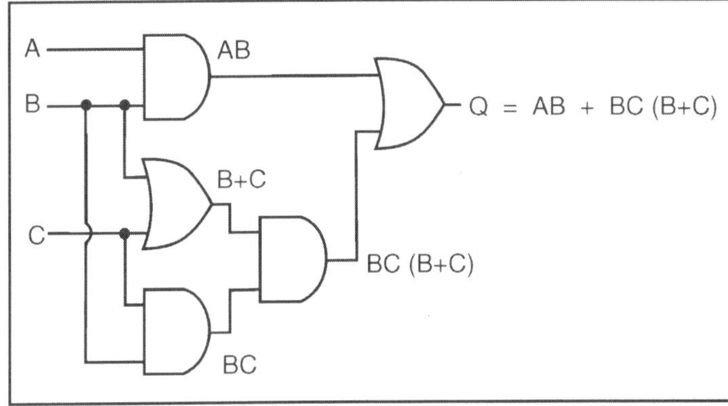

Now that we have a Boolean expression to work with, we need to apply the rules of Boolean algebra to reduce the expression to its simplest form (simplest defined as requiring the fewest gates to implement):

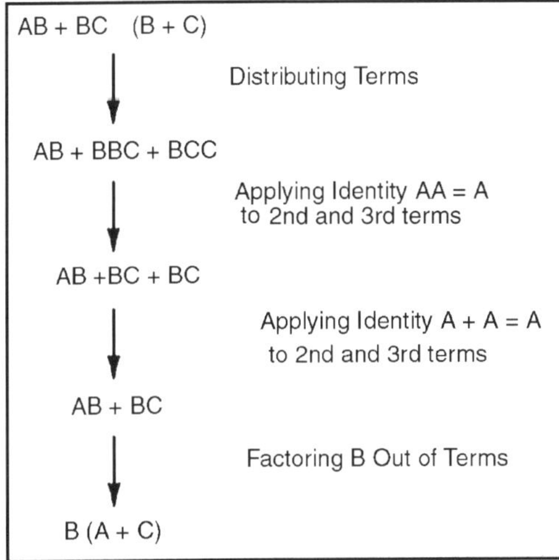

The final expression, B(A + C), is much simpler than the original, yet performs the same function. If you would like to verify this, you may generate a truth table for both expressions and determine Q's status (the circuits' output) for all eight logic-state combinations of A, B, and C, for both circuits. The two truth tables should be identical.

Now, we must generate a schematic diagram from this Boolean expression. To do this, evaluate the expression, following proper mathematical order of operations (multiplication before addition, operations inside parentheses before anything else), and draw gates for each step. Remember again that OR gates are equivalent to Boolean addition, while AND gates are equivalent to Boolean multiplication. In this case, we would begin with the sub-expression "A + C", which is an OR gate:

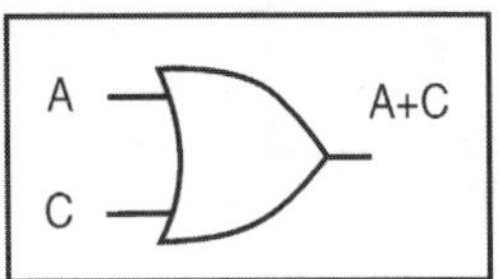

The next step in evaluating the expression "B(A + C)" is to multiply (AND gate) the signal B by the output of the previous gate (A + C):

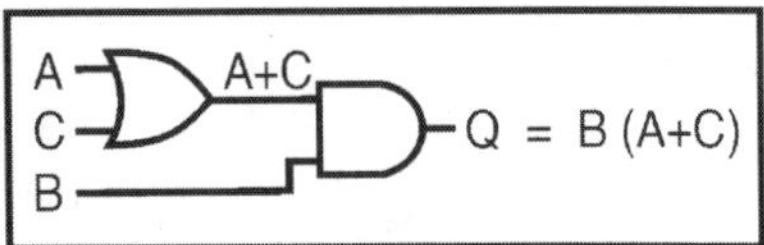

Obviously, this circuit is much simpler than the original, having only two logic gates instead of five. Such component reduction results in higher operating speed (less delay time from input signal transition to output signal transition), less power consumption, less cost, and greater reliability. Electromechanical relay circuits, typically being slower, consuming more electrical power to operate, costing more, and having a shorter average life than their semiconductor counterparts, benefit dramatically from Boolean simplification. Let's consider an example circuit:

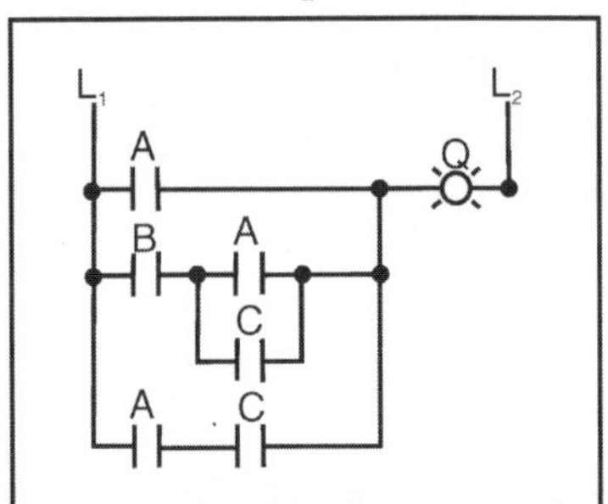

As before, our first step in reducing this circuit to its simplest form must be to develop a Boolean expression from the schematic.

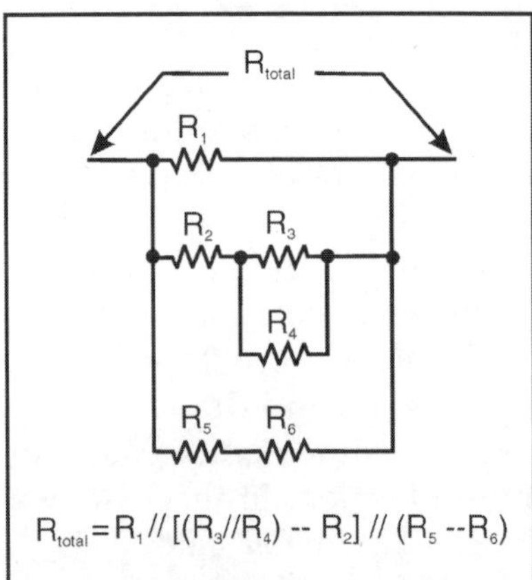

The easiest way I've found to do this is to follow the same steps I'd normally follow to reduce a series-parallel resistor network to a single, total resistance. For example, examine the following resistor network with its resistors arranged in the same connection pattern as the relay contacts in the former circuit, and corresponding total resistance formula:

Remember that parallel contacts are equivalent to Boolean addition, while series contacts are equivalent to Boolean multiplication. Write a Boolean expression for this relay contact circuit, following the same order of precedence that you would follow in reducing a series-parallel resistor network to a total resistance. It may be helpful to write a Boolean sub-expression to the left of each ladder "rung," to help organize your expression-writing:

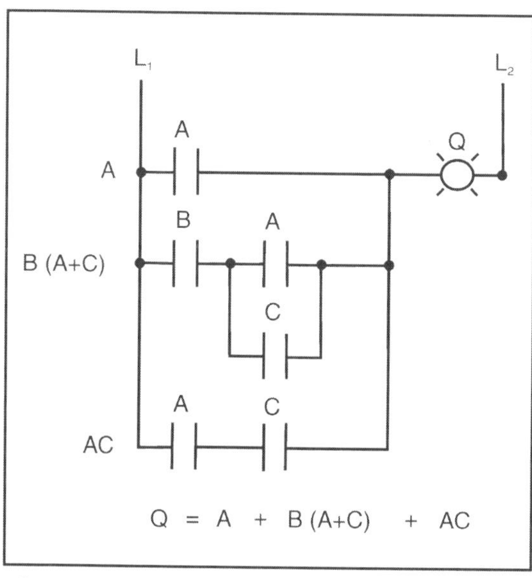

Now that we have a Boolean expression to work with, we need to apply the rules of Boolean algebra to reduce the expression to its simplest form (simplest defined as requiring the fewest relay contacts to implement):

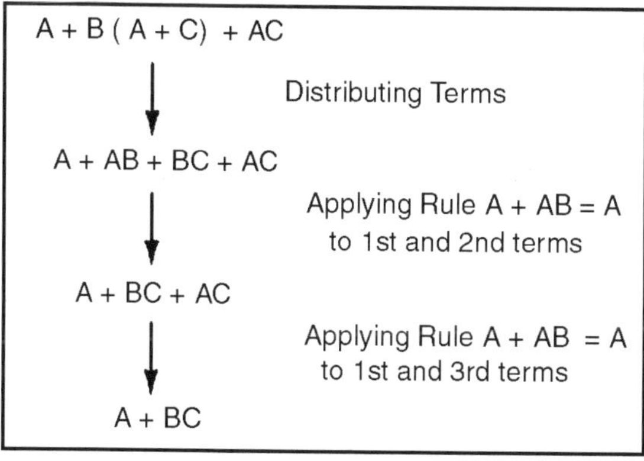

The more mathematically inclined should be able to see that the two steps employing the rule "A + AB = A" may be combined into a single step, the rule being expandable to:

"A + AB + AC + AD +... = A"

A + B (A + C) + AC

↓ Distributing Terms

A + AB + BC + AC

↓ Applying (Expanded) Rule A + AB = A to 1st, 2nd, and 4th terms

A + BC

As you can see, the reduced circuit is much simpler than the original, yet performs the same logical function:

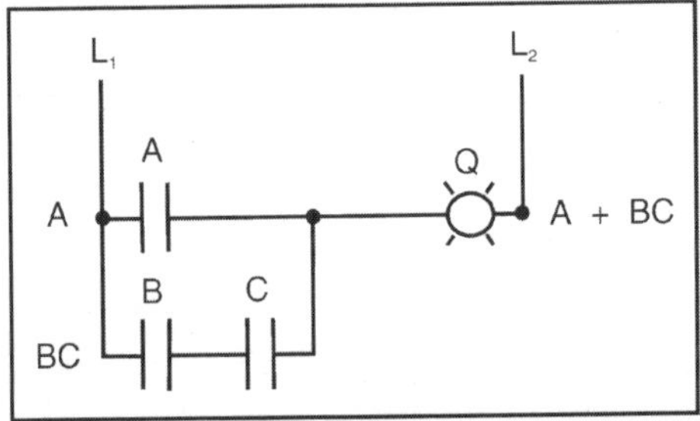

REVIEW

- To convert a gate circuit to a Boolean expression, label each gate output with a Boolean sub-expression corresponding to the gates' input signals, until a final expression is reached at the last gate.
- To convert a Boolean expression to a gate circuit, evaluate the expression using standard order of operations: multiplication before addition, and operations within parentheses before anything else.
- To convert a ladder logic circuit to a Boolean expression, label each rung with a Boolean sub-expression corresponding to the contacts' input signals, until a final expression is reached at the last coil or light. To determine proper order of evaluation, treat the contacts as though they were resistors, and as if you were determining total resistance of the series-parallel network formed by them. In other words, look for contacts that are either *directly* in series or *directly* in parallel with each other first, then "collapse" them into equivalent Boolean sub-expressions before proceeding to other contacts.
- To convert a Boolean expression to a ladder logic circuit, evaluate the expression using standard order of operations: multiplication before addition, and operations within parentheses before anything else.

EVALUATION OF BOOLEAN EXPRESSIONS

Here is another topic that this instructor normally forgets to mention, as it is so natural to one who has been in the "business" for many years. The question to be addressed now is: "What are the rules for evaluating Boolean expressions?"

OPERATOR PRECEDENCE

The main question to be addressed is the relative precedence of the basic Boolean operators: AND, OR, and NOT. This author is not aware of the relative precedence of the XOR operator and prefers to use parentheses around an XOR expression when there is any doubt.

The relative precedence of the operators is:

- NOT do this first
- AND
- OR do this last

Consider the Boolean expression A•B + C•D, often written as AB + CD. Without the precedence rules, there are two valid interpretations: either (A•B) + (C•D) or A•(B + C)•D. The precedence rules for the operators indicate that the first is the correct interpretation; in this Boolean algebra follows standard algebra as taught in high-school. Consider now the expression $\overline{A} \bullet B + C \bullet \overline{D}$; according to our rules, this is read as.

$$\left(\left(\overline{A}\right) \bullet B\right) + \left(C \bullet \left(\overline{D}\right)\right)$$

Note that parentheses and explicit extension of the NOT overbar can override the precedence rules, so that *A*·(*B* + *C*)·*D* is read as the logical AND of three terms: *A*, (*B* + *C*), and *D*. Note also that the two expressions $\overline{A \bullet B}$ and $\overline{A} \bullet \overline{B}$ are different. The first expression, better written as $\overline{(A \bullet B)}$, refers to the logical NOT of the logical AND of A and B; in a language such as LISP it would be written as NOT (AND A B). The second expression, due to the precedence rules, refers to the logical AND of the logical NOT of A and the logical NOT of B; in LISP this might be written as AND((NOT A) (NOT B)).

Evaluation of Boolean expressions implies giving values to the variables and following the precedence rules in applying the logical operators. Let *A* = 1, *B* = 0, *C* = 1, and *D* = 1.

$$A \bullet B + C \bullet D = 1 \bullet 0 + 1 \bullet 1 = 0 + 1 = 1$$

$$A \bullet (B + C) \bullet D = 1 \bullet (0 + 1) \bullet 1 = 1 \bullet 1 \bullet 1 = 1 \; \overline{A} \bullet B + C \bullet \overline{D}$$

$$= \overline{1} \bullet 0 + 1 \bullet \overline{1} = 0 \bullet 0 + 1 \bullet 0 = 0 + 0 = 0$$

$$\overline{A \bullet B} = \overline{1 \bullet 0} = \overline{0} = 0$$

$$\overline{A} \bullet \overline{B} = \overline{1} \bullet \overline{0} = 0 \bullet 1 = 0$$

Also

$$A \bullet (B + C \bullet D) = 1 \bullet (0 + 1 \bullet 1) = 1 \bullet (0 + 1) = 1 \bullet 1 = 1$$

$$(A \bullet B + C) \bullet D = (1 \bullet 0 + 1) \bullet 1 = (0 + 1) \bullet 1 = 1$$

The Basic Axioms and Postulates of Boolean Algebra

We close our discussion of Boolean algebra by giving a formal definition of the algebra along with a listing of its basic axioms and postulates.

Definition: A Boolean algebra is a closed algebraic system containing a set K of two or more elements and three operators, two binary and one unary. The binary operators are denoted "+" for OR and "•" for AND. The unary operator is NOT, denoted by a overbar placed over the variable. The system is closed, so that for all X and Y in K, we have $X + Y$ in K, $X \bullet Y$ in K and NOT(X) in K. The set K must contain two constants 0 and 1 (FALSE and TRUE), which obey the postulates below. In normal use, we say $K = \{0, 1\}$ – set notation.

The postulates of Boolean algebra are:

*P*1 Existence of 0 and 1 *a*) $X + 0 = X$ for all X
b) $X \bullet 1 = X$ for all X

*P*2 Commutativity *a*) $X + Y = Y + X$ for all X and Y
b) $X \bullet Y = Y \bullet X$ for all X and Y

*P*3 Associativity *a*) $X + (Y + Z) = (X + Y) + Z$, for all X, Y, and Z
b) $X \bullet (Y \bullet Z) = (X \bullet Y) \bullet Z$, for all X, Y, and Z

*P*4 Distributivity *a*) $X \bullet (Y + Z) = (X \bullet Y) + (X \bullet Z)$, for all X, Y,
b) $X + (Y \bullet Z) = (X + Y) \bullet (X + Z)$, for all X, Y, Z

*P*5 Complement *a*) $X + \overline{X} = 1$, for all X
b) $X \bullet \overline{X} = 0$, for all X

The theorems of Boolean algebra are:

*T*1 Idempotency *a*) $X + X = X$, for all X
b) $X \bullet X = X$, for all X

*T*2 1 and 0 Properties *a*) $X + 1 = 1$, for all X
b) $X \bullet 0 = 0$, for all X

*T*3 Absorption *a*) $X + (X \bullet Y) = X$, for all X and Y
b) $X \bullet (X + Y) = X$, for all X and Y

*T*4 Absorption *a*) $X + \overline{X} \bullet Y = X + Y$, for all X and Y
b) $X \bullet (\overline{X} + Y) = X \cdot Y$, for all X and Y

*T*5 DeMorgan's Laws a)= $\overline{(X+Y)}\ \overline{X} \bullet \overline{Y}$
b) $\overline{(X \bullet Y)} = \overline{X} + \overline{Y}$

*T*6 Consensus a) $X \bullet Y + \overline{X} \bullet Z + Y \bullet Z = X \bullet Y + \overline{X} \bullet Z$
b) $(X+Y) \bullet (\overline{X}+Z) \bullet (Y+Z) = (X+Y) \bullet (\overline{X}+Z)$

The observant student will note that most of the postulates, excepting only P4b, seem to be quite reasonable and based on experience in high-school algebra. Some of the theorems seem reasonable or at least not sufficiently different from high school algebra to cause concern, but others such as *T1* and *T2* are decidedly different. Most of the unsettling differences have to do with the properties of the Boolean OR function, which is quite different from standard addition, although commonly denoted with the same symbol.

The principle of duality is another property that is unique to Boolean algebra – there is nothing like it in standard algebra. This principle says that if a statement is true in Boolean algebra, so is its dual.The dual of a statement is obtained by changing ANDs to ORs, ORs to ANDs, 0s to 1s, and 1s to 0s. In the above, we can arrange the postulates as duals.

Postulate	**Dual**
$0\bullet X=0$	$1 + X=1$
$1\bullet X=X$	$0 + X=X$
$0 + X=X$	$1\bullet X=X$
$1 + X=1$	$0\bullet X=0$

These last two statements just repeat the first two if one reads correctly.

Both standard algebra and Boolean algebra have distributive postulates. Standard algebra has one distributive postulate; Boolean algebra must have two distributive postulates as a result of the principle of duality. In standard algebra, we have the equality $A\bullet(B + C) = A\bullet B + A\bullet C$ for all values of A, B, and C. This is the distributive postulate as seen in high school; we know it and expect it. In Boolean algebra we have the distributive postulate $A\cdot(B + C) = A\bullet B + A\bullet C$, which looks familiar. The principle of duality states that if $A\bullet(B + C) = A\bullet B + A\bullet C$ is true then the dual statement $A + B\bullet C = (A + B)\bullet(A + C)$ must also be true. We prove the second statement using a method unique to Boolean algebra. This method depends on the fact that there are only two possible values for A: $A = 0$ and $A = 1$. We consider both cases using a proof technique much favoured by this instructor: consider both possibilities for one variable. If $A = 1$, the statement becomes $1 + B\bullet C = (1 + B)\bullet(1 + C)$, or $1 = 1\bullet 1$, obviously true. If $A = 0$, the statement becomes $0 + B\bullet C = (0 + B)\bullet(0 + C)$, or $B\bullet C = B\bullet C$.

Just for fun, we offer a truth-table proof of the second distributive postulate.

Table. 8.3 $A + B\bullet C = (A + B)\bullet(A + C)$

A	B	C	B·C	A + B·C	(A + B)	(A + C)	(A + B)•(A + C)
0	0	0	0	0	0	0	0
0	0	1	0	0	0	1	0
0	1	0	0	0	1	0	0
0	1	1	1	1	1	1	1
1	0	0	0	1	1	1	1
1	0	1	0	1	1	1	1
1	1	0	0	1	1	1	1
1	1	1	1	1	1	1	1

Note that to complete the proof, one must construct the truth table, showing columns for each of the two functions $A + B \bullet C$ and $(A + B) \bullet (A + C)$, then note that the contents of the two columns are identical for each row.

A Few More Proofs

At this point in the course, we note two facts:

1. That some of the theorems above look a bit strange.
2. That we need more work on proof of Boolean theorems.

Towards that end, let's look at a few variants of the Absorption Theorem. We prove that,

$$X + (X \bullet Y) = X, \text{ for all } X \text{ and } Y$$

The first proof will use a truth table. Remember that the truth table will have four rows, corresponding to the four possible combinations of values for X and Y:

$X = 0$ and $Y = 0$, $X = 0$ and $Y = 1$, $X = 1$ and $Y = 0$, $X = 1$ and $Y = 1$.

X	**Y**	**X • Y**	**X + (X•Y)**
0	0	0	0
0	1	0	0
1	0	0	1
1	1	1	1

Having computed the value of $X + (X \bullet Y)$ for all possible combinations of X and Y, we note that in each case we have $X + (X \bullet Y) = X$. Thus, we conclude that the theorem is true.

Here is another proof method much favoured by this author. It depends on the fact that there are only two possible values of X: $X = 0$ and $X = 1$. A theorem involving the variable X is true if and only if it is true for both $X = 0$ and $X = 1$. Consider the above theorem, with the claim that $X + (X \bullet Y) = X$. We look at the cases $X = 0$ and $X = 1$, computing both the LHS (Left Hand Side of the Equality) and RHS (Right Hand Side of the Equality).

If $X = 0$, then $X + (X \bullet Y) = 0 + (0 \bullet Y) = 0 + 0 = 0$,

and LHS = RHS.

If $X = 1$, then $X + (X \bullet Y) = 1 + (1 \bullet Y) = 1 + Y = 1$,

and LHS = RHS.

Consider now a trivial variant of the absorption theorem. $\overline{X} + (\overline{X} \bullet Y) = \overline{X}$, for all X and Y

There are two ways to prove this variant of the theorem., given that the above standard statement of the absorption theorem is true. An experienced mathematician would claim that the proof is obvious. We shall make it obvious.

The absorption theorem, as proved, is $X + (X \bullet Y) = X$, for all X and Y. We restate the theorem, substituting the variable W for X, getting $W + (W \bullet Y) = W$, for all X and Y. Now, we let $W = \overline{X}$, and the result is indeed obvious.

X	Y	$\overline{X} \bullet Y$	$X + (\overline{X} \bullet Y)$	X + Y
0	0	0	0	0
0	1	1	1	1
1	0	0	1	1
1	1	0	1	1

We close this section with a proof of the other standard variant of the absorption theorem, that $X + (\overline{X} \bullet Y) = X + Y$ for all X and Y. The theorem can also be proved using the two case method.

Yet Another Sample Proof

While this instructor seems to prefer the obscure "two-case" method of proof, most students prefer the truth table approach. We shall use the truth table approach to prove one of the variants of DeMorgan's laws: $= \overline{(X \bullet Y)} = \overline{X} + \overline{Y}$

X	Y	$X \bullet Y$	$\overline{(X \bullet Y)}$	$\overline{X}$	$\overline{Y}$	$\overline{X} + \overline{Y}$	**Comment**
0	0	0	1	1	1	1	Same
0	1	0	1	1	0	1	Same
1	0	0	1	0	1	1	Same
1	1	1	0	0	0	0	Same

In using a truth table to show that two expressions are equal, one generates a column for each of the functions and shows that the values of the two functions are the same for every possible combination of the input variables. Here, we have computed the values of both

$\overline{(X \bullet Y)}$ and $(\overline{X} + \overline{Y})$ for all possible combinations of the values of X and Y and shown that for every possible combination the two functions have the same value. Thus they are equal.

SOME BASIC FORMS: PRODUCT OF SUMS AND SUM OF PRODUCTS

We conclude our discussion of Boolean algebra by giving a formal definition to two notations commonly used to represent Boolean functions:

SOP	Sum of Products
POS	Product of Sums

We begin by assuming the definition of a variable. A literal is either a variable in its true form or its complemented form, examples are *A*, *C′*, and *D*.

A product term is the logical AND of one or more literals; a sum term is the logical OR of one or more literals. According to the strict definition the single literal *B′* is both a product term and sum term.

A sum of products (SOP) is the logical OR of one or more product terms.

A product of sums (POS) is the logical AND of one or more sum terms.

Sample SOP:

$$A + B \bullet C \quad A \bullet B + A \bullet C + B \bullet C \quad A \bullet B + A \bullet C + A \bullet B \bullet C$$

Sample POS:

$$A \bullet (B + C) \quad (A + B) \bullet (A + C) \bullet (A + B) \bullet (A + C) \bullet (A + B) \bullet (A + C)(A + B + C)$$

The student will note a few oddities in the above definition. These are present in order to avoid special cases in some of the more general theorems. The first oddity is the mention of the logical AND of one term and the logical OR of one term; each of these operators is a binary operator and requires two input variables. What we are saying here is that if we take a single literal as either a sum term or a product term, our notation is facilitated. Consider the expression $(A + B + C)$, which is either a POS or SOP expression depending on its use. As a POS expression, it is a single sum term comprising three literals. As an SOP expression, it is the logical OR of three product terms, each of which comprising a single literal. For cases such as these the only rule is to have a consistent interpretation.

We now consider the concept of normal and canonical forms.

These forms apply to both Sum of Products and Produce of Sums expressions, so we may have quite a variety of expression types including the following.

- Not in any form at all.
- Sum of Products, but not normal.
- Product of Sums, but not normal.
- Normal Sum of Products.
- Normal Product of Sums.
- Canonical Sum of Products.
- Canonical Product of Sums.

In order to define the concept of a normal form, we must consider the idea of inclusion.

A product term X is included in another product term Y if every literal that is in X is also in Y. A sum term X is included in another sum term Y if every literal that is in X is also in Y. For inclusion, both terms must be product terms or both must be sum terms.

Consider the SOP formula $A \bullet B + A \bullet C + A \bullet B \bullet C$. Note that the first term is included in the third term as is the second term. The third term $A \bullet B \bullet C$ contains the first term $A \bullet B$, etc.

Consider the POS formula $(A + B) \bullet (A + C) \bullet (A + B + C)$. Again, the first term (A + B) is included in the third term (A + B + C), as is the second term.

An extreme form of inclusion is observed when the expression has identical terms. Examples of this would be the SOP expression $A \bullet B + A \bullet C + A \bullet B$ and the POS expression $(A + B) \bullet (A + C) \bullet (A + C)$. Each of these has duplicate terms, so that the inclusion is 2-way. The basic idea is that an expression with included (or duplicate) terms is not written in the simplest possible form. The idea of simplifying such expressions arises from the theorems of Boolean algebra, specifically the following two.

T1 Idempotency *a*) $X + X = X$, for all X

b) $X \bullet X = X$, for all X

T3 Absorption *a*) $X + (X \bullet Y) = X$, for all X and Y

b) $X \bullet (X + Y) = X$, for all X and Y

As a direct consequence of these theorems, we can perform the following simplifications.

$A \bullet B + A \bullet C + A \bullet B = A \bullet B + A \bullet C$

$(A + B) \bullet (A + C) \bullet (A + C) = (A + B)(A + C)$

$A \bullet B + A \bullet C + A \bullet B \bullet C = A \bullet B + A \bullet C$

$(A + B) \bullet (A + C) \bullet (A + B + C) = (A + B) \bullet (A + C)$

We now consider these two formulae with slight alterations:A•B+A•C+A'•B•Cand(A + B)•(A + C)•(A' + B + C). Since the literal must be included exactly, neither of formulae in this second set contains included terms.

We now can produce the definitions of normal forms. A formula is in a normal form only if it contains no included terms; thus, a normal SOP form is a SOP form with no included terms and a normal POS form is a POS form with no included terms.

Sample Normal SOP:

$A + B \bullet C$ $\quad A \bullet B + A \bullet C + B \bullet C$ $\quad A' \bullet B + A \bullet C + B \bullet C'$

Sample Normal POS:

$A \bullet (B + C)$ $\quad (A + B) \bullet (A + C) \bullet (B + C)$ $\quad (A' + B) \bullet (A + C')$

We now can define the canonical forms. A normal form over a number of variables is in canonical form if every term contains each variable in either the true or complemented form. A canonical SOP form is a normal SOP form in which every product term contains a literal for every variable. A canonical POS form is a normal POS form in which every sum term in which every sum term contains a literal for every variable.

Note that all canonical forms are also normal forms. Canonical forms correspond directly to truth tables and can be considered as one-to-one translations of truth tables. Here are the rules for converting truth tables to canonical forms.

To produce the Sum of Products representation from a truth table, follow this rule.

- Generate a product term for each row where the value of the function is 1.
- The variable is complemented if its value in the row is 0, otherwise it is not.

To produce the Product of Sums representation from a truth table, follow this rule.

- Generate a sum term for each row where the value of the function is 0.
- The variable is complemented if its value in the row is 1, otherwise it is not.

As an example of conversion from a truth table to the normal forms, consider the two Boolean functions *F*1 and *F*2, each of three Boolean variables, denoted *A*, *B*, and *C*. Note that a truth table for a three-variable function requires $2^3 = 8$ rows.

Recall that the truth table forms a complete specification of both *F*1 and *F*2. We may elect to represent each of *F*1 and F2 in either normal or canonical form, but that is not required.

Row	A	B	C	F1	F2
0	0	0	0	0	0
1	0	0	1	1	0
2	0	1	0	1	0
3	0	1	1	0	1
4	1	0	0	1	0
5	1	0	1	0	1
6	1	1	0	0	1
7	1	1	1	1	1

There are two ways to represent the canonical forms. We first present a pair of forms that this author calls the Σ–list and Π–list. The Σ–list is used to represent canonical SOP and the Π–list is used to represent canonical POS forms.

To generate the Σ–list, we just list the rows of the truth table for which the function has a value of 1. In the truth table, we have the following.

*F*1 has value 1 for rows 1, 2, 4, and 7; so $F1 = \Pi(1, 2, 4, 7)$.

*F*2 has value 1 for rows 3, 5, 6, and 7; so $F2 = \Pi(3, 5, 6, 7)$.

To generate the Π–list, we just list the rows of the truth table for which the function has a value of 0. In the truth table, we have the following.

*F*1 has value 1 for rows 0, 3, 5, and 6; so $F1 = \Pi(0, 3, 5, 6)$.

*F*2 has value 1 for rows 0, 1, 2, and 4; so $F2 = \Pi(0, 1, 2, 4)$.

Note that conversion directly between the Σ–list and Π–list forms is easy if one knows how many variables are involved. Here we have 3 variables, with 8 rows numbered 0 through 7. Thus, if $F1 = \Sigma(1, 2, 4, 7)$, then $F1 = \Pi(0, 3, 5, 6)$ because the numbers 0, 3, 5, and 6 are the only numbers in the range from 0 to 7 inclusive that are not in the Σ–list. The conversion rule works both ways. If $F2 = \Pi(0, 1, 2, 4)$, then $F2 = \Sigma(3, 5, 6, 7)$ because the numbers 3, 5, 6, and 7 are the only numbers in the range from 0 to 7 inclusive not in the Π–list.

We now address the generation of the canonical SOP form of *F*1 from the truth table. The rule is to generate a product term for each row for which the function value is 1. Reading from the top of the truth table, the first row of interest is row 1.

A	B	C	F1
0	0	1	1

We have $A = 0$ for this row, so the corresponding literal in the product term is A'.

We have $B = 0$ for this row, so the corresponding literal in the product term is B'.

We have $C = 1$ for this row, so the corresponding literal in the product term is C.

The product term generated for this row in the truth table is $A' \bullet B' \bullet C$.

We now address the generation of the canonical POSP form of $F1$ from the truth table.The rule is to generate a sum term for each row for which the function value is 0. Reading from the top of the truth table, the first row of interest is row 1.

A	B	C	F1
0	0	0	0

We have $A = 0$ for this row, so the corresponding literal in the product term is A.

We have $B = 0$ for this row, so the corresponding literal in the product term is B.

We have $C = 1$ for this row, so the corresponding literal in the product term is C.

The product term generated for this row in the truth table is $A + B + C$.

Thus we have the following representation as Sum of Products,

$$F1 = A' \bullet B' \bullet C + A' \bullet B \bullet C + A \bullet B' \bullet C' + A \bullet B \bullet C$$

$$F2 = A' \bullet B \bullet C + A \bullet B' \bullet C + A \bullet B \bullet C' + A \bullet B \bullet C$$

We have the following representation as Product of Sums

$$F1 = (A + B + C) \bullet (A + B' + C') \bullet (A' + B + C') \bullet (A' + B' + C)$$

$$F2 = (A + B + C) \bullet (A + B + C') \bullet (A + B' + C) \bullet (A' + B + C)$$

The choice of representations depends on the number of terms. For N Boolean variables

Let C_{SOP} be the number of terms in the Sum of Products representation. Let C_{POS} be the number of terms in the Product of Sums representation.

Then $C_{SOP} + C_{POS} = 2^N$. In the above example $C_{SOP} = 4$ and $C_{POS} = 4$, so either choice is equally good. However, if the Sum of Products representation has 6 terms, then the Product of Sums representation has only 2 terms, which might recommend it.

NON-CANONICAL FORMS

The basic definition of a canonical form (both SOP and POS) is that every term contain each variable, either in the negated or non-negated state. In the example above, note that every product term in the SOP form contains all three variables *A*, *B*, and C in some form. The same holds for the POS forms.

It is often the case that a non-canonical form is simpler. For example, one can easily show that $F2(A, B, C) = A \bullet B + A \bullet C + B \bullet C$. Note that in this simplified form, not all variables appear in each of the terms.Thus, this is a normal SOP form, but not a canonical form.

The simplification can be done by one of three ways: algebraic methods, Karnaugh Maps (*K*-Maps), or the Quine-McCluskey procedure.Here we present a simplification of *F*2(*A*, *B*, *C*) by the algebraic method with notes to the appropriate postulates and theorems.

The idempotency theorem, states that for any Boolean expression *X*, we have

$X + X = X$. Thus $X = X + X = X + X + X$ and $A \bullet B \bullet C = A \bullet B \bullet C + AB \bullet C + A \bullet B \bullet C$, as $A \bullet B \bullet C$ is a valid Boolean expression for any values of *A*, *B*, and *C*.

$F2 = A' \bullet B \bullet C + A \bullet B' \bullet C + A \bullet B \bullet C' + A \bullet B \bullet C$ Original form

$= A' \bullet B \bullet C + A \bullet \mathbf{B'} \bullet C + A \bullet B \bullet C' + A \bullet B \bullet C$

$+ A \bullet B \bullet C + A \bullet B \bullet C$ Idempotence

$= A' \bullet B \bullet C + A \bullet B \bullet C + A \bullet B' \bullet C + A \bullet B \bullet C$

$+ A \bullet B \bullet C' + A \bullet B \bullet C$ Commutativity

$= (A' + A) \bullet B \bullet C + A \bullet (B' + B) \bullet C$

$+ A \bullet B \bullet (C' + C)$ Distributivity

$= 1 \bullet B \bullet C + A \bullet 1 \bullet C + A \bullet B \bullet 1$

$= B \bullet C + A \bullet C + A \bullet B$

$= A \bullet B + A \bullet C + B \bullet C$ Commutativity

The main reason to simplify Boolean expressions is to reduce the number of logic gates required to implement the expressions. This was very important in the early days of computer engineering when the gates themselves were costly and prone to failure. It is less of a concern these days.

More On Conversion Between Truth Tables and Boolean Expressions

We now give a brief discussion on the methods used by this author to derive Boolean expressions from truth tables and truth tables from canonical form Boolean expressions. We shall do this be example.Consider the truth table expression for our function *F*1.

The truth table for F1 is shown below.

Row	A	B	C	F1
0	0	0	0	0
1	0	0	1	1
2	0	1	0	1
3	0	1	1	0
4	1	0	0	1
5	1	0	1	0
6	1	1	0	0
7	1	1	1	1

The rows for which the function F1 has value 1 are shown in bold font. As noted above, we can write *F*1 in the Σ-list form as $F1 = \Sigma(1, 2, 4, 7)$. To convert this form to canonical SOP, we just replace the decimal numbers by binary; thus $F1 = \Sigma(001, 010, 100, 111)$. To work from the truth tables, we just note the values of A, B, and C in the selected rows. First write the Boolean expression in a form that cannot be correct, writing one identical Boolean product term for each of the four rows for which $F1 = 1$.

$$F1(A, B, C) = A \bullet B \bullet C + A \bullet B \bullet C + A \bullet B \bullet C + A \bullet B \bullet C$$

Then write under each term the 0's and 1's for the corresponding row.

$$F1(A, B, C) = A \bullet B \bullet C + A \bullet B \bullet C + A \bullet B \bullet C + A \bullet B \bullet C$$

0 0 1 0 1 0 1 0 0 1 1 1

Wherever one sees a 0, complement the corresponding variable, to get.

$$F1(A, B, C) = A' \bullet B' \bullet C + A' \bullet B \bullet C' + A \bullet B' \bullet C' + A \bullet B \bullet C$$

To produce the truth-table from the canonical form SOP expression, just write a 0 under every complemented variable and a 1 under each variable that is not complemented.

$$F2(A, B, C) = A' \bullet B \bullet C + A \bullet B' \bullet C + A \bullet B \bullet C' + A \bullet B \bullet C$$

0 1 1 1 0 1 1 1 0 1 1 1

This function can be written as $F2 = S(011, 101, 110, 111)$ in binary or $F2 = \Sigma(3, 5, 6, 7)$. To create the truth table for this, just make 8 rows and fill rows 3, 5, 6, and 7 with 1's and the other rows with 0's.

It is also possible to generate the canonical POS expressions using a similar procedure.Consider again the truth table for $F1$, this time with the rows with 0 values highlighted.

Row	A	B	C	F1
0	0	0	0	0
1	0	0	1	1
2	0	1	0	1
3	0	1	1	0
4	1	0	0	1
5	1	0	1	0
6	1	1	0	0
7	1	1	1	1

The rows for which the function $F1$ has value 0 are shown in bold font. As noted above, we can write $F1$ in the Π-list form as $F1 = P(0, 3, 5, 6)$. To convert this form to canonical POS, we just replace the decimal numbers by binary; thus $F1 = P(000, 011, 101, 110)$. To work from the truth tables, we just note the values of A, B, and C in the selected rows.

Again we write a Boolean expression with one sum term for each of the four rows for which the function has value 0. At the start, this is not correct.

$$F1 = (A + B + C) \bullet (A + B + C) \bullet (A + B + C) \bullet (A + B + C)$$

Now write the expression with 0's and 1's for the corresponding row numbers.

$$F1 = (A + B + C) \bullet (A + B + C) \bullet (A + B + C) \bullet (A + B + C)$$

0 0 0 0 1 1 1 0 1 1 1 0

Wherever one sees a 1, complement the corresponding variable, to get.

$F1 = (A + B + C)\bullet(A + B' + C')\bullet(A' + B + C')\bullet(A' + B' + C)$

To produce the truth-table from the canonical form POS expression, just write a 1 under every complemented variable and a 0 under each variable that is not complemented.

$F2 = (A + B + C)\bullet(A + B + C')\bullet(A + B' + C)\bullet(A' + B + C)$

0 0 0 0 0 1 0 1 0 1 0 0

This function can be written as **F2** = Π(000, 001, 010, 100) in binary or **F2** = Π(0, 1, 2, 4). To create the truth table for this, just make 8 rows and fill rows 0, 1, 2, and 4 with 0's and the other rows with 1's.

CONVERTING TRUTH TABLES INTO BOOLEAN EXPRESSIONS

In designing digital circuits, the designer often begins with a truth table describing what the circuit should do. The design task is largely to determine what type of circuit will perform the function described in the truth table.

While some people seem to have a natural ability to look at a truth table and immediately envision the necessary logic gate or relay logic circuitry for the task, there are procedural techniques available for the rest of us. Here, Boolean algebra proves its utility in a most dramatic way. To illustrate this procedural method, we should begin with a realistic design problem. Suppose we were given the task of designing a flame detection circuit for a toxic waste incinerator.

The intense heat of the fire is intended to neutralize the toxicity of the waste introduced into the incinerator. Such combustion-based techniques are commonly used to neutralize medical waste, which may be infected with deadly viruses or bacteria:

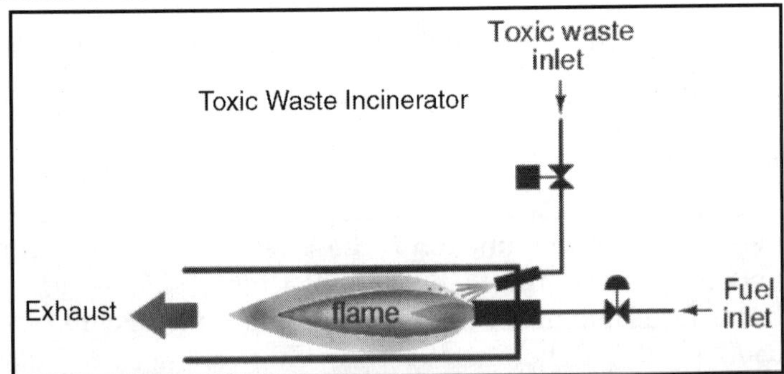

So long as a flame is maintained in the incinerator, it is safe to inject waste into it to be neutralized. If the flame were to be extinguished, however, it would be unsafe to continue to inject waste into the combustion chamber, as it would exit the exhaust un-neutralized, and pose a health threat to anyone in close proximity to the exhaust. What we need in this system is a sure way of detecting the presence of a flame, and permitting waste to be injected only if a flame is "proven" by the flame detection system.

Several different flame-detection technologies exist: optical (detection of light), thermal (detection of high temperature), and electrical conduction (detection of ionized particles in the flame path), each one with its unique advantages and disadvantages.

Suppose that due to the high degree of hazard involved with potentially passing un-neutralized waste out the exhaust of this incinerator, it is decided that the flame detection system be made redundant (multiple sensors), so that failure of a single sensor does not lead to an emission of toxins out the exhaust. Each sensor comes equipped with a normally-open contact (open if no flame, closed if flame detected) which we will use to activate the inputs of a logic system:

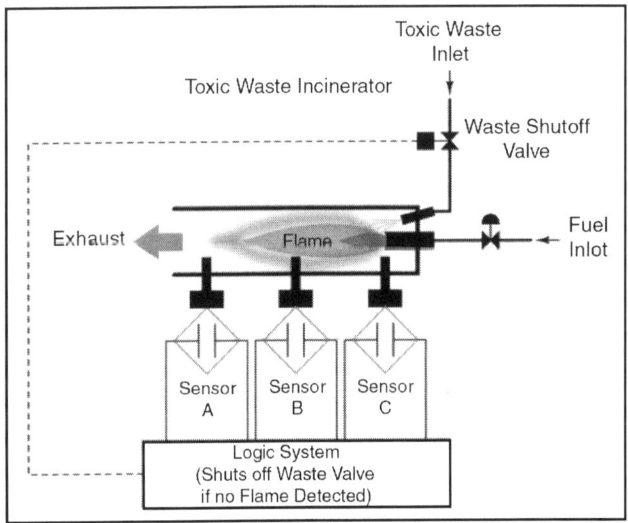

Our task, now, is to design the circuitry of the logic system to open the waste valve if and only if there is good flame proven by the sensors. First, though, we must decide what the logical behaviour of this control system should be. Do we want the valve to be opened if only one out of the three sensors detects flame? Probably not, because this would defeat the purpose of having multiple sensors.

If any one of the sensors were to fail in such a way as to falsely indicate the presence of flame when there was none, a logic system based on the principle of "any one out of three sensors showing flame" would give the same output that a single-sensor system would with the same failure. A far better solution would be to design the system so that the valve is commanded to open if any only if *all three sensors* detect a good flame. This way, any single, failed sensor falsely showing flame could not keep the valve in the open position; rather, it would require all three sensors to be failed in the same manner — a highly improbable scenario — for this dangerous condition to occur.

Thus, our truth table would look like this:

Sensor Inputs

A	B	C	Output
0	0	0	0
0	0	1	0
0	1	0	0
0	1	1	0
1	0	0	0
1	0	1	0
1	1	0	0
1	1	1	1

Output = 0 (Close Valve)

Output = 1 (Open Valve)

It does not require much insight to realise that this functionality could be generated with a three-input AND gate: the output of the circuit will be "high" if and only if input A *AND* input B *AND* input C are all "high:"

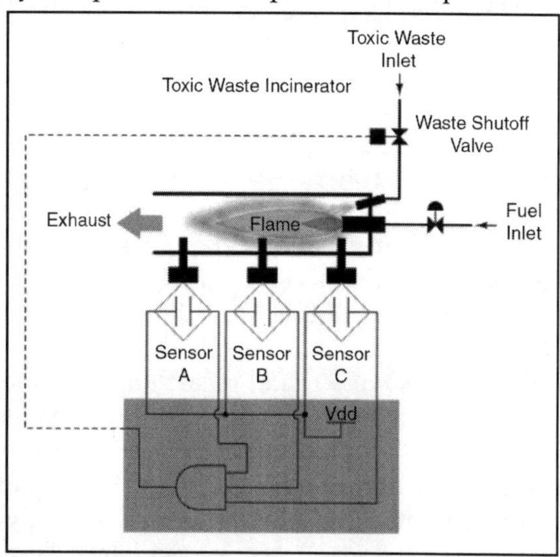

If using relay circuitry, we could create this AND function by wiring three relay contacts in series, or simply by wiring the three sensor contacts in series, so that the only way electrical power could be sent to open the waste valve is if all three sensors indicate flame:

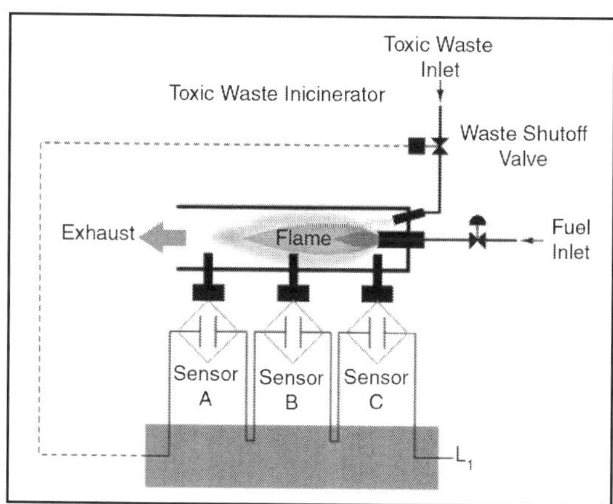

While this design strategy maximizes safety, it makes the system very susceptible to sensor failures of the opposite kind. Suppose that one of the three sensors were to fail in such a way that it indicated no flame when there really was a good flame in the incinerator's combustion chamber. That single failure would shut off the waste valve unnecessarily, resulting in lost production time and wasted fuel (feeding a fire that wasn't being used to incinerate waste). It would be nice to have a logic system that allowed for this kind of failure without shutting the system down unnecessarily, yet still provide sensor redundancy so as to maintain safety in the event that any single sensor failed "high" (showing flame at all times, whether or not there was one to detect). A strategy that would meet both needs would be a "two out of three" sensor logic, whereby the waste valve is opened if at least two out of the three sensors show good flame.

The truth table for such a system would look like this:

Sensor Inputs

A	B	C	Output
0	0	0	0
0	0	1	0
0	1	0	0
0	1	1	1
1	0	0	0
1	0	1	1
1	1	0	1
1	1	1	1

Output = 0 (Close Valve)

Output = 1 (Open Valve)

Here, it is not necessarily obvious what kind of logic circuit would satisfy the truth table. However, a simple method for designing such a circuit is found in a standard form of Boolean expression called the *Sum-Of-Products*, or *SOP*, form. As you might suspect, a Sum-Of-Products Boolean expression is literally a set of Boolean terms added (*summed*) together, each term being a multiplicative (*product*) combination of Boolean variables.

An example of an SOP expression would be something like this: ABC + BC + DF, the sum of products "ABC," "BC," and "DF."

Sum-Of-Products expressions are easy to generate from truth tables. All we have to do is examine the truth table for any rows where the output is "high" (1), and write a Boolean product term that would equal a value of 1 given those input conditions. For instance, in the fourth row down in the truth table for our two-out-of-three logic system, where A=0, B=1, and C=1, the product term would be A'BC, since that term would have a value of 1 if and only if A=0, B=1, and C=1:

Sensor Inputs

A	B	C	Output	
0	0	0	0	
0	0	1	0	
0	1	0	0	
0	1	1	1	$\bar{A}BC = 1$
1	0	0	0	
1	0	1	1	
1	1	0	1	
1	1	1	1	

Sensor Inputs

A	B	C	Output	
0	0	0	0	
0	0	1	0	
0	1	0	0	
0	1	1	1	$\bar{A}BC = 1$
1	0	0	0	
1	0	1	1	$A\bar{B}C = 1$
1	1	0	1	$AB\bar{C} = 1$
1	1	1	1	$ABC = 1$

Three other rows of the truth table have an output value of 1, so those rows also need Boolean product expressions to represent them: Finally, we join these four Boolean product expressions together by addition, to create a single Boolean expression describing the truth table as a whole:

Sensor Inputs

A	B	C	Output	
0	0	0	0	
0	0	1	0	
0	1	0	0	
0	1	1	1	$\bar{A}BC = 1$
1	0	0	0	
1	0	1	1	$A\bar{B}C = 1$
1	1	0	1	$AB\bar{C} = 1$
1	1	1	1	$ABC = 1$

Output = $\bar{A}BC + A\bar{B}C + AB\bar{C} + ABC$

Now that we have a Boolean Sum-Of-Products expression for the truth table's function, we can easily design a logic gate or relay logic circuit based on that expression:

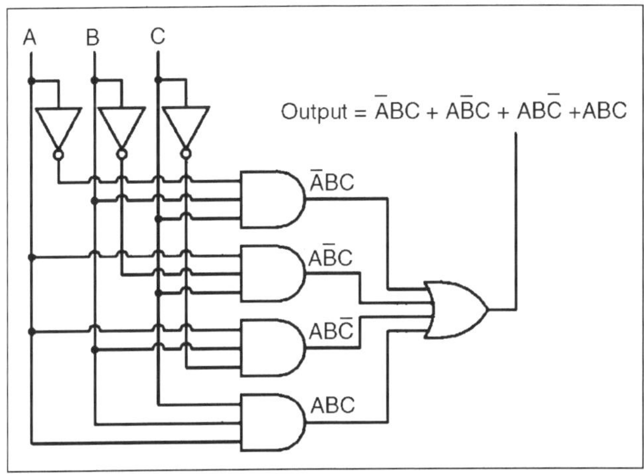

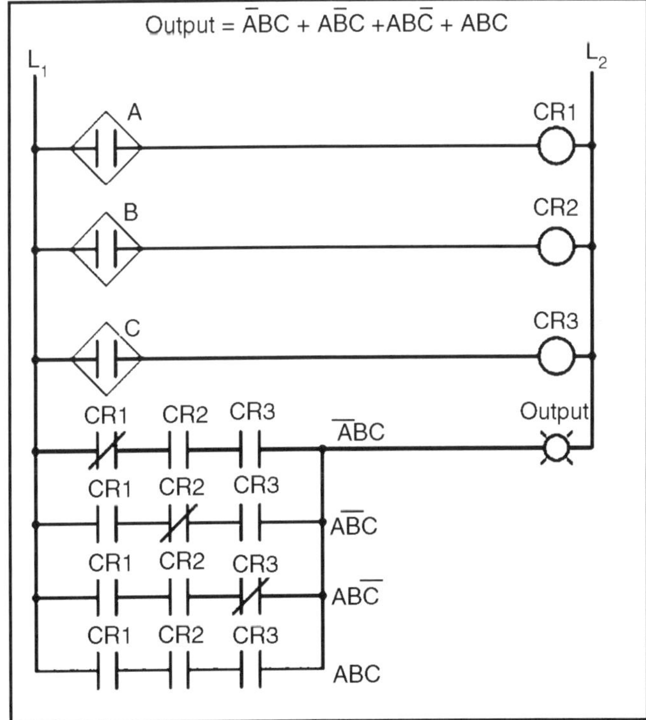

Unfortunately, both of these circuits are quite complex, and could benefit from simplification.

The result of the simplification, we can now build much simpler logic circuits performing the same function, in either gate or relay form:

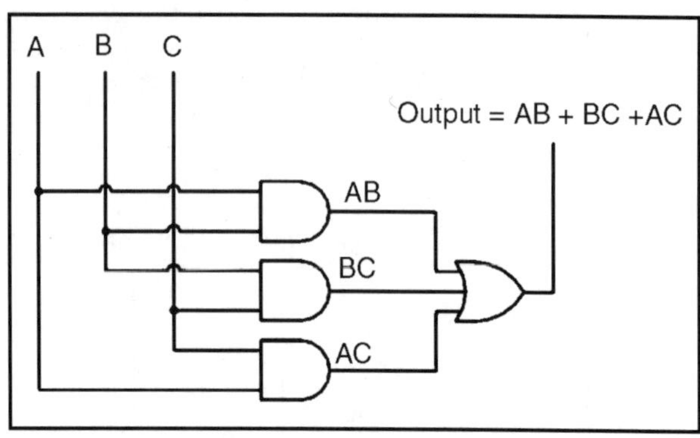

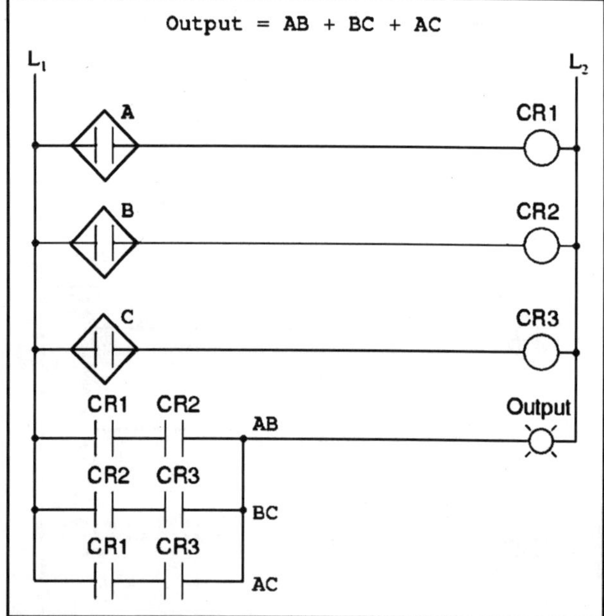

Either one of these circuits will adequately perform the task of operating the incinerator waste valve based on a flame verification from two out of the three flame sensors. At minimum, this is what we need to have a safe incinerator system. We can, however, extend the functionality of the system by adding to it logic circuitry designed to detect if any one of the sensors does not agree with the other two.

If all three sensors are operating properly, they should detect flame with equal accuracy. Thus, they should either all register "low" (000: no flame) or all register "high" (111: good flame). Any other output combination (001, 010, 011, 100, 101, or 110) constitutes a disagreement between sensors, and may therefore serve as an indicator of a potential sensor failure. If we added circuitry to detect any one of the six "sensor disagreement" conditions, we could use the output of that circuitry to activate an alarm.

Whoever is monitoring the incinerator would then exercise judgment in either continuing to operate with a possible failed sensor (inputs: 011, 101, or 110), or shut the incinerator down to be absolutely safe. Also, if the incinerator is shut down (no flame), and one or more of the sensors still indicates flame (001, 010, 011, 100, 101, or 110) while the other(s) indicate(s) no flame, it will be known that a definite sensor problem exists.

The first step in designing this "sensor disagreement" detection circuit is to write a truth table describing its behaviour.

Since we already have a truth table describing the output of the "good flame" logic circuit, we can simply add another output column to the table to represent the second circuit, and make a table representing the entire logic system:

Output = 0 (close Value)

Output = 1 (Open Valve)

Output = 0 (Sensors Agree)

Output = 1 (Aensors disagree)

Sensor Inputs			Good Flame	Sensor Disagreement
A	B	C	Output	Output
0	0	0	0	0
0	0	1	0	1
0	1	0	0	1
0	1	1	1	1
1	0	0	0	1
1	0	1	1	1
1	1	0	1	1
1	1	1	1	0

While it is possible to generate a Sum-Of-Products expression for this new truth table column, it would require six terms, of three variables each! Such a Boolean expression would require many steps to simplify, with a large potential for making algebraic errors:

An alternative to generating a Sum-Of-Products expression to account for all the "high" (1) output conditions in the truth table is to generate a *Product-Of-Sums*, or *POS*, expression, to account for all the "low" (0) output conditions instead.

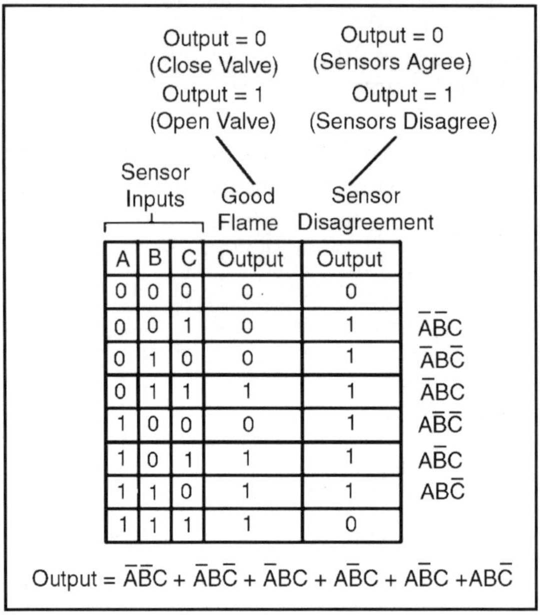

Output = 0 (Close Valve)
Output = 1 (Open Valve)

Output = 0 (Sensors Agree)
Output = 1 (Sensors Disagree)

Sensor Inputs — Good Flame — Sensor Disagreement

A	B	C	Output	Output	
0	0	0	0	0	
0	0	1	0	1	$\bar{A}\bar{B}C$
0	1	0	0	1	$\bar{A}B\bar{C}$
0	1	1	1	1	$\bar{A}BC$
1	0	0	0	1	$A\bar{B}\bar{C}$
1	0	1	1	1	$A\bar{B}C$
1	1	0	1	1	$AB\bar{C}$
1	1	1	1	0	

$$\text{Output} = \bar{A}\bar{B}C + \bar{A}B\bar{C} + \bar{A}BC + A\bar{B}\bar{C} + A\bar{B}C + AB\bar{C}$$

Being that there are much fewer instances of a "low" output in the last truth table column, the resulting Product-Of-Sums expression should contain fewer terms. As its name suggests, a Product-Of-Sums expression is a set of added terms (*sums*), which are multiplied (*product*) together. An example of a POS expression would be (A + B)(C + D), the product of the sums "A + B" and "C + D".

To begin, we identify which rows in the last truth table column have "low" (0) outputs, and write a Boolean sum term that would equal 0 for that row's input conditions. For instance, in the first row of the truth table, where A=0, B=0, and C=0, the sum term would be (A + B + C), since that term would have a value of 0 if and only if A=0, B=0, and C=0:

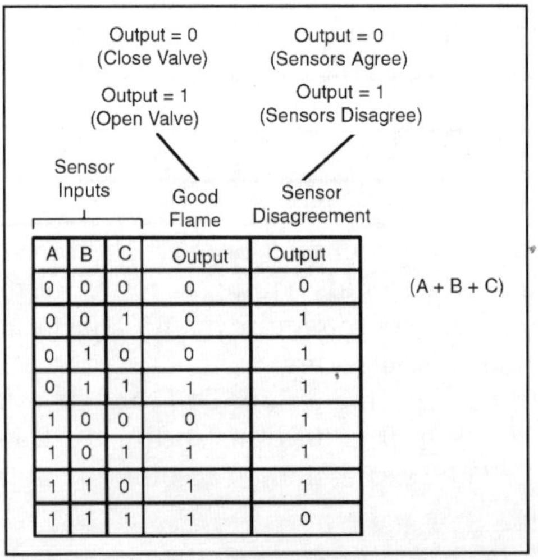

Output = 0 (Close Valve)
Output = 1 (Open Valve)

Output = 0 (Sensors Agree)
Output = 1 (Sensors Disagree)

Sensor Inputs — Good Flame — Sensor Disagreement

A	B	C	Output	Output	
0	0	0	0	0	(A + B + C)
0	0	1	0	1	
0	1	0	0	1	
0	1	1	1	1	
1	0	0	0	1	
1	0	1	1	1	
1	1	0	1	1	
1	1	1	1	0	

Only one other row in the last truth table column has a "low" (0) output, so all we need is one more sum term to complete our Product-Of-Sums expression. This last sum term represents a 0 output for an input condition of A=1, B=1 and C=1. Therefore, the term must be written as (A′ + B′+ C′), because only the sum of the *complemented* input variables would equal 0 for that condition only:

Output = 0 (Close Valve)
Output = 1 (Open Valve)
Output = 0 (Sensors Agree)
Output = 1 (Sensors Disagree)

Sensor Inputs			Good Flame	Sensor Disagreement	
A	B	C	Output	Output	
0	0	0	0	0	(A + B + C)
0	0	1	0	1	
0	1	0	0	1	
0	1	1	1	1	
1	0	0	0	1	
1	0	1	1	1	
1	1	0	1	1	
1	1	1	1	0	$(\bar{A} + \bar{B} + \bar{C})$

The completed Product-Of-Sums expression, of course, is the multiplicative combination of these two sum terms:

Output = 0 (Close Valve)
Output = 1 (Open Valve)
Output = 0 (Sensors Agree)
Output = 1 (Sensors Disagree)

Sensor Inputs			Good Flame	Sensor Disagreement	
A	B	C	Output	Output	
0	0	0	0	0	(A + B + C)
0	0	1	0	1	
0	1	0	0	1	
0	1	1	1	1	
1	0	0	0	1	
1	0	1	1	1	
1	1	0	1	1	
1	1	1	1	0	$(\bar{A} + \bar{B} + \bar{C})$

Output = $(A + B + C)(\bar{A} + \bar{B} + C)$

Whereas a Sum-Of-Products expression could be implemented in the form of a set of AND gates with their outputs connecting to a single OR gate, a

Product-Of-Sums expression can be implemented as a set of OR gates feeding into a single AND gate:

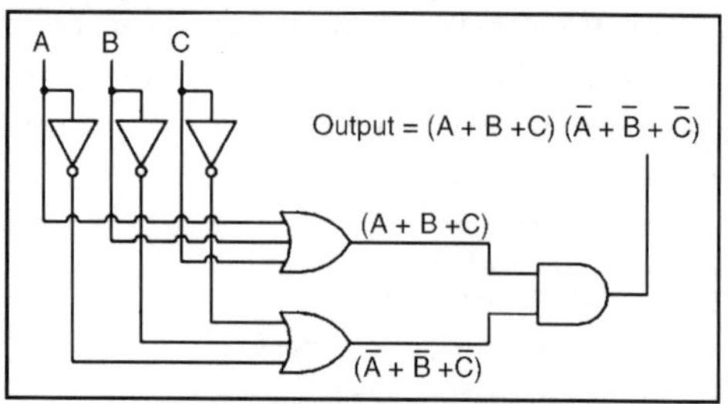

Correspondingly, whereas a Sum-Of-Products expression could be implemented as a parallel collection of series-connected relay contacts, a Product-Of-Sums expression can be implemented as a series collection of parallel-connected relay contacts:

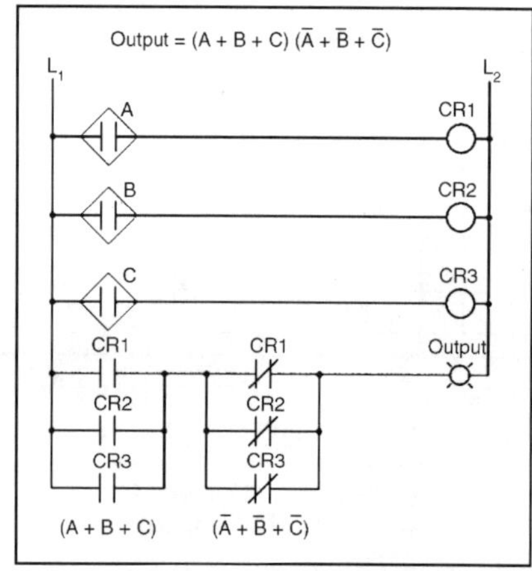

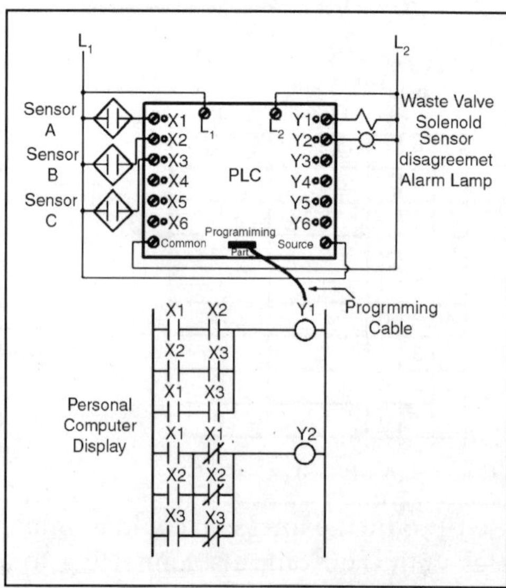

The previous two circuits represent different versions of the "sensor disagreement" logic circuit only, not the "good flame" detection circuit(s). The entire logic system would be the combination of both "good flame" and "sensor disagreement" circuits, shown on the same diagram.

Implemented in a Programmable Logic Controller (PLC), the entire logic system might resemble something like this:

As you can see, both the Sum-Of-Products and Products-Of-Sums standard Boolean forms are powerful tools when applied to truth tables. They allow us to derive a Boolean expression — and ultimately, an actual logic circuit — from nothing but a truth table, which is a written specification for what we want a logic circuit to do. To be able to go from a written specification to an actual circuit using simple, deterministic procedures means that it is possible to automate the design process for a digital circuit.

In other words, a computer could be programmed to design a custom logic circuit from a truth table specification! The steps to take from a truth table to the final circuit are so unambiguous and direct that it requires little, if any, creativity or other original thought to execute them.

EXCLUSIVE AND FUNCTION OF BOOLEN OPERATIONS

One element conspicuously missing from the set of Boolean operations is that of Exclusive-OR. Whereas the OR function is equivalent to Boolean addition, the AND function to Boolean multiplication, and the NOT function (inverter) to Boolean complementation, there is no direct Boolean equivalent for Exclusive-OR. This hasn't stopped people from developing a symbol to represent it, though:

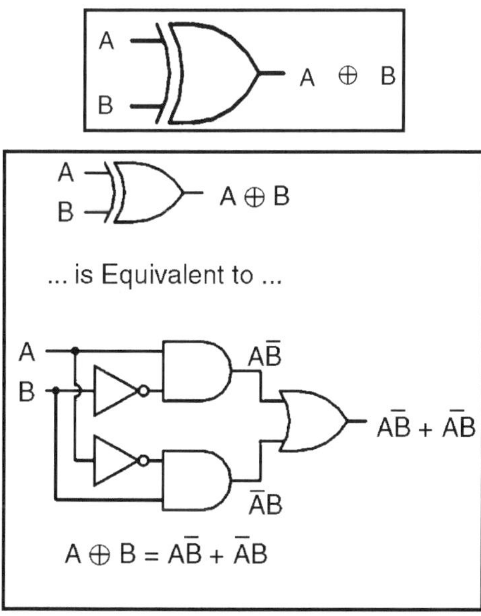

This symbol is seldom used in Boolean expressions because the identities, laws, and rules of simplification involving addition, multiplication, and complementation do not apply to it.

However, there is a way to represent the Exclusive-OR function in terms of OR and AND, as has been shown in previous chapters:

$$AB' + A'B$$

As a Boolean equivalency, this rule may be helpful in simplifying some Boolean expressions. Any expression following the AB′ + A′B form (two AND gates and an OR gate) may be replaced by a single Exclusive-OR gate.

Index

G

I

O

P

Q

R

S

T

U

V

W